CYBER OPERATIONS AN
OF FORCE IN INTERNAT

Cyber Operations and the Use of Force in International Law

MARCO ROSCINI

The Leverhulme Trust

OXFORD
UNIVERSITY PRESS

OXFORD
UNIVERSITY PRESS

Great Clarendon Street, Oxford, OX2 6DP,
United Kingdom

Oxford University Press is a department of the University of Oxford.
It furthers the University's objective of excellence in research, scholarship,
and education by publishing worldwide. Oxford is a registered trade mark of
Oxford University Press in the UK and in certain other countries

© Marco Roscini 2014

The moral rights of the author have been asserted

First published 2014
First published in paperback 2016

All rights reserved. No part of this publication may be reproduced, stored in
a retrieval system, or transmitted, in any form or by any means, without the
prior permission in writing of Oxford University Press, or as expressly permitted
by law, by licence or under terms agreed with the appropriate reprographics
rights organization. Enquiries concerning reproduction outside the scope of the
above should be sent to the Rights Department, Oxford University Press, at the
address above

You must not circulate this work in any other form
and you must impose this same condition on any acquirer

Published in the United States of America by Oxford University Press
198 Madison Avenue, New York, NY 10016, United States of America

British Library Cataloguing in Publication Data
Data available

Library of Congress Cataloging in Publication Data
Data available

ISBN 978–0–19–965501–4 (Hbk.)
ISBN 978–0–19–879071–6 (Pbk.)

Links to third party websites are provided by Oxford in good faith and
for information only. Oxford disclaims any responsibility for the materials
contained in any third party website referenced in this work.

*For Ludovica, Federico and Margherita,
children of the Information Age*

[E]ach period has had its own peculiar forms of War,
its own restrictive conditions, and its own prejudices.

Carl von Clausewitz, *On War,* Book VIII, Chapter III.B
(London: Kegan Paul, Trench, Trübner & Co, 1940), vol III, p 103

Foreword

Innovative weaponry always intrigues military planners whose task is to prognosticate what the battlefield of the future will look like. How to handle a new weapon and maximize its effect? Will it perhaps become a 'game changer'? As a rule, debates of this nature—confined as they are to eventual (hence, hypothetical) armed conflicts—tend to take place in-house, certainly away from the limelight. The public at large is not necessarily aware of them. But, even if it is, the typical posture is to regard such deliberations as moot (until the floodgates of an armed conflict actually open). Still, every once in a long while, a new weapon arrives upon the scene attracting exceptional lay interest. Once engaged, the public is loath to leave the matter entirely in the hands of military professionals. The civil society—in the broadest sense of the term—deems itself fully entitled to enquire, express views, offer guidance and lay down the law.

There are a handful of historical illustrations for public fascination with select novel weapons. None is comparable to the profound present allure of cyber weapons. Conceivably, the reason why so many people are so enthusiastically engrossed in cyber is that nowadays almost everybody (from an astonishingly early age and in every quarter of the world) has access to the internet and to the social media. Millions of people have suffered from or heard first-hand about phenomena such as 'hacking', 'phishing', or the malicious implant of a computer virus. The resultant trauma for the victim of a (peacetime) cyber attack makes him or her feel qualified to draw lessons and arrive at far-reaching conclusions (germane even in wartime).

When the average person seems to grasp the nature of the topic, to be aware of what the stakes are, and to be in a position to offer a valid opinion, pressures on international lawyers to join the fray begin to mount. International lawyers are expected by public opinion—indeed, morally compelled—to investigate the repercussions of the use of the new weapon and to speculate about the resolution of problems that are anticipated. This, of course, is not what international lawyers ordinarily do in the sphere of armed conflicts. Generally speaking, lawyers do not speculate: they react to needs that have already become manifest in the world of reality. The international law of war, either at the preliminary stage of dissection or in the final phase of consolidation, usually comes in the wake—and not in advance—of the facts.

It is necessary to bear in mind that—notwithstanding a number of notable cyber attacks reported in peacetime—there has not yet been a single armed conflict featuring such attacks in any meaningful sense. Nevertheless, before the first cyber 'shot' has been 'fired', we already have—for instance—the detailed (albeit non-binding) *Tallinn Manual on the International Law Applicable to Cyber Warfare*, published in 2013 under the aegis of NATO. The preparation of an elaborate international legal Manual with respect to a new weapon, antecedent to its actual use, is entirely unprecedented. The danger in putting the legal cart before the horse of warfare is that, when the moment of truth arrives, the two are liable to go in two separate directions. It may be worthwhile to remember what transpired when the Hague Peace Conference of 1899—a time when balloons were the only available platforms—deemed it appropriate to introduce (quite laconically) a ban on the discharge of projectiles and explosives from the air. The ban was repeated in the second Hague Conference of 1907. Yet, once put to the test of actual air warfare—subsequent to both Conferences—the outcome was a total fiasco.

This is not to say that international legal examination of cyber warfare should stop at this point in time. Indeed, now that the legal genie is anyhow out of the bottle, the reverse is true. There appears to be no alternative at present but to consider dispassionately the host of doctrinal legal propositions already put forward, and to ask whether—when the time comes, i.e. when they will be assayed in an ordeal of hostilities—they are likely to pass muster.

The present volume by Dr Marco Roscini is a systematic, up-to-date and well-informed analysis of the legal discourse that has taken place thus far. The author identifies the issues that have given rise to much discussion, marshals the evidence and provides a clear picture of where cyber operations stand in the overall scheme of the international law of armed conflict. This gives him an opportunity to delve into many controversial aspects of that law, irrespective of their kinetic/cyber application.

The book is basically divided into *jus ad bellum* and *jus in bello* sections (with a short supplementary chapter on neutrality). As far as the *jus ad bellum* is concerned, the principal issue is one of reconciling cyber attacks with a well-entrenched law that seems made-to-measure for kinetic warfare.

The pivotal challenge that has to be overcome is getting through the portals of the relevant provisions of the Charter of the United Nations. But surely the Security Council's vast discretion in determining the existence of a threat to the peace, breach of the peace or act of aggression (under Article 39) is not diminished by the inability of the framers of the Charter—in 1945—to foresee cyber attacks. Equally, all armed attacks (justifying individual and collective self-defence in response, pursuant to Article 51) must be subject to the same criteria, whatever weapon is resorted to.

Many people (lawyers as well as lay persons) are unable to equate in their minds a cyber attack with the image of a classical kinetic armed attack like Pearl Harbor. But that is only because they are thinking of cyber attacks in the peacetime context of 'hacking' and ignore the potential effects that can ensue from a war-inducing takeover of an enemy computer controlling installations, dams, aircraft, etc., causing large-scale human fatalities and devastation of property.

When it comes to the *jus in bello*, there are those who entertain the notion that (once it is upon us) cyber warfare will revolutionize international law, eliciting a whole new set of rules befitting a previously unknown phenomenon. This is an ahistorical point of view. In the past, *grosso modo*, all new means and methods of warfare have been absorbed into the pre-existing system of *jus in bello*. Of course, the process of absorption entails adaptation and some modification. Air and missile warfare is a good illustration: a number of special exigencies have led to a revised roster of custom-made rules, yet none of the general principles of the *jus in bello* as we know it (distinction, unnecessary suffering, proportionality in collateral damage, etc.) have undergone any transformation. There is no reason to believe that cyber warfare will resist more vigorously the magnetic field of the *jus in bello* in force.

The point of departure of any serious legal study of future cyber warfare must therefore be that, at bottom, it will be subordinated to the general *jus in bello*. Once this is properly perceived, there is no escape from getting into the thicket of current discords about the meaning of direct participation in hostilities (engendering civilian loss of protection from attack), the degree of organization required in non-international armed conflicts (and the consequences of belonging to an armed organized group), and so forth.

The present volume brings to the fore a string of quandaries accompanying the present application of both the *jus ad bellum* and the *jus in bello*. It is not required to agree with the author on every thesis presented by him (e.g. as regards the controversial issue of anticipatory self-defence). What really counts is that intricate problems are scrutinized in a sober fashion, and that the legal investigation displays erudition as well as insight. This is accomplished here without fail. The book surely sets the stage for the future encounter between law and reality.

Professor Yoram Dinstein
January 2014

Acknowledgements

I have been fortunate enough to have benefited from the help of many people throughout the writing of this book.

Valentina Azarov, Giulio Bartolini, Bill Boothby, Emin Çalışkan and his colleagues from the Cyber Security Institute of Tubitak (Ünal Tatar, Bahtiyar Bircan, Fatih Karayumak, Abdülkerim Demir, Abdurrahim Özel), Yoram Dinstein, Matt Evans (who first suggested that I should write a book on cyber operations), Dieter Fleck, Charles Garraway, Daniel Joyner, Erik Koppe, Simon Olleson, Natalino Ronzitti, Joseph Savirimuthu, Attila Tanzi, and Matthew Waxman kindly read previous versions of the manuscript in whole or in part and offered precious comments from a technical or legal perspective. Alicia de la Cour Venning, Andraz Kastelic, Barbara Sonczyk and Paolo Turrini provided excellent research assistance. Merel Alstein and Anthony Hinton at Oxford University Press have been supportive and patient editors throughout the preparation of the book.

Last but not least, I acknowledge the generous financial support of the Leverhulme Trust, that awarded me a Research Fellowship for the academic year 2012–13 so that I could work full time on the present book.

My deep gratitude goes to all the above. All remaining errors and omissions are, of course, my sole responsibility.

Marco Roscini
London, 30 September 2013

Contents

Table of Cases	xv
Table of Legislation and Other Documents	xix
List of Abbreviations	xxvii

1. Identifying the Problem and the Applicable Law — 1

I. The Emergence of the Cyber Threat to International Security — 1
II. The Taxonomy of Military Cyber Operations: Definitions and Classification — 10
III. The Applicable Law: *Inter (Cyber) Arma Enim Silent Leges*? — 19
IV. Identification and Attribution Problems — 33
V. The Book's Scope and Purpose — 40

2. Cyber Operations and the *jus ad bellum* — 43

I. Introduction — 43
II. Cyber Operations and the Prohibition of the Threat and Use of Force in International Relations — 44
III. Cyber Operations and the Law of Self-Defence — 69
IV. Remedies Against Cyber Operations Short of Armed Attack — 104
V. Chapter VII of the United Nations Charter and the Role of the Security Council — 110
VI. Conclusions — 115

3. The Applicability of the *jus in bello* to Cyber Operations — 117

I. Introduction — 117
II. Cyber Operations in and as International Armed Conflicts — 119
III. Cyber Operations During Partial or Total Belligerent Occupation — 141
IV. Cyber Operations in and as Non-International Armed Conflicts — 148
V. Cyber Operations as 'Internal Disturbances and Tensions' — 159
VI. Conclusions — 161

4. Cyber Operations and the Conduct of Hostilities — 164

I. Introduction — 164
II. The Legality of Means and Methods of Cyber Warfare — 168
III. The Law of Targeting — 176
IV. Cyber Operations Short of 'Attack' — 239
V. Cyber Operations as Remedies Against Violations of the Law of Armed Conflict — 242
VI. Conclusions — 245

5. Cyber Operations and the Law of Neutrality		246
I.	Introduction	246
II.	When Does the Law of Neutrality Apply?	248
III.	The Law of Neutrality and its Consequences on the Conduct of Cyber Operations	253
IV.	Non-Belligerency	267
V.	The Law of Neutrality and the UN Charter	269
VI.	Remedies Against the Violations of the Law of Neutrality	272
VII.	Conclusions	277
General Conclusions		280
Select Bibliography		289
Index		301

Table of Cases

INTERNATIONAL COURTS AND TRIBUNALS

Permanent Court of International Justice
The Case of the S.S. 'Lotus' (France v Turkey), Judgment, 7 September 1927 19

Nuremberg International Military Tribunal
Judgment, 1 October 1946. 276

International Court of Justice
Case Concerning Application of the Convention on the Prevention and Punishment of the Crime of Genocide (Bosnia and Herzegovina v Serbia and Montenegro), Judgment (Merits), 26 February 2007 . . . . 35, 37–8, 87, 100, 103, 113, 138–9, 196
Case Concerning Armed Activities on the Territory of the Congo (Democratic Republic of the Congo v Uganda), Judgment, 19 December 2005 . . . . 77, 82, 83–4, 90, 99, 109, 144
Case Concerning the Land and Maritime Boundary between Cameroon and Nigeria (Cameroon v Nigeria: Equatorial Guinea intervening), Judgment (Merits), 10 October 2002 109
Case Concerning the Military and Paramilitary Activities in and against Nicaragua (Nicaragua v United States of America), Judgment (Jurisdiction), 26 November 1984 102
Case Concerning the Military and Paramilitary Activities in and against Nicaragua (Nicaragua v United States of America), Judgment (Merits), 27 June 1986 . . . . 25, 29, 33, 35, 37, 38, 44, 46, 48, 53, 63, 66, 68, 71–2, 75, 77, 84, 87, 92, 98, 99, 100, 103–5, 109, 128, 137, 138, 149, 195, 274, 282
Case Concerning Oil Platforms (Iran v United States), Judgment (Merits), 6 November 2003 . . . . 72–3, 77, 90, 98, 99–100, 104–5, 109, 271
Case Concerning United States Diplomatic and Consular Staff in Tehran (United States of America v Iran), Judgment, 24 May 1980 . . . . 39, 82
Corfu Channel Case (United Kingdom of Great Britain and Northern Ireland v Albania), Judgment (Merits), 9 April 1949 . . . . 40, 63, 68, 82, 87, 99, 100, 102–3, 274, 282
Dispute Regarding Navigational and Related Rights (Costa Rica v Nicaragua), Judgment, 13 July 2009. . . . . 20–1, 59, 280
Fisheries Jurisdiction Case (Spain v Canada), Judgment, 4 December 1998 54
Legal Consequences of the Construction of a Wall in the Occupied Palestinian Territory, Advisory Opinion, 9 July 2004. . . . . 44, 83, 118, 144, 168, 236, 243
Legality of the Threat or Use of Nuclear Weapons, Advisory Opinion, 8 July 1996 22
North Sea Continental Shelf Cases (Federal Republic of Germany v Denmark/The Netherlands), Judgment, 20 February 1969 . . . . 25, 28

International Criminal Tribunal for the Former Yugoslavia
Prosecutor v Dragoljub Kunarać, Radomir Kovać and Zoran Voković, Case No IT–96–23 and IT–96–23/1–A, Appeals Chamber Judgment, 12 June 2002 . . . . 118, 124, 125
Prosecutor v Duško Tadić, Case No IT–94–1, Decision on the Defence Motion for Interlocutory Appeal on Jurisdiction, 2 October 1995 . . . . 26–7, 112–3, 118–9, 127, 132, 150–1, 161, 164–5, 177, 218, 251
Prosecutor v Duško Tadić, Case No IT–94–1–A, Appeals Chamber Judgment, 15 July 1999. . . . . .37–8, 128–9, 138, 250

Prosecutor v Duško Tadić, Case No IT–94–1–T, Trial Chamber Opinion and Judgment,
   7 May 1997 . . . . 125, 151, 199
Prosecutor v Fatmir Limaj, Haradin Bala and Isak Musliu, Case No IT–03–66–T, Trial
   Chamber II Judgment, 30 November 2005 . . . . 125, 151, 152, 154
Prosecutor v Ivica Rajić aka Viktor Andrić, Case No IT–95–12–R61, Review of the
   Indictment pursuant to Rule 61 of the Rules of Procedure and Evidence,
   13 September 1996 129
Prosecutor v Ljube Boškoski and Johan Tarčulovski, Case No IT–04–82–T, Trial Chamber II
   Judgment, 10 July 2008 . . . . 152, 154, 158
Prosecutor v Miodrag Jokić, Case No IT–01–42/1–S, Trial Chamber I Sentencing Judgment,
   18 March 2004 124
Prosecutor v Mladen Naletilić, aka 'Tuta', and Vinko Martinović, aka "Štela",
   Case No IT–98–34–T, Trial Chamber Judgment, 31 March 2003 142–4
Prosecutor v Pavle Strugar, Case No IT–0–42–A, Appeals Chamber Judgment,
   17 July 2008 . . . . 207, 241
Prosecutor v Ramush Haradinaj, Case No IT–04–84–T, Trial Chamber I Judgment,
   3 April 2008 . . . . 152, 154
Prosecutor v Stanislav Galić, Case No IT–98–29–T, Trial Chamber I Judgment,
   5 December 2003 . . . . 221, 228
Prosecutor v Vidoje Blagojević and Dragan Jokić, Case No IT–02–60–T, Trial Chamber
   I Judgment, 17 January 2005 124
Prosecutor v Zejnil Delalić, Zdravko Mucić also known as "Pavo", Hazim Delić and Esad
   Landžo aka 'Zenga', Case No IT–96–21–T, Trial Chamber Judgment, 16 November 1998   132
Prosecutor v Zoran Kupreškić, Mirjan Kupreškić, Vlatko Kupreškić, Drago Josipović,
   Dragan Papić and Vladimir Šantić, also known as "Vlado", Case No IT–96–16–T,
   Trial Chamber Judgment, 14 January 2000 . . . . 29, 177, 224, 232

International Criminal Tribunal for Rwanda
Georges Anderson Nderubumwe Rutaganda v Prosecutor, Case No ICTR–96–3–A,
   Appeals Chamber Judgment, 26 May 2003 124
Prosecutor v Jean-Paul Akayesu, Case No ICTR–96–4, Appeals Chamber Judgment,
   1 June 2001 125
Prosecutor v Jean-Paul Akayesu, Case No ICTR–96–4–T, Trial Chamber I Judgment,
   2 September 1998 . . . . 118, 203

International Criminal Court
Prosecutor v Germain Katanga and Mathieu Ngudjolo Chui, ICC–01/04–01/07–717,
   Pre-Trial Chamber I Decision on the Confirmation of Charges, 30 September 2008 146
Prosecutor v Thomas Lubanga Dyilo, Case No ICC–01/04–01/06–803, Pre-Trial Chamber
   I Decision on the Confirmation of Charges, 29 January 2007 146
Prosecutor v Thomas Lubanga Dyilo, Case No ICC–01/04–01/06–2842, Trial Chamber
   I Judgment, 14 March 2012 . . . . 139–40, 147

Inter-American Commission on Human Rights
Juan Carlos Abella v Argentina, Case 11.137, Report No 55/97,
   18 November 1997 . . . . 152, 154, 160

Eritrea–Ethiopia Claims Commission
Partial Award, Central Front–Eritrea's Claims 2, 4, 6, 7, 8, and 22, 28 April 2004 143
Partial Award, Central Front–Ethiopia's Claim 2, 28 April 2004 229
Partial Award, *Jus ad bellum*, Ethiopia's Claims 1–8, 19 December 2005 . . . . . . 73, 99, 104, 121–2
Partial Award, Western Front, Aerial Bombardment and Related Claims, Eritrea's Claims 1,
   3, 5, 9–13, 14, 21, 25, and 26, 19 December 2005 177

Other Arbitral Tribunals

Dalmia Cement Ltd v National Bank of Pakistan, International Chamber of Commerce,
　Arbitration Tribunal, 18 December 1967 128
Guyana and Suriname, Arbitral Award, 17 September 2007 .... 68, 69
Tinoco Claims Arbitration (Great Britain v Costa Rica), Arbitral Award, 18 October 1923 249

NATIONAL COURTS AND TRIBUNALS

Germany

Case No 2 WD 12.04, Judgment, Bundesverfassungsgericht, 21 June 2005 268

Israel

Beit Sourik Village v The Government of Israel, HCJ 2056/04, Supreme Court,
　30 June 2004 228
Public Committee Against Torture in Israel et al v The Government of Israel et al, HCJ
　769/02, Supreme Court, 11 December 2005 .... 129, 139, 146, 181–2, 195, 203–5,
　　　　　　　　　　　　　　　　　　　　　　　　207, 210, 214, 215, 219, 225, 236, 241
Tsemel et al v Minister of Defence et al, HCJ 102/82, Supreme Court, 13 July 1983 143

Japan

Ryuichi Shimoda et al v The State, District Court of Tokyo, 7 December 1963 172

United States

Hamdan v Rumsfeld et al, 548 US 557, 126 S Ct 2749, Supreme Court, 29 June 2006 140
United States v List (Hostages Trial), Case No 7, United States Military Tribunal at
　Nuremberg, 19 February 1948 142

Table of Legislation and Other Documents

TREATIES

1856 Declaration Respecting Maritime
Law ('Declaration of Paris') 247
1868 Declaration Renouncing the
Use, in Time of War, of Explosive
Projectiles Under 400 Grammes
Weight ('Saint Petersburg Declaration')
Preamble 172
1871 Treaty between Her Majesty and
the United States of America for
the Amicable Settlement of all
Causes of Difference between
the Two Countries, including
the Rules Defining Duties of a
Neutral Government during War
("Washington Rules"). 247
1899 Hague Convention II with Respect
to the Laws and Customs of War
on Land
Preamble 126
1899 Regulations Concerning the Laws
and Customs of the War on Land
annexed to the 1899 Hague
Convention II
Article 54 247
Articles 57–60 247
1907 Hague Convention III Relative to
the Opening of Hostilities
Article 1 121
1907 Hague Convention IV on the Laws
and Customs of War on Land
Article 3 . . . . 36, 124, 244
1907 Regulations Respecting the Laws
and Customs of War on Land
annexed to the 1907 Hague
Convention IV. 144
Article 1 196
Article 2 212
Article 3 192
Article 23(b). 214
Article 23(e). 171, 172
Article 24 240
Article 27 233
Article 42 . . . . 119, 141, 143
Article 53 145
1907 Hague Convention V Respecting
the Rights and Duties of Neutral
Powers and Persons in Case of War
on Land 247, 248, 253, 269, 277
Article 1 . . . . 254, 263, 286
Article 2 . . . . . 254, 256, 259, 260, 263, 286
Article 3 . . . . 263, 264, 286
Article 4 . . . . 258, 265
Article 5 . . . . 256, 258, 259, 265, 266
Article 6 266
Article 7 . . . . 259, 266
Article 8 . . . . 256, 263, 264, 286
Article 9 . . . . 264, 266
Article 10 273
Article 16 259
Article 17 259
1907 Hague Convention VIII Relating
to the Laying of Automatic
Submarine Contact Mines 247
1907 Hague Convention IX Concerning
Bombardment by Naval Forces in
Time of War
Article 2 182
1907 Hague Convention XI Relative to
Certain Restrictions With Regard
to the Exercise of the Right of
Capture in Naval War. 247
1907 Hague Convention XII Relative to
the Creation of the International
Prize Court. 247
1907 Hague Convention XIII
Respecting the Rights and Duties
of Neutral Powers in
Naval War 247, 269, 277
Article 1 . . . . 254, 255
Article 5 . . . . 255, 263
Article 6 266
Article 8 257
Article 10 260
Article 25 257
Article 26 253
1909 London Declaration Concerning
the Laws of Naval Warfare 247
1919 Covenant of the League of Nations . . . 43
1928 Havana Convention on Maritime
Neutrality. 247
1928 General Treaty for Renunciation of
War as an Instrument of National
Policy ("Pact of Paris") 43
Article 1 118

1933 Montevideo Convention on the
  Rights and Duties of States
  Article 8 64
1945 Charter of the United
  Nations 20, 21, 22, 33, 45, 64,
      73, 80, 111, 115, 117, 231,
      243, 246, 248, 262, 268–72
  Preamble 45
  Article 2(4)...... 41, 43–8, 53, 55, 58–62,
      65–8, 71, 83, 104, 115, 128,
      131, 243, 250, 273, 281
  Article 2(5)...................... 269
  Article 25....................... 269
  Article 35....................... 110
  Article 36(1).................... 110
  Article 39............... 67, 112, 269
  Article 41................. 45, 49, 270
  Article 42............. 49, 50, 114, 271
  Article 43................. 259, 271
  Article 44.................. 45, 271
  Article 46.................. 45, 271
  Article 48(1).................... 271
  Article 51....... 17, 41, 50, 70, 71,
      74, 76–8, 83, 84, 88, 93, 96, 97,
      103–6, 110, 115, 271, 281
  Article 53(1)............... 114, 115
  Article 54....................... 115
  Article 103........ 96, 269, 270, 271, 276
1945 Statute of the International Court
  of Justice 98
  Article 38..................... 20, 25
1947 Inter-American Treaty of
  Reciprocal Assistance ("Rio Treaty") ... 84
  Article 3(1)....................... 93
  Article 6......................... 71
1948 Charter of the Organization of
  American States ....
  Article 19..................... 46, 64
1948 Convention on the Prevention and
  Punishment of the Crime of Genocide
  Article 3 39
1949 North Atlantic Treaty
  ("NATO Treaty")
  Article 4...................... 93, 94
  Article 5....................... 93–7
  Article 6...................... 96, 97
  Article 7......................... 96
1949 Geneva Convention I for the
  Amelioration of the Condition of
  the Wounded and Sick in Armed
  Forces in the Field
  Article 1................... 242, 243
  Article 2........... 119, 120, 126, 134,
      136, 141, 251
  Article 3 ...... 23, 29, 140, 149–61, 164,
      177, 203, 242, 283
  Article 4......................... 182
  Article 6......................... 23
  Article 19(1)..................... 230
  Article 19(2)..................... 182
  Article 24....................... 230
  Article 25................. 194, 230
  Article 35....................... 230
  Article 36....................... 230
1949 Geneva Convention II for the
  Amelioration of the Condition of
  Wounded, Sick and Shipwrecked
  Members of Armed Forces at Sea
  Article 1................... 242, 243
  Article 2........... 119, 120, 126, 134,
      136, 141, 251
  Article 3 ...... 23, 29, 140, 149–61, 164,
      177, 203, 242, 283
  Article 6......................... 23
  Article 22(1)..................... 230
  Article 36....................... 230
  Article 39....................... 230
  Article 122..................... 252
1949 Geneva Convention III Relative to
  the Treatment of Prisoners of War ... 268
  Article 1................... 242, 243
  Article 2........... 119, 120, 126, 134,
      136, 141, 251
  Article 3 .... 23, 29, 140, 149–61,
      164, 177, 203, 242, 283
  Article 4......................... 198
  Article 4(A)(1) 194
  Article 4(A)(2) 146, 195, 196
  Article 4(A)(4) 211
  Article 4(A)(6) 144, 212, 213
  Article 4(B)(2) 267
  Article 6......................... 23
  Article 122..................... 252
1949 Geneva Convention IV Relative to
  the Protection of Civilian Persons
  in Time of War
  Article 1................... 242, 243
  Article 2........... 119, 120, 126, 134,
      136, 141, 251
  Article 3 .... 23, 29, 140, 149–61,
      164, 177, 203, 242, 283
  Article 4......................... 182
  Article 5......................... 203
  Article 6(1)..................... 143
  Article 6(3)..................... 119
  Article 7......................... 23
  Article 14....................... 231
  Article 15....................... 231
  Article 18(1)..................... 230
  Article 18(5)..................... 182

Table of Legislation and Other Documents

Article 20(1)....................... 230
Article 21.......................... 230
Article 22(1)....................... 230
Article 27.......................... 203
Article 40(2)....................... 210
Articles 41–3 203
Article 51(1).................. 144, 210
Article 53.......................... 145
Article 56(1)....................... 145
Article 78.......................... 203
Article 147......................... 210
Article 154......................... 144
1951 Security Treaty between Australia,
  New Zealand, and the United
  States ("ANZUS Treaty") 92
1954 Hague Convention for the
  Protection of Cultural Property in
  the Event of Armed Conflict ... 164, 229
Article 18(1)....................... 119
1954 First Hague Protocol for the
  Protection of Cultural Property in
  the Event of Armed Conflict 229
1969 Vienna Convention on the
  Law of Treaties
Article 2(1)(a)..................... 20
Article 30(2)....................... 96
Article 31...................... 78, 134
Article 31(1)....................... 45
Article 31(3)(b) 20
Article 31(3)(c).................... 96
Article 32................... 45, 54, 78
1977 Protocol I Additional to the
  Geneva Conventions of 12
  August 1949, and Relating to
  the Protection of Victims of
  International Armed Conflicts 117,
      164, 167, 175, 181,
      225, 230, 247, 268
Preamble 118
Article 1 ...... 22, 23, 126, 134, 140, 141,
      199, 224, 242, 249, 281
Article 2(c)................... 252, 267
Article 3(b)........................ 119
Article 9(2)(a)..................... 252
Article 12(1)....................... 230
Article 13(1)....................... 194
Article 15(1)....................... 230
Article 15(5)....................... 230
Article 19.......................... 252
Article 20.......................... 244
Articles 21ff 230
Article 22(2)(a).................... 252
Article 31.......................... 252
Article 35(2)................... 172, 173
Article 35(3).................. 174, 229
Article 36............... 23, 51, 60, 62,
      170, 171, 281
Article 37................... 215–7, 256
Article 38.......................... 217
Article 39................. 217, 252, 256
Article 41(2)....................... 194
Article 43.............. 36, 192–4, 207
Article 44..................... 197, 198
Article 46.......................... 240
Article 47.......................... 199
Article 48..................... 177, 178
Article 49........ 17, 50, 166, 167, 178,
      179, 181, 184, 188, 210,
      214, 218, 222, 239, 241,
      245, 248, 254, 255, 284
Article 50........ 186, 195, 203, 211, 225
Article 51(1)................ 176–8, 188
Article 51(2).......... 178, 188, 192, 241
Article 51(3).......... 129, 188, 203, 204
Article 51(4)............. 173, 188, 217–8
Article 51(5)............. 218, 219, 222,
      224, 235, 263
Article 51(6)....................... 244
Article 52(1).................. 178, 244
Article 52(2)....... 180, 182–8, 191, 194,
      222, 224, 285
Article 53(a)....................... 229
Article 53(c)....................... 244
Article 54........... 221, 230, 231, 244
Article 55(1).................. 174, 229
Article 55(2)....................... 244
Article 56........... 221, 230, 232, 244
Article 57...... 178, 188, 219, 226, 229,
      232, 233, 235, 239
Article 58................ 232, 237, 285
Article 59(1)....................... 231
Article 60(1)....................... 231
Article 61.......................... 212
Article 62.......................... 212
Article 64.......................... 252
Article 65.......................... 212
Article 66.......................... 212
Article 67.......................... 212
Article 75.......................... 204
Article 89.......................... 243
Article 91................ 36, 123–4, 244
Article 96.......................... 141
1977 Protocol II Additional to the
  Geneva Conventions of 12 August
  1949, and relating to the Protection
  of Victims of Non-international
  Armed Conflicts...... 117, 157–9, 161,
      164, 219, 242, 283

Preamble 157
Article 1 .... 141, 153, 157
Article 9(1)....................... 230
Article 12......................... 217
Article 13....... 129, 177, 178, 192, 199,
                  203, 207, 219, 232
Article 14......................... 230
Article 15......................... 230
1978 Economic Community of
   West African States Protocol on
   Non-aggression 43
1980 Convention on Prohibitions or
   Restrictions on the Use of Certain
   Conventional Weapons Which
   May Be Deemed to Be Excessively
   Injurious or to Have Indiscriminate
   Effects (as amended in 2001) 164
1980 Protocol II on Prohibitions or
   Restrictions on the Use of Mines,
   Booby-Traps and Other Devices
   (as amended in 1996)
   Article 1(2)...................... 164
   Article 2(6)...................... 183
   Article 3(4)...................... 234
   Article 3(10)..................... 234
1980 Protocol III on Prohibitions
   or Restrictions on the Use of
   Incendiary Weapons
   Article 1(5)...................... 234
1981 Economic Community of West
   African States Protocol Relating to
   Mutual Assistance of Defence........ 43
1989 United States–Soviet Union
   Agreement on the Prevention of
   Dangerous Military Activities
   Article 2(1)(d) 64
1992 Treaty on the European Union
   Article 42(7)...................... 97
1993 Convention on the Prohibition
   of the Development, Production,
   Stockpiling and Use of Chemical
   Weapons and on Their Destruction .....
   Article 1(1)...................... 164
1993 The Republic of Estonia and the
   Russian Federation Agreement
   on Legal Assistance and Legal
   Relationship in Civil, Family and
   Criminal Matters 40
1994 Convention on the Safety of UN
   and Associated Personnel ....
   Article 2(2)...................... 231
   Article 7(1)...................... 231
1997 Ottawa Convention on
   Anti-Personnel Mines

Article 1 164
1998 Statute of the International
   Criminal Court
   Article 8(2)(b)(iii)................. 231
   Article 8(2)(b)(xix)(3).............. 173
   Article 8(2)(e)......... 150, 159, 177, 231
   Article 8(2)(f).................... 159
1999 Second Hague Protocol for
   the Protection of Cultural
   Property in the Event of
   Armed Conflict 164, 229
   Article 1(f)...................... 183
2000 Constitutive Act of the African Union
   Article 4......................... 64
2000 Economic Community of Central
   African States Mutual Assistance
   Pact 43
2001 Budapest Convention on
   Cyber Crime 19
   Chapter II, Section 1, Title 1 4
2002 Charter of the Shanghai
   Cooperation Organization
   Article 2......................... 64
2004 Central African Economic and
   Monetary Union Non-aggression Pact . 43
2005 African Union Non-aggression and
   Common Defence Pact 43
   Article 1(c)...................... 85
   Article 4(b)..................... 85, 93
2005 Protocol III Additional to the
   Geneva Conventions of 12 August
   1949, and Relating to the Adoption
   of an Additional Distinctive
   Emblem 117
2006 Great Lakes Protocol on
   Non-aggression and Mutual
   Defence 43, 85
2007 Charter of the Association of
   Southeast Asian Nations
   Article 2(2)(e).................... 64
2007 Treaty on the Functioning of the
   European Union
   Article 222....................... 97
2008 Charter of the Organisation of the
   Islamic Conference
   Article 2(4)...................... 64
2008 Convention on Cluster Munitions
   Article 1(1)...................... 164
2009 Agreement between the
   Governments of the Member
   States of the Shanghai Cooperation
   Organization on Cooperation
   in the Field of International
   Information Security 130

OTHER INTERNATIONAL INSTRUMENTS

1938 Declaration for the Purpose of Establishing Similar Rules of Neutrality.................... 147
1948 UN General Assembly's Universal Declaration of Human Rights 64
1965 UN General Assembly's Declaration on the Inadmissibility of Intervention in the Domestic Affairs of States and the Protection of their Independence and Sovereignty.................. 64, 82
1970 UN General Assembly's Declaration on Principles of International Law concerning Friendly Relations and Co-operation among States in accordance with the Charter of the United Nations...... 46, 64, 65, 82, 250
1974 UN General Assembly's Declaration on the Definition of Aggression 46, 48, 71, 111, 129, 250
1975 Helsinki Final Act of the Conference on Security and Co-operation in Europe
 Principle VI 64
1976 UN General Assembly's Declaration on Non-interference in the Internal Affairs of States 64, 65
1981 UN General Assembly's Declaration on the Inadmissibility of Intervention and Interference in the Internal Affairs of States 64, 65
1987 UN General Assembly's Declaration on the Enhancement of the Effectiveness of the Principle of Refraining from the Threat or Use of Force in International Relations 46
1994 Annex to Security Council Resolution 955 ("Statute of the International Criminal Tribunal for Rwanda")
 Article 1........................ 139
 Article 7........................ 139
1995 UN Commission on Human Rights' Declaration of Minimum Humanitarian Standards ("Turku Declaration") 160–1
1998 Security Council Resolution 1189........................ 82
2001 Security Council Resolution 1368........................ 71, 84
2001 Security Council Resolution 1373.................... 71, 82, 84
2001 International Law Commission's Articles on the Responsibility of States for Internationally Wrongful Acts 34, 123
 Article 4....................... 34, 35
 Article 5........................... 35
 Article 7........................... 36
 Article 8................. 35, 37, 39, 81
 Article 11......................... 39
 Article 14........................ 209
 Article 15........................ 109
 Article 25........................ 273
 Article 42....... 105, 242, 268, 272, 273
 Article 48........................ 243
 Article 49................... 106, 268
 Article 50............ 63, 105, 106, 272
 Article 51........................ 106
 Article 52........................ 106
 Article 54................... 243, 268
 Article 55......................... 34
 Part Three, Chapter II 105
2003 Security Council Resolution 1483... 141
2004 African Union's Solemn Declaration on a Common African Defence and Security Policy
 Chapter III, para 13(t) 43
2010 Organization for Security and Co-operation in Europe's Astana Commemorative Declaration towards a Security Community 3
2011 International Law Commission's Draft Articles on the Effects of Armed Conflict on Treaties................ 127, 150, 151
2011 International Law Commission's Draft Articles on the Responsibility of International Organizations 34

PRIVATE CODIFICATIONS

1923 Hague Rules of Aerial Warfare...... 247
 Article 21(1)..................... 241
 Article 24(1)..................... 182
 Article 24(2)..................... 182
 Article 39....................... 255
 Article 40................... 254, 260
 Article 42....................... 257
 Article 44....................... 266
 Article 45....................... 266
 Article 46....................... 257
 Article 47................... 255, 257

Article 48 273
1939 Harvard Draft Convention on the
  Rights and Duties of Neutral States
  in Naval and Aerial War
  Article 22 263
  Article 24 273
1956 International Committee of the
  Red Cross' New Delhi Draft Rules
  for the Limitation of the Dangers
  Incurred by the Civilian Population
  in Time of War 182, 189, 203
2005 International Committee of the
  Red Cross' Study on Customary
  International Humanitarian Law 26,
    28, 30, 32, 49, 117, 158,
    165, 168, 171, 172, 173,
    174, 177, 178, 183, 186,
    192–4, 198, 203, 217,
    218, 219, 229, 230,
    231, 232, 238
2009 International Committee of
  the Red Cross' Interpretive
  Guidance on the Notion of Direct
  Participation in Hostilities 123–4,
    130, 165, 182, 194,
    195, 196, 199–212,
    215, 241–2

MILITARY MANUALS

International Military Manuals

1994 International Institute of
  Humanitarian Law's San Remo
  Manual on International Law
  Applicable to Armed Conflict
  at Sea 31, 32, 166, 246, 247,
    257, 260, 268, 272, 275
2006 International Institute of
  Humanitarian Law's Manual on the
  Law of Non-International Armed
  Conflict 31–2, 174, 183, 193,
    199, 219, 225, 232
2009 HPCR Manual on International
  Law Applicable to Air and Missile
  Warfare 12, 13, 31, 49, 52, 130,
    166, 170, 179–80, 205, 207,
    240, 247, 253, 255, 256,
    257, 260, 261, 264, 267,
    268, 269, 273
2013 Tallinn Manual on the
  International Law Applicable to
  Cyber Warfare, Prepared by the
  International Group of Experts
  at the Invitation of The NATO
  Cooperative Cyber Defence Centre
  of Excellence 1, 11, 13, 14,
    24, 25, 30–2, 35, 40, 41, 45,
    49, 55, 63, 65, 66, 69, 72,
    75–6, 77, 79, 82, 85, 88,
    90, 91, 92, 100, 102, 104,
    114, 123, 127, 129–30, 131,
    133, 144, 145, 151, 152,
    153, 155–6, 165, 168–9,
    170, 172, 173, 175, 176,
    178, 179, 180, 182–6, 191,
    192, 195, 197, 198, 203,
    209, 212, 216, 217, 218,
    220–4, 226, 229–31, 233–6,
    238–40, 244, 247, 248, 254,
    256–8, 260–3, 268–70,
    272–6, 284

National Military Manuals

Australia
2006 Law of Armed Conflict 137

Canada
2001 Law of Armed Conflict Manual:
  At the Operational and Tactical
  Levels 137, 225

France
2001 Manuel de droit des conflits
  armés 133, 219

Germany
1992 Humanitarian Law in Armed
  Conflicts 127, 129, 137, 144,
    151, 153, 172, 175, 191,
    192, 219, 233, 241, 251,
    252, 255, 260, 263–4, 266

Italy
1941 Military Penal Code of War
  (as amended in 2002) 133

United Kingdom
2004 Manual of the Law of Armed
  Conflict 27, 61, 122, 126, 136,
    170, 172, 183, 189, 191,
    198, 216, 219, 226,
    232, 233, 235, 255, 267,
    269–70, 276

United States
1956 The Law of Land Warfare 275

2007 The Commander's Handbook on
the Law of Naval Operations ... 27, 186,
221, 226, 230, 275

OTHER DOCUMENTS, INCLUDING OPERATIONAL HANDBOOKS, NATIONAL SECURITY STRATEGIES AND DOCTRINES

Australia
2004 Operations Law for RAAF
Commanders 134
2006 Australia's Defence Doctrine 137

Finland
2009 Finnish Security and Defence
Policy 2009 283

France
2013 Livre blanc. Défense et sécurité
nationale 220

Georgia
2011 National Security Concept of Georgia.... 8

Germany
2006 White Paper 2006 on German
Security Policy and the Future of
the Bundeswehr 75

North Atlantic Treaty Organization
1999 The Alliance's Strategic Concept 93
1999 Policy on Non-Lethal Weapons 61
2010 Active Engagement, Modern
Defence. Strategic Concept for
the Defence and Security of the
Members of the North Atlantic
Treaty Organisation 3, 94
2013 Glossary of Terms and
Definitions................... 12, 13

Poland
2008 Vision of the Polish Armed
Forces 2030 .... 9, 10

United Kingdom
2009 The National Security Strategy
of the United Kingdom: Update
2009—Security for the Next
Generation..................... 95

2010 A Strong Britain in an Age of
Uncertainty: The National Security
Strategy 3, 51

United States
1976 International Law — The Conduct
of Armed Conflict and Air
Operations.................... 187
1995 Doctrine for Military Operations
Other Than War................. 133
1998 Intelligence Targeting Guide ... 187, 191
2000 Joint Vision 2020—America's
Military. Preparing for Tomorrow..... 51
2002 Joint Doctrine for Targeting 191,
219, 230
2002 The National Security Strategy of
the United States of America 78
2003 National Strategy for the
Physical Protection of Critical
Infrastructures and Key Assets 56
2004 The National Military Strategy of
the United States of America 60
2006 Counterinsurgency Field
Manual................... 221, 225
2008 National Defense Strategy.......... 50
2010 Dictionary of Military and
Associated Terms 9, 11, 12, 15,
55, 69, 169
2010 National Security
Strategy 1, 50–1, 78
2010 Psychological Operations.......... 240
2013 Joint Targeting.................. 176

CYBER SECURITY STRATEGIES AND DOCTRINES AND OTHER CYBER-RELATED DOCUMENTS

Australia
2009 Cyber Security Strategy.......... 2, 57

Canada
2010 Canada's Cyber Security Strategy..... 89

Estonia
2008 Cyber Security Strategy............ 55

European Union
2013 Cybersecurity Strategy of the
European Union.................. 97

Germany
2011 Cyber Security Strategy for
Germany 12, 57

India
2011 National Cyber Security Policy (discussion draft) 56–7

Italy
2010 Relazione sulle possibili implicazioni e minacce per la sicurezza nazionale derivanti dall'utilizzo dello spazio cibernetico 2, 12, 95, 101, 122
2012 La posizione italiana sui principi fondamentali di Internet 22, 70

Lithuania
2011 Programme for the Development of Information Security (Cyber-Security) for 2011-2019 92

Netherlands
2012 Dutch Government response to the Advisory Council on International Affairs/Advisory Committee on Issues of Public International Law's Report on Cyber Warfare 10, 22, 84–5, 256

North Atlantic Treaty Organization
2008 Policy on Cyber Defence 93
2011 Defending the Networks—The NATO Policy on Cyber Defence 94

Organization of American States
2004 Inter-American Integral Strategy to Combat Threats to Cyber Security 93

Russian Federation
2000 Conceptual Views Regarding the Activities of the Armed Forces of the Russian Federation in the Information Space 22, 70, 111
2011 Convention on International Information Security (Concept)............. 58, 82, 112, 287

United Kingdom
2011 The UK Cyber Security Strategy. Protecting and Promoting the UK in a Digital World 56
2012 Defence Committee's Defence and Cyber-Security 34, 90, 194, 202

United States
1997 Cornerstones of Information Warfare 4, 50, 186
1999 Assessment of International Legal Issues in Information Operations........ 3, 12, 41, 47, 58–9, 60, 74, 77, 86, 100–1, 113, 237–8, 260, 262, 271, 274, 275
2003 National Strategy to Secure Cyberspace............... 1, 4, 56, 70
2006 National Military Strategy for Cyberspace Operations........ 3, 9, 11, 12, 13, 14, 263
2009 Cyberspace Policy Review 4
2010 Air Force's Doctrine on Cyberspace Operations................ 4, 14, 101
2010 Joint Terminology for Cyberspace Operations........ 9, 11, 12, 13, 14–5, 16, 24, 56, 207, 223
2011 Cyberspace Policy Report 21, 33, 60–1, 74, 86, 247, 259–60, 275
2011 International Strategy for Cyberspace.......... 4, 12, 21, 32, 56, 70, 87, 89, 92, 111, 282
2011 Legal Reviews of Weapons and Cyber Capabilities (Air Force Instruction 51-402) 51, 60, 169, 171
2011 Strategy for Operating in Cyberspace............... 3–4, 12, 92
2012 Joint Doctrine for Information Operations................ 4, 11, 12
2012 Presidential Policy Directive 20 ..... 12, 15–6, 21, 55, 61, 70, 89, 234–5, 248

List of Abbreviations

AF	Air Force
AIV/CAVV	Advisory Council on International Affairs/Advisory Committee on Issues of Public International Law
AU	African Union
BGP	Border Gateway Protocol
CBMs	Confidence Building Measures
CC	Cyber Collection
CCDCOE	NATO Cooperative Cyber Defence Centre of Excellence
CERT	Computer Emergency Readiness Team
CIA	Central Intelligence Agency
CNA	Computer Network Attack
CND	Computer Network Defense
CNE	Computer Network Exploitation
CNO	Computer Network Operations
COPASIR	Comitato parlamentare per la Sicurezza della Repubblica
CYBERCOM	US Cyber Command
DCEO	Defensive Cyber Effects Operations
DCO-RA	Defensive Cyberspace Operations Response Action
DDoS	Distributed Denial of Service
DoD	US Department of Defense
DoS	Denial of Service
DRC	Democratic Republic of the Congo
EECC	Eritrea-Ethiopia Claims Commission
EU	European Union
EW	Electronic Warfare
GC	1949 Geneva Conventions on the protection of Victims of War
GGE	Group of Governmental Experts
GOSCC	Global Operations and Security Control Centre
HPCR	Program on Humanitarian Policy and Conflict Research
IAC	International armed conflict
IAEA	International Atomic Energy Agency
ICC	International Criminal Court
ICJ	International Court of Justice
ICRC	International Committee of the Red Cross

ICT	Information and Communications Technologies
ICTR	International Criminal Tribunal for Rwanda
ICTY	International Criminal Tribunal for the former Yugoslavia
IHL	International humanitarian law
ILA	International Law Association
ILC	International Law Commission
IP	Internet Protocol
ISP	Internet Service Provider
ITU	International Telecommunications Union
JWICS	Joint Worldwide Intelligence Communications System
LOAC	Law of armed conflict
LOIAC	Law of international armed conflict
MoD	UK Ministry of Defence
MoU	Memorandum of understanding
NAC	North Atlantic Council
NATO	North Atlantic Treaty Organization
NCI	National critical infrastructure
NDCM	Nonintrusive Defensive Countermeasures
NIAC	Non-international armed conflict
OCC	Offensive Counter-Cyber Operations
OCEO	Offensive Cyber Effects Operations
OCO	Offensive Cyberspace operations
OSCE	Organization for Security and Co-operation in Europe
PCIJ	Permanent Court of International Justice
POW	Prisoner of war
RAAF	Royal Australian Air Force
RBN	Russian Business Network
RIAA	Reports of International Arbitral Awards
SCADA	Supervisory Control and Data Acquisition
SIPRNet	Secret Internet Protocol Router Network
UAV	Unmanned aerial vehicle
UN	United Nations
UNTS	United Nations Treaty Series
USB	Universal Serial Bus

1
Identifying the Problem and the Applicable Law

I. The Emergence of the Cyber Threat to International Security

Modern societies have become increasingly dependent on computers, computer systems, and networks, with vital services now relying on the internet.[1] This 'digital revolution' has involved not only civilian infrastructures, but also the armed forces: as *The Economist* noted in a famous article, '[b]ombs are guided by GPS satellites; drones are piloted remotely from across the world; fighter planes and warships are now huge data-processing centres; even the ordinary foot-soldier is being wired up'.[2] Digitalization, however, is a double-edged sword: as the US Deputy Secretary of Defense has emphasized, '[i]n the 21st Century, bits and bytes can be as threatening as bullets and bombs'.[3] In fact, the more digitally reliant a state is, the more vulnerable to cyber attacks: if computer networks become the society's 'nerve system', incapacitating them may mean paralysing the country.[4]

Cyber security is likely to acquire increasing importance in the next few years.[5] The threat no longer comes exclusively from the proverbial teenage hacker, but also from ideologically motivated individuals ('hacktivists'), states, and criminal

---

[1] While a computer is '[a] device that processes data', a computer system is '[o]ne or more interconnected computers with associated software and peripheral devices' (*Tallinn Manual on the International Law Applicable to Cyber Warfare* (Cambridge: Cambridge University Press, 2013), p 258). A computer network links two or more computers or computer systems (also known as 'network nodes') to exchange data by using wired, wireless, or mixed technology.

[2] 'War in the fifth domain', *The Economist*, 1 July 2010, <http://www.economist.com/node/16478792>. The US *National Strategy to Secure Cyberspace* acknowledges that '[b]y 2003, our economy and national security became fully dependent upon information technology and the information infrastructure' (*The National Strategy to Secure Cyberspace*, February 2003, p 6, <http://www.us-cert.gov/reading_room/cyberspace_strategy.pdf>).

[3] *Remarks on the Department of Defense Cyber Strategy, As Delivered by Deputy Secretary of Defense William J. Lynn, III*, 14 July 2011, <http://www.defense.gov/speeches/speech.aspx?speechid=1593>.

[4] The 2010 US *National Security Strategy* recalls that '[t]he very technologies that empower us to lead and create also empower those who would disrupt and destroy' (*National Security Strategy*, May 2010, p 27, <http://www.whitehouse.gov/sites/default/files/rss_viewer/national_security_strategy.pdf>).

[5] As noted by the US National Strategy to Secure Cyberspace in 2003, 'the attack tools and methodologies are becoming widely available, and the technical capability and sophistication of users bent on causing havoc or disruption is improving' (*The National Strategy to Secure Cyberspace*, p 6). The views, however, are not unanimous: see Thomas Rid, *Cyber War Will Not Take Place* (London: Hurst, 2013).

and terrorist organizations: cyber technologies and expertise are relatively easy and cheap to acquire, which allows weaker states and even non-state actors to potentially cause considerable damage to countries with superior conventional military power.[6] Indeed, cyber operations may not only be used for industrial espionage or intelligence collection, but also to delete, alter, or corrupt software and data resident in computers, with possible negative repercussions on the functionality of computer-operated physical infrastructures. Even though extreme scenarios have not occurred yet, a cyber operation could go as far as to disable power generators, cut off the military command, control, and communication systems, cause trains to derail and aeroplanes to crash, nuclear reactors to melt down, pipelines to explode, weapons to malfunction, banking systems to cripple. Geographical distance and frontiers also become irrelevant in the cyber context, as a target could be hit on the other side of the world in a matter of seconds. The advent of cloud computing, with software and data stored in remote servers instead of resident computers, further complicates the matter and increases potential security risks: breaking the defences of the remote server means having access to the information of all users.[7]

It is, therefore, hardly surprising that cyber threats have become a concern of the international community, with the UN General Assembly adopting a series of annual resolutions on information security since 1998 emphasizing that 'the dissemination and use of information technologies and means affect the interests of the entire international community',[8] that 'the criminal misuse of information technologies may have a grave impact on all States'[9] and that these technologies 'can potentially be used for purposes that are inconsistent with the objectives of maintaining international stability and security'.[10] The resolutions called for the views of the UN member states on information security and established three Groups of

---

[6] As noted in the Australian *Cyber Security Strategy*, '[t]he distinction between traditional threat actors—hackers, terrorists, organised criminal networks, industrial spies and foreign intelligence services—is increasingly blurred' (Australian Government, *Cyber Security Strategy*, 2009, p 3, <http://www.ag.gov.au/RightsAndProtections/CyberSecurity/Documents/AG%20Cyber%20Security%20Strategy%20-%20for%20website.pdf>). It does not seem, however, that 'terrorist' groups have been particularly active so far in conducting cyber operations, with the possible exception of Al-Qaeda: see Gregory J Rattray and Jason Healey, 'Non-State Actors and Cyber Conflict', in *America's Cyber Future. Security and Prosperity in the Information Age*, edited by Kristin M Lord and Travis Sharp (Center for a New American Security, June 2011), Vol II, p 72, <http://www.cnas.org/files/documents/publications/CNAS_Cyber_Volume%20II_2.pdf>; Richard Garnett and Paul Clarke, 'Cyberterrorism: A New Challenge for International Law', in *Enforcing International Law Norms Against Terrorism*, edited by Andrea Bianchi (Oxford and Portland: Hart, 2004), p 467; Susan W Brenner, *Cyberthreats: The Emerging Fault Lines of the Nation State* (New York: Oxford University Press, 2009), p 43.

[7] Comitato Parlamentare per la Sicurezza della Repubblica (COPASIR), *Relazione sulle possibili implicazioni e minacce per la sicurezza nazionale derivanti dall'utilizzo dello spazio cibernetico*, Doc XXXIV, no 4, 7 July 2010, p 47.

[8] See eg the Preambles to GA Resolutions 55/28 of 20 November 2000; 56/19 of 29 November 2001; 59/61 of 3 December 2004; 60/45 of 8 December 2005; 61/54 of 6 December 2006; 62/17 of 5 December 2007; 63/37 of 2 December 2008; 64/25 of 2 December 2009; 65/41 of 8 December 2010; 66/24 of 2 December 2011; 67/27 of 3 December 2012.

[9] See eg the Preambles to GA Resolutions 55/63 of 4 December 2000; 56/121 of 19 December 2001.

[10] See eg the Preambles to GA Resolutions 58/32 of 8 December 2003; 59/61 of 3 December 2004; 60/45 of 8 December 2005; 61/54 of 6 December 2006; 62/17 of 5 December 2007; 63/37 of 2 December 2008; 64/25 of 2 December 2009; 65/41 of 8 December 2010; 66/24 of 2 December 2011; 67/27 of 3 December 2012.

Governmental Experts (GGE) that examined threats in cyberspace and discussed cooperative measures to address them.[11] The General Assembly also endorsed the holding of the World Summit on the Information Society, that took place, in two phases, in Geneva in 2003 and Tunis in 2005.[12] It is not only the United Nations, however, that has become concerned with cyber security. The 2010 Organization for the Security and Cooperation in Europe (OSCE)'s Astana Commemorative Declaration also mentioned cyber threats as one of the 'emerging transnational threats'.[13] NATO's New *Strategic Concept*, adopted in November 2010, recognizes the new security environment and emphasizes that, if 'the threat of a conventional attack against NATO territory is low', '[c]yber attacks are becoming more frequent, more organized and more costly in the damage that they inflict [and] can reach a threshold that threatens national and Euro-Atlantic prosperity, security and stability'.[14] In September 2011, China, the Russian Federation, Tajikistan, and Uzbekistan submitted a draft resolution to the UN General Assembly on an international Code of Conduct for Information Security.[15] The United Kingdom's *National Security Strategy*, published in October 2010, highlights that cyber attacks by states and non-state actors are one of the four high-priority risks for the UK's national security.[16] In particular, the document claims that '[a]ctivity in cyberspace will continue to evolve as a direct national security and economic threat, as it is refined as a means of espionage and crime, and continues to grow as a terrorist enabler, as well as a military weapon for use by states and possibly others'.[17] The United Kingdom also adopted a Cyber Security Strategy, as did several other states and international organizations.[18] The United States has been a particularly prolific issuer of documents on cyber security issues: apart from commissioning a study on information operations as early as 1999,[19] the Department of Defense (DoD) adopted a partly declassified *National Military Strategy for Cyberspace Operations* (2006)[20]

---

[11] While the first Group, established in 2004, did not produce a substantial report, the second, created in 2009, issued a report in 2010 (UN Doc A/65/201, 30 July 2010). A third Group met between 2012 and 2013 and also adopted a final report containing a set of recommendations (UN Doc A/68/98, 24 June 2013).

[12] For the documents adopted at the Summit, see <http://www.itu.int/wsis/index.html>.

[13] OSCE, *Astana Commemorative Declaration—Towards a Security Community*, SUM.DOC/1/10/Corr.1, 3 December 2010, para 9, <http://www.osce.org/cio/74985?download=true>.

[14] NATO, *Active Engagement, Modern Defence. Strategic Concept for the Defence and Security of the Members of the North Atlantic Treaty Organisation*, November 2010, paras 7, 12, <http://www.nato.int/lisbon2010/strategic-concept-2010-eng.pdf>.

[15] UN Doc A/66/359, 14 September 2011.

[16] *A Strong Britain in an Age of Uncertainty: The National Security Strategy*, October 2010, pp 29–30, <http://www.direct.gov.uk/prod_consum_dg/groups/dg_digitalassets/@dg/@en/documents/digitalasset/dg_191639.pdf?CID=PDF&PLA=furl&CRE=nationalsecuritystrategy>.

[17] *A Strong Britain*, p 29.

[18] See the documents on the NATO Cooperative Cyber Defence Centre of Excellence (CCDCOE)'s website: <http://www.ccdcoe.org/328.html>.

[19] US Department of Defense, *An Assessment of International Legal Issues in Information Operations*, May 1999, <http://www.au.af.mil/au/awc/awcgate/dod-io-legal/dod-io-legal.pdf>.

[20] Chairman of the Joint Chiefs of Staff, *The National Military Strategy for Cyberspace Operations*, December 2006, <http://www.dod.mil/pubs/foi/joint_staff/jointStaff_joint Operations/ 07-F-2105 doc1.pdf>.

and a *Strategy for Operating in Cyberspace* (2011).[21] The Air Force published the pioneering *Cornerstones of Information Warfare* in 1997[22] and subsequently a comprehensive doctrine of cyber operations.[23] The Joint Chiefs of Staff also released, among others, a *Joint Doctrine for Information Operations*,[24] while the Bush and Obama Presidencies adopted a *National Strategy to Secure Cyberspace* in 2003[25] and a *Cyberspace Policy Review* in 2009,[26] followed by the adoption of an *International Strategy for Cyberspace* in 2011, respectively.[27]

If 'cyber crime', ie the offences against the confidentiality, integrity, and availability of computer data and systems committed by individuals or private entities for personal gain,[28] is essentially a domestic law matter, cyber activities conducted by states against other states fall under the remit of international law. The applicable legal paradigm, then, depends first and foremost on whether or not the operation is attributable to a subject of international law. Several states have in fact been the object of cyber attacks of which other states were suspected. As early as June 1982, a logic bomb installed in the computer control system of a Soviet gas pipeline by the US Central Intelligence Agency (CIA) allegedly caused a major explosion in Siberia.[29] Predictably, however, there was no official confirmation of the incident by either the United States or the Soviet Union and it is still uncertain whether the attack actually occurred. Fast forward 25 years and, in 2007, a three-week Distributed Denial of Service (DDoS) attack targeted Estonia, one of the most wired countries in the world, shutting down government websites first and then extending to newspapers, TV stations, banks, and other targets.[30] The attack, which, at least in its second phase, involved more than one million computers based in

---

[21] *Department of Defense Strategy for Operating in Cyberspace*, July 2011, <http://www.defense.gov/news/d20110714cyber.pdf>.

[22] US Department of the Air Force, *Cornerstones of Information Warfare*, 17 April 1997, <http://www.dtic.mil/cgi-bin/GetTRDoc?AD=ADA323807>.

[23] *Cyberspace Operations*, Air Force Doctrine Document 3-12, 15 July 2010, p 49, <http://www2.gwu.edu/~nsarchiv/NSAEBB/NSAEBB424/docs/Cyber-060.pdf>.

[24] White House, *Information Operations*, Joint Publication 3-13, 27 November 2012, <http://www.dtic.mil/doctrine/new_pubs/jp3_13.pdf> ('Joint Doctrine for Information Operations'). Previous versions dated to 1998 and 2006.

[25] *The National Strategy to Secure Cyberspace*.

[26] *Cyberspace Policy Review. Assuring a Trusted and Resilient Information and Communications Infrastructure*, May 2009, <http://www.whitehouse.gov/assets/documents/Cyberspace_Policy_Review_final.pdf>.

[27] *International Strategy for Cyberspace. Prosperity, Security and Openness in a Networked World*, May 2011, <http://www.whitehouse.gov/sites/default/files/rss_viewer/international_strategy_for_cyberspace.pdf>. On the international law aspects of the Strategy, see David P Fidler, 'International Law and the Future of Cyberspace: The Obama Administration's *International Strategy for Cyberspace*', *ASIL Insights*, Vol 15, issue 15 (8 June 2011), <http://www.asil.org/insights/volume/15/issue/15/international-law-and-future-cyberspace-obama-administration%E2%80%99s>.

[28] The language is borrowed from Chapter II, Section 1, Title 1 of the 2001 Budapest Convention on Cyber Crime. The text of the Convention is in *International Legal Materials* 41 (2002), pp 282 ff.

[29] Heather Harrison Dinniss, *Cyber Warfare and the Laws of War* (Cambridge: Cambridge University Press, 2012), p 6.

[30] On DDoS attacks, see below, Section II, p 18 of this Chapter. For the facts of the case, see Eneken Tikk, Kadri Kaska, and Liis Vihul, *International Cyber Incidents. Legal Considerations* (CCDCOE, 2010), pp 18 ff, <http://www.ccdcoe.org/publications/books/legalconsiderations.pdf>.

over 100 countries hijacked and linked through the use of botnets, followed the decision of the Estonian government to remove a Soviet war memorial from Tallinn's city centre and, overall, lasted almost a month. The attack caused some limited economic and communication disruption, but no material damage, injuries, or loss of life.[31] Websites were also defaced and their content replaced with pro-Russia propaganda. Because of the political context in which the operation occurred and the fact that Russian Internet Protocol (IP) addresses were involved, fingers were pointed at Russia, which, however, firmly denied any involvement. In addition to Estonia, cyber operations also hit, among others, Azerbaijan,[32] Kyrgyzstan,[33] Lithuania,[34] Montenegro,[35] South Korea,[36] Switzerland,[37] Taiwan,[38] the United Kingdom,[39] and the United States.[40] In August 2012, a virus, dubbed 'Shamoon' from a word contained in its computer code, destroyed the data of about 30,000 company computers of Saudi Aramco, the world's largest oil producer, and, according to Saudi Arabia, targeted the country's economy with the purpose of stopping pumping oil into domestic and international markets.[41] The deleted data were replaced with a burning American flag.

---

[31] Sean M Watts, 'Low-Intensity Computer Network Attack and Self-Defense', *International Law Studies* 87 (2011), p 70.

[32] Letter dated 6 September 2012 from the Chargé d'affaires a.i. of the Permanent Mission of Azerbaijan to the United Nations addressed to the Secretary-General, UN Doc A/66/897–S/2012/687.

[33] 'The fog of cyberwar', *The Guardian, Technology Supplement*, 5 February 2009, p 1; Fred Schreier, *On Cyber Warfare*, DCAF Horizon 2015 Working Paper no 7, p 113, <http://www.dcaf.ch/Publications/On-Cyberwarfare>, 7 September 2012.

[34] In June 2008, after the Lithuanian Parliament adopted a law prohibiting the public display of Soviet symbols, political and private websites were defaced and their content replaced with pro-Soviet propaganda (Tikk, Kaska, and Vihul, *International Cyber Incidents*, pp 63 ff).

[35] A cyber attack forced the shutdown of more than 150 websites, including the postal service and several banks' websites in March 2010. The attack apparently originated in Kosovo ('Cyber-attack shut 150 Montenegrin websites', *The Sydney Morning Herald*, 12 March 2010, <http://news.smh.com.au/breaking-news-technology/cyberattack-shut-150-montenegrin-websites-20100312-q1xo.html>).

[36] Matthew Weaver, 'Cyber attackers target South Korea and US', *The Guardian*, 8 July 2009, <http://www.guardian.co.uk/world/2009/jul/08/south-korea-cyber-attack>; Schreier, *On Cyber Warfare*, p 114.

[37] Michael Barkoviak, 'Swiss Ministry Suffers Cyber Attack', *Daily Tech*, 28 October 2009, <http://www.dailytech.com/Swiss+Ministry+Suffers+Cyber+Attack/article16629.htm>.

[38] Susan W Brenner, '"At Light Speed": Attribution and Response to Cybercrime/Terrorism/Warfare', *Journal of Criminal Law and Criminology* 97 (2006–07), p 402.

[39] Jonathan Richards, 'Thousands of cyber attacks each day on key utilities', *The Times*, 23 August 2008, <http://www.thetimes.co.uk/tto/news/uk/crime/article1874881.ece>. According to the *Annual Report 2009–2010* of the UK Intelligence and Security Committee, the greatest threat of electronic attacks to the United Kingdom comes from states, in particular from Russia and China (Intelligence and Security Committee, *Annual Report 2009–2010*, March 2010), p 16, <http://www.gov.uk/government/uploads/system/uploads/attachment_data/file/61295/isc-annualreport0910.pdf>.

[40] See, for instance, the 2003 'Titan Rain' operation, that infiltrated governmental computer networks in the United States for four years through the installation of back door programs to steal information (Scott J Shackelford, 'From Nuclear War to Net War: Analogizing Cyber Attacks in International Law', *Berkeley Journal of International Law* 27 (2009), p 204). See also the other incidents reported in Schreier, *On Cyber Warfare*, pp 107, 114.

[41] 'Saudi Aramco says cyber attack targeted kingdom's economy', *Al Arabiya News*, 9 December 2012, <http://www.alarabiya.net/articles/2012/12/09/254162.html>. Oil production, however, remained uninterrupted.

But what has been epitomized as a 'game changer' was first discovered in September 2010, when it was reported that a computer worm, dubbed Stuxnet, had attacked Iran's industrial infrastructure with the alleged ultimate purpose of sabotaging the gas centrifuges at the Natanz uranium enrichment facility, one of the sites where the Islamic Republic is developing a nuclear programme.[42] Even though an earlier version had already been released in 2007,[43] the worm—which presumably infiltrated the Natanz system, which is not usually connected to the internet for security reasons, through laptops and USB drives—mainly operated in three waves between June 2009 and May 2010.[44] Unlike other worms, Stuxnet did not limit itself to self-replicate, but also contained a 'weaponized' payload, designed to give instructions to other programs[45] and (if one excludes the above-mentioned almost legendary case of the Siberian pipeline) is, in fact, the first and so far only known use of malicious software designed to cause material damage by attacking the Supervisory Control and Data Acquisition (SCADA) system of a national critical infrastructure (NCI).[46] Stuxnet had two components: one designed to force a change in the centrifuges' rotor speed, inducing excessive vibrations or distortions, and one that recorded the normal operations of the plant and then sent them back to plant operators to make it look as if everything was functioning normally.[47] Although the exact consequences of the incident are still the object of debate in 2010, the International Atomic Energy Agency (IAEA) reported that Iran stopped feeding uranium into a significant number of gas centrifuges at Natanz.[48] In October 2011, another worm, dubbed DuQu, was discovered: its code had striking similarities with Stuxnet although its payload was not designed to cause physical damage, but to obtain information that could be used to attack industrial control systems.[49] Malware, known as

---

[42] For a comprehensive technical analysis of Stuxnet, see Symantec's Nicolas Falliere, Liam O Murchu, and Eric Chien, *W32.Stuxnet Dossier*, version 1.4, February 2011, <http://www.symantec.com/content/en/us/enterprise/media/security_response/whitepapers/w32_stuxnet_dossier.pdf>. Iran claims that its uranium enrichment programme is for purely civilian purposes.

[43] Ivanka Barzashka, 'Are Cyber-Weapons Effective? Assessing Stuxnet's Impact on the Iranian Enrichment Programme', *RUSI Journal* 158, no 2 (April 2013), pp 50, 55.

[44] It was also reported that, in December 2012, the worm reappeared and targeted companies in southern Iran ('US general warns over Iranian cyber-soldiers', *BBC News Technology*, 18 January 2013, <http://www.bbc.co.uk/news/technology-21075781>).

[45] Jeremy Richmond, 'Evolving Battlefields: Does Stuxnet Demonstrate a Need for Modifications to the Law of Armed Conflict?', *Fordham International Law Journal* 35 (2012), p 849.

[46] SCADA systems are computer-controlled industrial control systems that monitor and control industrial processes of physical infrastructures. On NCIs, see Chapter 2, Section II.1.2.

[47] William J Broad, John Markoff, and David E Sanger, 'Israeli Test on Worm Called Crucial in Iran Nuclear Delay', *The New York Times*, 15 January 2011, <http://www.cfr.org/iran/nyt-israeli-test-worm-called-crucial-iran-nuclear-delay/p23850>.

[48] William J Broad, 'Report Suggests Problems with Iran's Nuclear Effort', *The New York Times*, 23 November 2010, <http://www.nytimes.com/2010/11/24/world/middleeast/24nuke.html>. It is, however, unconfirmed whether this was due to Stuxnet or to technical malfunctions inherent to the equipment used (Katharina Ziolkowski, *Stuxnet—Legal Considerations* (CCDCOE, 2012), p 5; Barzashka, 'Are Cyber-Weapons Effective?', p 52).

[49] Symantec, *W32.DuQu—The Precursor to the Next Stuxnet*, 23 November 2011, <http://www.symantec.com/content/en/us/enterprise/media/security_response/whitepapers/w32_duqu_the_precursor_to_the_next_stuxnet.pdf>. For a discussion of the legal aspects of DuQu, see David P Fidler,

Flame, was also found in May 2012 to have penetrated the computers of senior Iranian officials with the alleged goal of stealing sensitive data. Disguised as a routine Microsoft update, Flame collected intelligence from a variety of sources and sent it back to its controllers, but, unlike Stuxnet, did not cause material damage.[50] It is entirely possible that DuQu and Flame worked together with Stuxnet for the same purpose: slowing down Iran's nuclear programme, which is allegedly aimed at developing nuclear weapons. Although the evidence is at best circumstantial,[51] the sophistication of Flame and DuQu and, in the case of Stuxnet, also its consequences on the Natanz facility have raised claims that states could be behind the incidents, in particular Israel and the United States: it has been reported that cyber efforts to disrupt the Iranian nuclear programme, codenamed 'Operation Olympic Games', were started in 2006 by the Bush Administration with Israel's cooperation and were expanded by President Barack Obama.[52]

Cyber operations have also been used in connection with a military operation or an armed conflict. It appears, for instance, that, during Operation Allied Force in 1999, the United States considered launching a cyber attack against Yugoslavia's air defence command network to disrupt its ability to target NATO aircraft, but eventually cancelled the plan because of doubts on its legality and of the risks for civilian aviation.[53] Pro-Serbian hacking groups such as the 'Black Hand', however, attacked NATO internet infrastructure during the armed conflict: although it is unknown whether their actions were attributable to Yugoslavia, their stated goal was to disrupt NATO's military operations.[54] In the second Chechen war (1999–2000), Russia disabled the insurgents' websites in order to prevent them from delivering anti-Russian propaganda: the Chechen insurgents are in fact considered pioneers in the use of the internet as a war propaganda tool.[55] It also seems that the 2007 bombing by Israel of a nuclear facility in Syria (codenamed 'Operation Orchard') was preceded by a cyber attack that neutralized ground radars and anti-aircraft batteries.[56] The cyber operations against Georgia of July–August 2008, that occurred before and during the armed conflict with the Russian

---

'Tinker, Tailor, Soldier, Duqu: Why cyberespionage is more dangerous than you think', *International Journal of Critical Infrastructure Protection* 5 (2012), pp 28–9.

[50] Ellen Nakashima, Greg Miller and Julie Tate, 'U.S., Israel developed Flame computer virus to slow down Iranian nuclear efforts, officials say', *The Washington Post*, 19 June 2012, <http://articles.washingtonpost.com/2012-06-19/world/35460741_1_stuxnet-computer-virus-malware>.

[51] On the standard of proof required for attribution of cyber operations amounting to a use of force, see below, Chapter 2, Section III.6.

[52] David E Sanger, 'Obama Order Sped Up Wave of Cyberattacks Against Iran', *The New York Times*, 1 June 2012, <http://www.nytimes.com/2012/06/01/world/middleeast/obama-ordered-wave-of-cyberattacks-against-iran.html?_r=0>.

[53] Jeffrey TG Kelsey, 'Hacking into International Humanitarian Law: The Principles of Distinction and Neutrality in the Age of Cyber Warfare', *Michigan Law Review* 106 (2008), pp 1434–5.

[54] Schreier, *On Cyber Warfare*, p 108.

[55] Eneken Tikk, Kadri Kaska, Kristel Rünnimeri, Mari Kert, Anna-Maria Talihärm, and Liis Vihul, *Cyber Attacks Against Georgia: Legal Lessons Identified* (CCDCOE, November 2008), p 5. On the use of the internet by armed groups for propaganda and communication purposes, see Wael Adhami, 'The Strategic Importance of the Internet for Armed Insurgent Groups in Modern Warfare', *International Review of the Red Cross* 89 (2007), pp 867–70.

[56] Harrison Dinniss, *Cyber Warfare*, p 7; Schreier, *On Cyber Warfare*, pp 110–11.

Federation, caused the governmental websites to go off-line and slowed down internet services.[57] In particular, immediately before and after Russian troops entered the secessionist Georgian province of South Ossetia, several governmental websites were defaced and their content replaced with anti-Georgian propaganda, while DDoS attacks crippled the Caucasian nation's ability to disseminate information. Georgia accused the Russian Federation of carrying out the cyber attacks,[58] but Russia denied its involvement and claimed that the attacks were the responsibility of private citizens that voluntarily decided to take action. The cyber operations were mentioned in the 2009 Report of the Independent Fact-Finding Mission on the Conflict in Georgia, which, however, did not reach any conclusion on their attribution or legality but noted that '[i]f these attacks were directed by a government or governments, it is likely that this form of warfare was used for the first time in an inter-state armed conflict'.[59] Since 2000, the 'cyber war' in the Middle East has accompanied traditional hostilities. In October 2000, after the kidnapping of three Israeli soldiers, a Hezbollah website was defaced and its content replaced with Israel's flags and a sound file with the Israeli national anthem. Pro-Israeli hackers also attacked the official websites of military and political organizations such as the Palestinian National Authority, Hamas, and Iran. In response, hackers hit Israeli political, economic and military targets, including the Bank of Israel and the Tel Aviv Stock Exchange, as well as telecommunications, media, and universities.[60] In 2006, in the midst of another crisis between Israel and Gaza, some 700 Israeli internet domains were shut down by hackers.[61] Unusually severe cyber operations also targeted several of Israel's governmental websites during the 2008–09 Operation Cast Lead in the Gaza Strip, mainly for defacement purposes.[62] Israeli governmental and defence-related websites were also attacked by 'Anonymous' and other hacking groups in response to Israel's air raids and internet disruption in Gaza during the 2012 Operation Pillar of Defense.[63] Israel's chief information officer was quoted as saying that '[t]he war is taking place on three fronts. The first is physical, the second is on the world of social networks and the third is cyber'.[64]

---

[57] See the facts of the case and their legal anaylsis in Tikk, Kaska, Rünnimeri, Kert, Talihärm, and Vihul, *Cyber Attacks Against Georgia*, pp 4 ff. For the technical aspects of the cyber operations against Georgia, see *Russia/Georgia Cyber War—Findings and Analysis*, Project Grey Goose: Phase I Report, 17 October 2008, <http://www.scribd.com/doc/6967393/Project-Grey-Goose-Phase-I-Report>.

[58] *National Security Concept of Georgia*, 2011, p 9, <http://www.nsc.gov.ge/files/files/National%20Security%20Concept.pdf>.

[59] *Report of the Independent Fact-Finding Mission on the Conflict in Georgia*, September 2009, Vol II, pp 217–19, <http://www.ceiig.ch/Report.html>.

[60] Kenneth Geers, 'Cyberspace and the changing nature of warfare', *SC Magazine*, 28 August 2008, <http://www.scmagazine.com/cyberspace-and-the-changing-nature-of-warfare/article/115929/#>.

[61] Geers, 'Cyberspace'.

[62] Stefan Kirchner, 'Distributed Denial-of-Service Attacks Under Public International Law: State Responsibility in Cyberwar', *The IUP Journal of Cyber Law* 8, no 3–4 (2009), p 14.

[63] Maya Epstein, 'The Fight for Public Opinion and Warfare on the Web', *Haaretz*, 19 November 2012, <http://www.haaretz.com/news/features/the-fight-for-public-opinion-and-warfare-on-the-web.premium-1.478993>.

[64] 'Mass cyber-war on Israel over Gaza raids', *Aljazeera*, 19 November 2012, <http://www.aljazeera.com/news/middleeast/2012/11/20121119973111746137.html>. On the use of new media to influence public opinion in the Palestinian–Israeli conflict, see Diana Allan and Curtis Brown, 'The *Mavi Marmara* at the Frontlines of Web 2.0', *Journal of Palestine Studies* 40 (2010), pp 63 ff.

During the 2011 armed conflict in Libya, the United States considered the use of cyber operations to disrupt Ghaddafi's air defence systems, although it eventually backed down.[65] Finally, the Syrian government has apparently used cyber operations through the self-styled 'Syrian Electronic Army' as part of its counterinsurgency campaign, while the opposition forces and 'Anonymous' have engaged in defacement operations against the Assad regime.[66]

The above list of incidents is by no means intended to be exhaustive but should sufficiently explain why the armed forces have become increasingly concerned with cyber security to the point that 'cyberspace', defined by the US DoD as '[a] global domain within the information environment consisting of the interdependent network of information technology infrastructures and resident data, including the Internet, telecommunications networks, computer systems, and embedded processors and controllers',[67] is now considered a fifth domain of warfare in addition to land, sea, air, and space.[68] As the 2010 Report of the GGE found, there is 'increased reporting that States are developing information and communications technologies as instruments of warfare and intelligence, and for political purposes'.[69] The Vision of the Polish Armed Forces 2030 expressly states that '[a]part from traditional geo-spaces, such as land, sea, air (including outer space), spheres unprovided with geographical parameters, immeasurable and unlimited, such as virtual cyberspace or information sphere, will be used as a battleground'.[70] This new battlefield 'will have no classical, linear nature, there will be no points of contact between fighting units nor delimitation lines. The future battlefield will be space in

---

[65] Thomas Rid and Peter McBurney, 'Cyber-Weapons', *RUSI Journal* 157, no 1 (February 2012), p 6.

[66] Justin Salhani, 'In Syria, the Cyberwar Intensifies', *Defense News*, 18 January 2013, <http://www.defensenews.com/article/20130118/C4ISR01/301180018/In-Syria-Cyberwar-Intensifies>.

[67] US DoD, *Dictionary of Military and Associated Terms*, Joint Publication 1–02, 8 November 2010 (As Amended Through 16 July 2013), p 70, <http://www.dtic.mil/doctrine/new_pubs/jp1_02.pdf>. Cyberspace, then, goes beyond the internet and includes all networked digital activities. A slightly different definition is contained in the 2006 US *National Military Strategy for Cyberspace Operations* and in the *Joint Terminology for Cyberspace Operations*: 'a domain characterized by the use of electronics and the electromagnetic spectrum to store, modify, and exchange data via networked systems and associated physical infrastructures' (US, *The National Military Strategy for Cyberspace Operations*, p 3; Memorandum for Chiefs of the Military Services, Commanders of the Combatant Commands, Directors of the Joint Staff Directorates, *Joint Terminology for Cyberspace Operations*, November 2010, p 7, <http://www.nsci-va.org/CyberReferenceLib/2010-11-Joint%20Terminology%20for%20Cyberspace%20Operations.pdf>).

[68] 'War in the fifth domain'. Unlike the traditional domains of warfare, however, cyberspace is man-made and has no specific boundaries. See the UNIDIR report on certain states that have included cyber warfare in their doctrine: Center for Strategic and International Studies, *Cybersecurity and Cyberwarfare—Preliminary Assessment of National Doctrine and Organization* (UNIDIR, 2011), <http://www.unidir.org/files/publications/pdfs/cybersecurity-and-cyberwarfare-preliminary-assessment-of-national-doctrine-and-organization-380.pdf>.

[69] UN Doc A/65/201, 30 July 2010, p 2. The Report was endorsed by the General Assembly in Resolution 65/41 of 8 December 2010.

[70] Ministry of National Defence, *Vision of the Polish Armed Forces 2030*, May 2008, p 13, <http://www.wp.mil.pl/pliki/File/vision_of_paf_2030.pdf>.

which combat operations and other actions of different nature and intensity will simultaneously take place.'[71]

The increasing militarization of cyberspace is reflected not only in the incorporation of cyber operations in military doctrines, but also in the creation of cyber units within national armies. Colombia has, for instance, established the Armed Forces Joint Cyber Command, which is mandated with preventing and countering cyber operations affecting national values and interests.[72] More famously, the United States has set up a military Cyber Command (a sub-unit of the Strategic Command).[73] China has also apparently created cyberspace battalions and regiments,[74] while North Korea's Unit 121, which at least partly operates from China because of the limited number of internet connections in North Korea, is believed to be responsible for disabling South Korea's military command, control, and communication networks.[75] Other states, including Argentina, Belgium, Brazil, Canada, Denmark, France, Germany, India, Iran, Israel, Japan, the Netherlands, South Korea, Switzerland, and the United Kingdom, have also either established military cyber units or plan to do so in the near future.[76]

II. The Taxonomy of Military Cyber Operations: Definitions and Classification

There are no consistent terminology or widely accepted definitions in this area. As is clear from its title, this book generally prefers to refer to 'cyber operations' instead of 'cyber war' to avoid using outdated notions and superficial and misleading analogies.[77] The expression 'cyber warfare' is also narrower than 'cyber operations'

---

[71] *Vision of the Polish Armed Forces*, p 13.   [72] UN Doc A/67/167, 23 July 2012, p 5.
[73] See the US Cyber Command's website: <http://www.arcyber.army.mil>.
[74] Sean M Condron, 'Getting It Right: Protecting American Critical Infrastructure in Cyberspace', *Harvard Journal of Law and Technology* 20 (2007), p 405; Eric Talbot Jensen, 'Computer Attacks on Critical National Infrastructure: A Use of Force Invoking the Right of Self-Defence', *Stanford Journal of International Law* 38 (2002), p 212; Sean Watts, 'Combatant Status and Computer Network Attack', *Virginia Journal of International Law* 50 (2010), p 405; Alexander Klimburg, 'Mobilising Cyber Power', *Survival* 53, no 1 (February–March 2011), p 45.
[75] Richard A Clarke and Robert K Knake, *Cyber War. The Next Threat to National Security and What to Do About It* (New York: Harpercollins, 2010), pp 27–8.
[76] John Goetz, Marcel Rosenbach, and Alexander Szandar, 'War of the Future: National Defense in Cyberspace', *Der Spiegel*, 11 February 2009, <http://www.spiegel.de/international/germany/war-of-the-future-national-defense-in-cyberspace-a-606987.html>; Elad Benari, 'Israel to Establish Cyber Warfare Administration', Israel National News, 13 January 2012, <http://www.israelnationalnews.com/News/News.aspx/151713>; 'UK to create new cyber defence force', BBC News, 29 September 2013, <http://www.bbc.co.uk/news/uk-24321717>; Dutch Government Response to the AIV/CAVV Report on Cyber Warfare, <http://www.rijksoverheid.nl/bestanden/documenten-en-publicaties/rapporten/2012/04/26/cavv-advies-nr-22-bijlage-regeringsreactie-en/cavv-advies-22-bijlage-regeringsreactie-en.pdf>, pp 3–4; Ziolkowski, *Stuxnet*, pp 51–2; Center for Strategic and International Studies, *Cybersecurity and Cyberwarfare*, p 4; Li Zhang, 'A Chinese Perspective on Cyber War', *International Review of the Red Cross* 94 (2012), p 805.
[77] See eg Michael Rundle, '"Anonymous" Hackers Declare Cyberwar on North Korea, Claim Internal Mail System Hacked', *The Huffington Post*, 4 April 2013, <http://www.huffingtonpost.co.uk/2013/04/04/anonymous-hackers-declare-war-north-korea_n_3012451.html>. As has been

and technically refers only to the conduct of hostilities in armed conflict using cyber technologies: it will therefore only be employed in the Chapters dealing with the law of armed conflict.[78]

In military doctrine, states' 'cyber operations' fall within the broader category of information operations.[79] 'Information operations' have been defined as the 'integrated employment of the core capabilities of electronic warfare, computer network operations, psychological operations, military deception, and operations security in concert with specified supporting and related capabilities, to influence, disrupt, corrupt, or usurp adversarial human and automated decision making while protecting our own'.[80] What characterizes cyber operations and makes them unique, however, is that information can also be used to inflict disruption or damage on an adversary.[81] The US DoD *Dictionary of Military and Associated Terms* defines 'cyberspace operations' as '[t]he employment of cyberspace capabilities where the primary purpose is to achieve objectives in or through cyberspace'.[82] The *Tallinn Manual on the International Law Applicable to Cyber Warfare*, published in 2013 by a Group of Experts at the invitation of NATO's CCDCOE,[83] slightly modifies this language and defines cyber operations as 'the employment of cyber capabilities with the primary purpose of achieving objectives in or by the use of cyberspace'.[84] More descriptively, the International Committee of the Red Cross (ICRC)'s definition refers to 'operations against or via a computer or a computer system through a data stream. Such operations can aim to do different things, for instance to infiltrate a system and collect, export, destroy, change, or encrypt data or to trigger, alter or otherwise manipulate processes controlled by the infiltrated computer system.'[85] All the above definitions suggest that cyberspace

---

observed, '[r]hetoric that uses a terminology of war, like "cyber war" or "cyber attack," can create situations in which a State has fewer obstacles to an aggressive response to a non-State actor's cyber threats or cyber conduct, stretching or overstepping the relevant legal boundaries' (Laurie R Blank, 'International Law and Cyber Threats from Non-State Actors', *International Law Studies* 89 (2013), p 437). The 'ideology of militarism' applied to cyberspace is also criticized by Mary Ellen O'Connell, 'Cyber Security without Cyber War', *Journal of Conflict and Security Law* 17 (2012), pp 191 ff.

[78] See Chapters 3, 4, and 5 of the present book.

[79] According to the *Oxford English Dictionary*, 'cyber' means 'relating to information technology, the Internet, and virtual reality' (*The Oxford Compact English Dictionary* (Oxford: Oxford University Press, 2003), p 268).

[80] US, *National Military Strategy for Cyberspace Operations*, p GL–2. The updated version of the Joint Doctrine for Information Operations (2012) describes them as the 'integrated employment, during military operations, of information-related capabilities in concert with other lines of operation to influence, disrupt, corrupt, or usurp the decision making of adversaries and potential adversaries while protecting our own' (Joint Doctrine for Information Operations, p GL-3).

[81] Daniel J Ryan, Maeve Dion, Eneken Tikk, and Julie JCH Ryan, 'International Cyberlaw: A Normative Approach', *Georgetown Journal of International Law* 42 (2011), p 1179.

[82] US DoD, *Dictionary of Military and Associated Terms*, p 70. See also *Joint Terminology for Cyberspace Operations*, p 8; and Joint Doctrine for Information Operations, p II–9.

[83] The CCDCOE is a think-tank based in Tallinn that was created after the 2008 DDoS attacks against the Baltic state. It is not integrated into NATO's structure or funded by it. On the Manual, see Section III.3 of this Chapter.

[84] Tallinn Manual, p 258.

[85] ICRC, *International Humanitarian Law and the Challenges of Contemporary Armed Conflicts*, ICRC Doc 31IC/11/5.1.2, October 2011, p 36, <http://www.icrc.org/eng/assets/files/red-cross-crescent-movement/31st-international-conference/31-int-conference-ihl-challenges-report-11-5-1-2-en.pdf>.

can be at the same time the target and the medium through which an attack is delivered.[86]

The 1999 US DoD's *Assessment of International Legal Issues in Information Operations*, the 2006 US *National Military Strategy for Cyberspace Operations* and the *Manual on International Law Applicable to Air and Missile Warfare*, adopted by the Program on Humanitarian Policy and Conflict Research (HPCR) at Harvard University in 2009, do not refer to 'cyber operations' but to 'computer network' operations (CNO). In strict linguistic terms, this latter notion is ambiguous, as it may lead to the erroneous belief that only computer networks are the targets of a cyber operation, while they may also include individual and specific computers within a network, as well as websites.[87] Furthermore, cyber operations can be conducted not only remotely through networks, but also through local installation of malware by agents that have physical access to the system. More recent documents, such as the 2010 US *International Strategy for Cyberspace*, the 2011 US DoD's *Strategy for Operating in Cyberspace*, the 2012 US Presidential Policy Directive 20 and the 2013 *Tallinn Manual on Cyber Warfare* drop the use of 'CNO' and refer to 'cyberspace operations' (the first two) and 'cyber operations' (the latter two).[88] The expressions CNO and its offshoots were eventually approved for removal also from the DoD *Dictionary of Military and Associated Terms* and do not appear in the 2012 version of the *Joint Doctrine for Information Operations*.[89]

There are different classifications of cyber operations in the US documents. In 2006, the US DoD distinguished CNO in computer network attacks (CNA), computer network defense (CND), and 'related computer network exploitation enabling operations' (CNE).[90] CNE was defined as '[e]nabling operations and intelligence collection to gather data from target or adversary automated information systems or networks'.[91] The *Joint Terminology for Cyberspace Operations* adds that CNE must occur 'through the use of computer networks'.[92] More

---

[86] Robin Geiss and Henning Lahmann, 'Cyber Warfare: Applying the Principle of Distinction in an Interconnected Space', *Israel Law Review* 45 (2012), p 384.

[87] HPCR, *Manual on International Law Applicable to Air and Missile Warfare* (Cambridge; Cambridge University press, 2013) 21.

[88] But see NATO's 2013 *Glossary of Terms and Definitions*, that reintroduces the distinction between CNA and CNE (p 2–C–11). The *Glossary* qualifies a CNA as a type of 'cyber attack' without, however, defining this expression.

[89] Joint Doctrine for Information Operations, p GL–3.

[90] US *National Military Strategy for Cyberspace Operations*, p GL–1. An alternative classification is contained in Germany's *Cyber Security Strategy*, which defines a 'cyber attack' as 'an IT attack in cyberspace directed against one or several other IT systems and aimed at damaging IT security'. It includes cyber espionage, ie an attack against the confidentiality of systems conducted by foreign intelligence services, and cyber sabotage, that prejudices the integrity and availability of IT systems (Federal Ministry of the Interior, *Cyber Security Strategy for Germany*, February 2011, p 16, <http://www.cio.bund.de/SharedDocs/Publikationen/DE/Strategische-Themen/css_engl_download.pdf?__blob=publicationFile>). Referring at the same time to the author and the purpose of the action as classification criteria, the Italian Comitato parlamentare per la sicurezza della Repubblica distinguishes between cyber crime, cyber terrorism, cyber espionage, and cyber war (COPASIR, *Relazione sulle possibili implicazioni*, p 17).

[91] US *National Military Strategy for Cyberspace Operations*, p GL–1.

[92] *Joint Terminology for Cyberspace Operations*, p 4.

vaguely, NATO's Glossary of Terms defines CNE as '[a]ction taken to make use of a computer or computer network, as well as the information hosted therein, in order to gain advantage'.[93]

The US *National Military Strategy for Cyberspace Operations* defines CNAs as '[o]perations to disrupt, deny, degrade, or destroy information resident in computers and computer networks, or the computers and networks themselves'.[94] A very similar definition appears in NATO's Glossary of Terms and Definitions.[95] This often cited definition distinguishes between two types of CNA, those targeting the computer or computer network and those targeting the *information* contained in the computer or computer network. As such, it may include kinetic or electronic attacks on the physical components of the cyber infrastructure.[96] The HPCR Manual adjusts the DoD's definition of CNA to also cover operations that 'manipulate' computer information and that aim 'to gain control over the computer or computer network'.[97] While both the DoD and HPCR definitions focus on the computers and computer systems as targets and do not indicate by what means (cyber, electronic or kinetic) the attack must be conducted,[98] the 2010 *Joint Terminology for Cyberspace Operations* more accurately defines CNAs as 'actions...taken *through the use of computer networks* to disrupt, deny, degrade, manipulate, or destroy information resident in the target information system or computer networks, or the systems/networks themselves'.[99] CNA, then, is narrower than 'cyber attack', which can be conducted not only through computer networks, but also through close access to the system, and whose intended effects 'are not necessarily limited to the targeted computer system or data themselves—for instance, attacks on computer systems which are intended to degrade or destroy infrastructure or C2 [command and control] capability'.[100]

As to CND, the US *National Military Strategy for Cyberspace Operations* defines it as '[a]ctions taken to protect, monitor, analyze, detect, and respond to unauthorized activity within DOD information systems and computer networks'.[101]

---

[93] NATO's Glossary of Terms and Definitions, p 2–C–11.
[94] US *National Military Strategy for Cyberspace Operations*, p GL–1. The definition is criticized by Dinstein, who argues that '[h]ad [it been] legally binding—or had it factually mirrored the whole gamut of the technological capabilities of the computer—the likelihood of a CNA ever constituting a full-fledged armed attack would be scant' (Yoram Dinstein, 'Computer Network Attacks and Self-Defense', *International Law Studies* 76 (2002), p 102).
[95] NATO's Glossary of Terms and Definitions, p 2–C–11.
[96] 'Cyber infrastructure' includes 'communications, storage, and computing resources upon which information systems operate' (Tallinn Manual, p 258).
[97] Rule 1(m), HPCR Manual, p 20. On the Manual, see Jordan J Paust, 'A Critical Appraisal of the Air and Missile Warfare Manual', *Texas International Law Journal* 47 (2012), pp 277 ff.
[98] Daniel T Kuehl, 'Information Operations, Information Warfare, and Computer Network Attack—Their Relationship to National Security in the Information Age', *International Law Studies* 76 (2002), pp 44–5.
[99] *Joint Terminology for Cyberspace Operations*, p 3 (emphasis added).
[100] *Joint Terminology for Cyberspace Operations*, p 5.
[101] US *National Military Strategy for Cyberspace Operations*, p GL–1. NATO's Glossary of Terms only distinguishes between CNAs and CNE and does not include CND.

CND employs information assurance capabilities, intelligence, counterintelligence, law enforcement, and also military capabilities, and includes both active and passive cyber defences:[102] while the latter consist of defending the networks through the use of firewalls, honeypots, encryption, routers, intrusion detection and prevention devices, numerical identifiers for communication between genuine users, anti-virus systems, and other tools which do not involve coercion or unauthorized intrusion into computer systems, the former are in kind responses to a previous cyber attack and are in fact attacks themselves.[103] Active defences capabilities, which can range from benign to aggressive, can work in an automated manner or be operated manually, and their details are often classified.[104]

In addition to referring to CNE, whose definition is identical to that of the 2006 National Military Strategy apart from the added specification that they must be conducted 'through the use of computer networks',[105] the US Air Force's *Doctrine for Cyberspace Operations*, adopted in July 2010 and updated in 2011, drops the expression CND and refers to 'cyberspace defense', defined as '[t]he passive, active and dynamic employment of capabilities to respond to imminent or on-going actions against AF [Air Force] or AF-protected networks, AF's portion of the Global Information Grid (GIG) or expeditionary communications assigned to the AF'.[106] It also introduces the concept of 'cyberspace force application', ie '[c]ombat operations in, through, and from cyberspace to achieve military objectives and influence the course and outcome of conflict by taking decisive actions against approved targets'.[107] Counter cyberspace operations are distinguished in offensive and defensive: the former, that replace CNAs, are defined as '[t]he operational planning and employment of capabilities to disrupt, deny, degrade, divert, neutralize or destroy an adversary's use of cyberspace capability or other data and information infrastructures to conduct activities or freedom of action', while the latter correspond to active defences.[108]

In November 2010, the US Joint Chiefs of Staff developed a terminology for cyberspace operations common to all US military forces. The document defines 'cyber warfare' as '[a]n armed conflict conducted in whole or part by cyber means. Military operations conducted to deny an opposing force the effective use of cyberspace systems and weapons in a conflict'.[109] Cyber warfare is divided in

---

[102] US *National Military Strategy for Cyberspace Operations*, p GL–1.
[103] Matthew J Sklerov, 'Solving the Dilemma of State Responses to Cyberattacks: A Justification for the Use of Active Defenses Against States Who Neglect Their Duty to Prevent', *Military Law Review* 201 (2009), pp 21–6.
[104] Active cyber defence involves 'launching a pre-emptive, preventive, or cyber counter-operation against the source', while passive cyber defence does not involve a counter-operation against the source but uses tools like firewalls, honeypots, anti-virus software, and the like (Tallinn Manual, pp 257, 261). See a categorization of active defences in Richard E Overill, 'Reacting to Cyber-intrusions: The Technical, Legal and Ethical Dimensions', *Journal of Financial Crime* 11 (2003), pp 163–4.
[105] *US Air Force, Cyberspace Operations*, p 49.
[106] *US Air Force, Cyberspace Operations*, p 50.
[107] *US Air Force, Cyberspace Operations*, p 50.
[108] *US Air Force, Cyberspace Operations*, pp 52 and 50, respectively.
[109] *Joint Terminology for Cyberspace Operations*, p 8.

cyber attack, cyber defence and cyber enabling operations. Cyber enabling operations presumably correspond to CNE. Cyber attack is defined as '[a] hostile act using computer or related networks or systems, and intended to disrupt and/or destroy an adversary's critical cyber systems, assets, or functions'.[110] Cyber attacks are different from CNAs in that 'the action meets use-of-force levels or is specifically intended to disrupt, deny, degrade, manipulate, and/or destroy adversary computer systems or data'.[111] Cyber attacks are also different from Offensive Counter-Cyber (OCC) operations as they can affect non-cyber systems and are not necessarily associated with imminent or ongoing hostilities.[112] Cyber defence is '[t]he integrated application of DOD or US Government cyberspace capabilities and processes to synchronize in real-time the ability to detect, analyse and mitigate threats and vulnerabilities, and outmaneuver adversaries, in order to defend designated networks, protect critical missions, and enable US freedom of action'.[113] It includes Proactive Net Operations, Defensive Counter Cyber and Defensive Countermeasures. 'Countermeasures' is not used in a legal sense, but indicates merely technical devices and techniques that fall below the use of force threshold.[114]

The US DoD *Dictionary of Military and Associated Terms* uses an alternative classification and distinguishes 'cyberspace operations' according to their purpose in defensive cyberspace operations (DCO), ie '[p]assive and active cyberspace operations intended to preserve the ability to utilize friendly cyberspace capabilities and protect data, networks, net-centric capabilities, and other designated systems' and offensive cyberspace operations (OCO), which are those 'intended to project power by the application of force in or through cyberspace'.[115] Defensive cyberspace operation response action (DCO-RA) are a type of DCO that involve '[d]eliberate, authorized defensive measures or activities taken outside of the defended network to protect and defend Department of Defense cyberspace capabilities or other designated systems'.[116] Cyber counterintelligence includes '[m]easures to identify, penetrate, or neutralize foreign operations that use cyber means as the primary tradecraft methodology, as well as foreign intelligence service collection efforts that use traditional methods to gauge cyber capabilities and intentions'.[117]

Finally, the leaked 2012 US Presidential Policy Directive 20 distinguishes 'cyber operations' in Cyber Collection (CC) and 'Cyber Effects' Operations (CEO). The former, which basically correspond to CNE, are '[o]perations and related programs or activities conducted by or on behalf of the United States Government, in or through cyberspace, for the primary purpose of collecting intelligence—including information that can be used for future operations—from computers, information

---

[110] *Joint Terminology for Cyberspace Operations*, p 5.
[111] *Joint Terminology for Cyberspace Operations*, p 6.
[112] *Joint Terminology for Cyberspace Operations*, p 13.
[113] *Joint Terminology for Cyberspace Operations*, p 6
[114] *Joint Terminology for Cyberspace Operations*, pp 4–5.
[115] US DoD, *Dictionary of Military and Associated Terms*, pp 75, 204.
[116] US DoD, *Dictionary of Military and Associated Terms*, p 75.
[117] US DoD, *Dictionary of Military and Associated Terms*, pp 69–70.

or communications systems, or networks with the intent to remain undetected'.[118] The latter's aim is to achieve a 'cyber effect', defined as '[t]he manipulation, disruption, denial, degradation, or destruction of computers, information or communication systems, networks, physical or virtual infrastructure controlled by computers or information systems, or information resident thereon'.[119] CEO are further distinguished into Defensive Cyber Effects Operations (DCEO) and Offensive Cyber Effects Operations (OCEO) depending on whether they are conducted in offence or in defence.[120] DCEO include Nonintrusive Defensive Countermeasures (NDCM), which do not entail unauthorized access to computer systems and networks and only produce minimum cyber effects to mitigate threats, but not Network Defense, ie programs, activities and tools for protection of computer systems and networks that do not require unauthorized access to them.[121]

In spite of the multiplicity of terms employed, what all the classifications above have in common is ultimately the main distinction between cyber exploitation and cyber attack. Cyber exploitation is hereby intended as referring to the unauthorized access to computers, computer systems, or networks, in order to obtain information, but without affecting the functionality of the accessed system or amending/deleting the data resident therein. As has been observed, '[t]he primary technical difference between cyber attack and cyberexploitation is in the nature of the payload to be executed—a cyber attack payload is destructive whereas a cyberexploitation payload acquires information nondestructively'.[122] Although they are often labelled in the press as 'cyber attacks', then, cyber exploitation operations are different as they do not affect the system's operation. They focus on intelligence collection, surveillance, and reconnaissance rather than on system disruption and can be preliminary to a kinetic or cyber attack that they aim to enable, for instance by mapping the architecture of the network or operating system to be attacked or by identifying previously unknown vulnerabilities.[123] Stealing security data or intellectual property from governments and corporations could also be an aim in itself and is a major threat to national security and commerce.[124] 'Trapdoors' and

---

[118] US, Presidential Policy Directive/PPD–20, October 2012, p 2, <http://www.guardian.co.uk/world/interactive/2013/jun/07/obama-cyber-directive-full-text>.
[119] US, Presidential Policy Directive 20, p 2.   [120] Presidential Policy Directive 20, p 3.
[121] US, Presidential Policy Directive 20, pp 2–3.
[122] Herbert S Lin, 'Offensive Cyber Operations and the Use of Force', *Journal of National Security Law and Policy* 4 (2010), p 64.
[123] Intelligence is 'any information concerning enemy forces and activities, as well as information necessary to facilitate one's own operations'. Surveillance is 'the systematic observation of areas, places, persons, or things, by visual, aural, electronic, photographic, or other means'. Reconnaissance is 'a single mission undertaken to obtain—by visual observation or other detection methods—specific information about the activities and resources of an enemy' (HPCR Manual, pp 320–1). See also *Joint Terminology for Cyberspace Operations*, p 11, according to which intelligence, surveillance and reconnaissance are '[a]n activity that synchronizes and integrates the planning and operation of sensors, assets, and processing, exploitation, and dissemination systems in direct support of current and future operations'.
[124] As has been noted, 'the cyber context changes the scale and consequences of theft and espionage to a degree that can result in harm to the country at least as severe as a physical attack' (Jack Goldsmith, 'How Cyber Changes the Laws of War', 24 *European Journal of International Law* (2013), p 133).

'sniffers' are particularly useful tools to conduct this type of operations: the former allow an external user to access software at any time without the computer's owner being aware of it, while the latter are programs executed from a remote computer that intercept and record data passing over a network in order to steal user IDs and passwords.

On the other hand, cyber attacks are those cyber operations, whether in offence or in defence, intended to alter, delete, corrupt, or deny access to computer data or software for the purposes of (a) propaganda or deception; and/or (b) partly or totally disrupting the functioning of the targeted computer, computer system or network, and related computer-operated physical infrastructure (if any); and/or (c) producing physical damage extrinsic to the computer, computer system, or network. As will be seen,[125] a 'cyber attack' might be an 'armed attack' in the sense of Article 51 of the UN Charter or an 'attack' under Article 49(1) of Protocol I Additional to the 1949 Geneva Conventions on the Protection of Victims of War, but care should be taken not to see these expressions as coterminous. In a military context, cyber attacks could be standalone operations, or used in conjunction with a subsequent kinetic or cyber operation that they aim to enable or facilitate, or be employed in armed conflict. A cyber attack can go from relatively innocuous psychological operations, such as website defacement, to acts that cause havoc in military campaigns by generating misinformation, or even acts resulting in major disruption of services and, material damage to property and loss of lives. In all cases, a cyber 'attack' involves an *action*, in offence or in defence, that is delivered in or through cyberspace, although not necessarily via a network, and could target either information systems or infrastructure control systems.[126] The former contain information but do not operate physical infrastructures, hence an attack on them causes loss or corruption of data but does not result in loss of functionality or material damage. The latter, of which a common type is SCADA systems, operate infrastructures: if corrupted, the consequence may be malfunctions or even physical damage.[127] For security

---

In the Moonlight Maze and Titan Rain operations, for instance, Russian and Chinese hackers stole sensitive information from the US DoD and Army's computers (Arie J Schaap, 'Cyber Warfare Operations: Development and Use Under International Law', *Air Force Law Review* 64 (2009), pp 141–2). As a consequence of the cyber intrusions allegedly originating from China, the US government adopted a new strategy to combat intellectual property theft (White House, *Administration Strategy on Mitigating the Theft of U.S. Trade Secrets*, February 2013 <http://www.whitehouse.gov//sites/default/files/omb/IPEC/admin_strategy_on_mitigating_the_theft_of_u.s._trade_secrets.pdf>, on which see David P Fidler, 'Economic Cyber Espionage and International Law: Controversies Involving Government Acquisition of Trade Secrets through Cyber Technologies', *ASIL Insights*, Vol 17, issuc 10 (20 March 2013) <http://www.asil.org/insights/volume/17/issue/10/economic-cyber-espionage-and-international-law-controversies-involving>).

[125] See Chapter 2, Section III.1 and Chapter 4, Section III.1.1.
[126] John Ricou Heaton, 'Civilians at War: Reexamining the Status of Civilians Accompanying the Armed Forces', *Air Force Law Review* 57 (2005), p 161. While syntactic attacks target the operating system, ie the instructions contained in a software program, semantic attacks alter or delete information stored in a computer system to mislead those that rely on that information (for instance, geographical coordinates in navigation systems). Mixed attacks combine the two (Marco Benatar, 'The Use of Cyber Force: Need for Legal Justification?', *Goettingen Journal of International Law* 1 (2009), pp 378–9).
[127] Ricou Heaton, 'Civilians at War', p 161.

reasons, SCADAs are normally 'air gapped' from the internet and the attack can only be delivered from within the closed network or through local installation of malware by agents that have close access to the system.

The most used methods to conduct a cyber attack are the corruption of hardware ('chipping')[128] or software, or flooding the system with so much information to cause its collapse. Popular software tools designed to interfere with the normal functioning of a computer are Trojan horses, logic bombs, viruses, and worms, which can be installed in a computer through chipping, hacking, via a portable storage device, or by inadvertently downloading them from a website or an email attachment.[129] A virus is a self-replicating program that usually attaches itself to a legitimate program on the target computer, modifying it and subsequently affecting other programs and, if the computer is connected to a network, potentially other computers as well. A virus will normally carry a payload, which is the code that corrupts or deletes computer data on the affected computer. A worm replicates itself in its entirety into other computers but, unlike viruses, does not usually modify other programs: it captures the addresses of the target computer and resends messages throughout the system so to cause a general slowdown of the system and potentially a crash. Unlike a virus, a worm can spread without human intervention.[130] Viruses and worms can be hidden in Trojan horses, an apparently innocuous code fragment that actually conceals a harmful program or allows remote access to the computer by an external user. Time and logic bombs are a type of Trojan horse designed to execute at a specific time or by certain circumstances, respectively. Denial of Service (DoS) attacks, of which 'flood attacks' are an example, are different as they do not normally penetrate into the system but aim to inundate the target with excessive calls, messages, enquiries, or requests in order to overload it and force its shut down.[131] Permanent DoS attacks are particularly serious attacks that damage the system and cause its replacement or reinstallation of hardware.[132] When the DoS attack is carried out by a large number of computers organized in botnets, it is referred to as a DDoS attack.[133]

---

[128] 'Chipping' involves 'integrating computer chips with built-in weaknesses or flaws' (Todd A Morth, 'Considering Our Position: Viewing Information Warfare as a Use of Force Prohibited by Article 2(4) of the U.N. Charter', *Case Western Journal of International Law* 30 (1998), p 572).

[129] Stephen J Cox, 'Confronting Threats Through Unconventional Means: Offensive Information Warfare as a Covert Alternative to Preemptive War', *Houston Law Review* 42 (2005–06), pp 888–9.

[130] Harrison Dinniss, *Cyber Warfare*, p 296.

[131] Richard E. Overill, 'Denial of Service Attacks: Threats and Methodologies', *Journal of Financial Crime* 6 (1999), p 353. Worms are a form of DoS attack to the extent that, by replicating themselves in each network node, they render the targeted system incapable of performing its normal functions (p 351). Unlike 'flood attacks', however, worms imply an intrusion into the targeted system.

[132] Schaap, 'Cyber Warfare Operations', p 135.

[133] 'Botnets' (short for 'robot networks'), which are the source of most spam, are networks of infected computers hijacked from their unaware owners by external users: linked together, such networks can be used to mount massive DDoS attacks (Stewart Baker, Shaun Waterman, and George Ivanov, *In the Crossfire—Critical Infrastructure in the Age of Cyber War*, 2009, p 6, <http://www.mcafee.com/us/resources/reports/rp-in-crossfire-critical-infrastructure-cyber-war.pdf>). The Mariposa botnet, started in 2008, was one of the world's biggest with up to 12.7 million computers controlled (Charles Arthur, 'Alleged controllers of "Mariposa" botnet arrested in Spain', *The Guardian*, 3 March 2010,

III. The Applicable Law: *Inter (Cyber) Arma Enim Silent Leges?*

Cyber operations amount to internationally wrongful acts if they are inconsistent with a primary rule of international law and are attributed to a state under the secondary rules on state responsibility.[134] The latter will be discussed in Section IV of this Chapter. As to the primary rules, there is so far only one treaty that expressly and specifically addresses cyber activities. The 2001 Budapest Convention on Cybercrime, negotiated in the framework of the Council of Europe and entered into force on 1 July 2004, requires states parties to criminalize certain cyber offences in their domestic legislation, to extend their jurisdiction to offences originating from their territory or committed by their nationals, and to provide mutual assistance in investigations and prosecutions.[135] An Additional Protocol concerning the criminalization of acts of a racist and xenophobic nature committed through computer systems was also adopted in 2003 and entered into force on 1 March 2006. The Convention, however, excludes from its scope of application 'conduct undertaken pursuant to lawful governmental authority'[136] and therefore does not apply to cyber operations conducted by states.

The lack of ad hoc rules does not mean that cyber operations may be conducted by states without restrictions. As pointed out by Judge Simma, the view according to which the 'absence of a legal prohibition...constitute[s] the presence of a legal permission'[137] reflects 'an old, tired view of international law'.[138] It is this book's contention that existing treaty and customary norms can be extended to cyber operations by means of interpretation even though the relevant treaties and customs do not expressly contemplate them. It cannot also be excluded that specific customary international law provisions are in the process of developing in relation to at least certain aspects of the conduct of cyber operations by states. These arguments will be explored in turn in the next two Sections.

<http://www.guardian.co.uk/technology/2010/mar/03/mariposa-botnet-spain>). On botnets, see William A Owens, Kenneth W Dam, and Herbert S Lin, *Technology, Policy, Law, and Ethics Regarding U.S. Acquisition and Use of Cyberattack Capabilities* (Washington, DC: National Academies Press, 2009), pp 92–6; Liis Vihul, Christian Czosseck, Katharina Ziolkowski, Lauri Aasmann, Ivo A Ivanov, and Sebastian Brüggemann, *Legal Implications of Countering Botnets* (CCDCOE, 2012).

[134] While primary rules are rules about conduct, secondary rules regulate the creation, modification, interpretation, validity, termination of primary rules and the consequences of their violation. The distinction between primary and secondary rules in the context of the works on state responsibility was first used by Roberto Ago, Second Report on State Responsibility—The Origin of International Responsibility, *Yearbook of the International Law Commission*, 1970, Vol II, p 179.

[135] A Committee formed of the parties to the Convention meets twice a year in plenary to consult on matters related to the Convention.

[136] Cyber Crime Convention, Explanatory Report, para 38, <http://conventions.coe.int/Treaty/EN/Reports/html/185.htm>.

[137] Julius Stone, '*Non Liquet* and the Function of Law in the International Community', *British Year Book of International Law* 35 (1959), 136. The presumption was famously asserted by the Permanent Court of International Justice (PCIJ) in *The Case of the S.S. 'Lotus' (France v Turkey)*, Judgment No 9, 1927, PCIJ, Series A, No 10, p 18.

[138] *Accordance with international law of the unilateral declaration of independence in respect of Kosovo*, Advisory Opinion, 22 July 2010, ICJ Reports 2010, Declaration of Judge Simma, para 2.

1. The applicability of existing treaties to cyber operations conducted by states

Together with customary international law, treaties are one of the two sources of international law.[139] Rules on the creation, interpretation, termination, and invalidity of treaties have been codified in the 1969 Vienna Convention on the Law of Treaties (entered into force in 1980), whose Article 2(1)(a) defines a treaty as 'an international agreement concluded between States in written form and governed by international law, whether embodied in a single instrument or in two or more related instruments and whatever its particular designation'.[140]

Although treaties have been concluded in all areas of international relations, those cyber operations that amount to a use of force or to acts of hostilities would fall within the provinces of international law that regulate the right of states to use force (*jus ad bellum*) and the conduct of warfare once an armed conflict has broken out (*jus in bello*, or the law of armed conflict, or international humanitarian law).[141] In the absence of ad hoc treaty regulation, the question is whether existing treaties that apply to traditional uses of force can be extended to cyber operations. The key *jus ad bellum* and *jus in bello* treaties are the 1945 Charter of the United Nations, the Hague Conventions of 1899 and 1907, the four 1949 Geneva Conventions on the Protection of Victims of War and their two 1977 Additional Protocols. It goes without saying that, for obvious historical reasons, none of the above texts refers to cyber issues. In the Advisory Opinion on the *Legal Consequences for States of the Continued Presence of South Africa in Namibia (South West Africa) notwithstanding Security Council Resolution 276 (1970)*, however, the International Court of Justice (ICJ) found that 'an international instrument has to be interpreted and applied within the framework of the entire legal system prevailing at the time of the interpretation'.[142] The concept of dynamic, or evolutionary, interpretation, which is also implied in Article 31(3)(b) of the Vienna Convention on the Law of Treaties,[143] was employed again by the Court in a subsequent Judgment, where it held that

> where parties have used generic terms in a treaty, the parties necessarily having been aware that the meaning of the terms was likely to evolve over time, and where the treaty has been entered

---

[139] See Art 38 of the Statute of the ICJ.

[140] The text of the Convention is in UNTS, Vol 1155, pp 331 ff. On the law of treaties, see Anthony Aust, *Modern Treaty Law and Practice*, 3rd edn (Cambridge: Cambridge University Press, 2013); Malgosia Fitzmaurice and Olufemi Elias, *Contemporary Issues in the Law of Treaties* (Utrecht: Eleven Publishing, 2005).

[141] Although there are slight differences of meaning in these expressions, they will be used as synonymous.

[142] *Legal Consequences for States of the Continued Presence of South Africa in Namibia (South West Africa) notwithstanding Security Council Resolution 276 (1970)*, Advisory Opinion, 21 June 1971, ICJ Reports 1971, para 53.

[143] According to Art 31(3)(b) of the Vienna Convention, treaties shall be interpreted taking into account, inter alia, 'any subsequent practice in the application of the treaty which establishes the agreement of the parties regarding its interpretation' (text in UNTS, vol 1115, pp 331 ff). Such practice includes 'documents, arrangements, and actions that express a specific understanding of the treaty' (Matthias Herdegen, 'Interpretation in International Law', in *Max Planck Encyclopedia of Public International Law* (2012), Vol VI, p 263). See also Rudolf Bernhardt, 'Evolutive Treaty Interpretation, Especially of the European Convention on Human Rights', *German Yearbook of International Law* 42 (1999), p 15.

into for a very long period or is 'of continuing duration', the parties must be presumed, as a general rule, to have intended those terms to have an evolving meaning.[144]

An 'interpretive reorientation'[145] of existing *jus ad bellum* and *jus in bello* provisions to accommodate cyber technology finds support in the fact that many states have affirmed the application of existing laws, including the UN Charter and the law of armed conflict, to cyber operations, often without distinguishing between treaties and customary norms. In a speech at the US CYBERCOM, the then Legal Advisor of the US State Department, Harold Koh, emphasized that 'international law principles do apply in cyberspace', including (but not limited to) the *jus ad bellum* and the *jus in bello*.[146] The White House's *International Strategy for Cyberspace* explains that '[t]he development of norms for state conduct in cyberspace does not require a reinvention of customary international law, nor does it render existing international norms obsolete'.[147] When submitting its views to the UN Secretary-General on information security, the United States also declared that '[d]espite the unique attributes of information and communications technologies, existing principles of international law serve as the appropriate framework within which to identify and analyse the rules and norms of behaviour that should govern the use of cyberspace in connection with hostilities'.[148] The 2012 US National Defense Authorization Act clarified that offensive cyber operations in cyberspace are subject, inter alia, to 'the policy principles and legal regimes that the Department [of Defense] follows for kinetic capabilities, including the law of armed conflict'.[149] Other states and international organizations that have affirmed the applicability of the existing law on the use of force and the law of armed conflict to cyber operations include Australia,[150] China,[151] Cuba,[152] the European

---

[144] *Dispute Regarding Navigational and Related Rights (Costa Rica v Nicaragua)*, Judgment, 13 July 2009, ICJ Reports 2009, para 66.

[145] Matthew C Waxman, 'Cyber-Attacks and the Use of Force: Back to the Future of Article 2(4)', *Yale Journal of International Law* 36 (2011), p 437. A leading commentary of the UN Charter, for instance, suggests that 'the rules on treaty interpretation and on the sources of international law do not exclude the possibility that Art 51 is reinterpreted, including on the basis of subsequent practice' (Albrecht Randelzhofer and Georg Nolte, 'Article 51', in *The Charter of the United Nations—A Commentary*, edited by Bruno Simma, Daniel-Erasmus Kahn, Georg Nolte, and Andreas Paulus, 3rd edn, Vol 2 (Oxford: Oxford University Press, 2012), p 1400).

[146] CarrieLyn D Guymon (ed), *Digest of United States Practice in International Law*, 2012, p 594.

[147] International Strategy for Cyberspace, p 9. See also US DoD, *Cyberspace Policy Report. A Report to Congress Pursuant to the National Defense Authorization Act for Fiscal Year 2011*, Section 934, November 2011, p 9, <http://www.defense.gov/home/features/2011/0411_cyberstrategy/docs/NDAA%20Section%20934%20Report_For%20webpage.pdf> ('[i]nternational legal norms, such as those found in the UN Charter and the law of armed conflict, which apply to the physical domains (i.e. sea, air, land, and space), also apply to the cyberspace domain').

[148] UN Doc A/66/152, 15 July 2011, p 18.

[149] National Defense Authorization Act for Fiscal Year 2012, H.R. 1540, 5 January 2012, Section 954, p 254, <http://www.gpo.gov/fdsys/pkg/BILLS-112hr1540enr/pdf/BILLS-112hr1540enr.pdf>. See also US Presidential Policy Directive 20, p 4.

[150] UN Doc A/66/152, 15 July 2011, p 6. [151] Zhang, 'A Chinese Perspective', p 4.

[152] UN Doc A/57/166/Add.1, 29 August 2002, p 3.

Union,[153] Hungary,[154] Iran,[155] Italy,[156] Mali,[157] the Netherlands,[158] Qatar,[159] the Russian Federation,[160] the United Kingdom.[161] On the basis of the views submitted by the UN member states, the 2013 Report of the GGE set up by the UN General Assembly was able to find that '[i]nternational law, and in particular the Charter of the United Nations, is applicable and is essential to maintaining peace and stability and promoting an open, secure, peaceful and accessible ICT [Information and Communications Technologies] environment'.[162]

With specific regard to international humanitarian law, the so-called Martens Clause provides, in its latest codification, that

[i]n cases not covered by this Protocol or by other international agreements, civilians and combatants remain under the protection and authority of the principles of international law derived from established custom, from the principles of humanity and from the dictates of public conscience.[163]

The Clause may be invoked in the interpretation of international humanitarian law treaties both to rule out that what is not expressly prohibited is permitted and as a presumption that favours humanitarian considerations whenever doubts exist on the meaning of certain provisions.[164] As such, the Clause can be used to found the extension of existing principles and rules to new weaponry so to avoid gaps in legal regulation. In its *Nuclear Weapons* Advisory Opinion, the ICJ found that the Martens Clause is 'an effective means of addressing the rapid evolution of military technology'.[165] According to the ICRC Commentary of Additional Protocol I, the Clause 'prevents the assumption that anything which is not explicitly prohibited by

---

[153] *Cybersecurity Strategy of the European Union: An Open, Safe and Secure Cyberspace*, 7 February 2013, pp 15–16, <http://ec.europa.eu/information_society/newsroom/cf//document.cfm?doc_id=1667>. See also Speech by EU High Representative Catherine Ashton on Cyber security: An open, free and secure Internet, Budapest, 4 October 2012, p 3, <http://europa.eu/rapid/press-release_SPEECH-12-685_en.htm>.

[154] Budapest Conference on Cyberspace, Opening Session, 4 October 2012, Welcome speech by János Martonyi, Minister of Foreign Affairs of Hungary, <http://www.cyberbudapest2012.hu/welcome-speech-by-janos-martonyi-hungarian-minister-of-foreign-affairs/>.

[155] Alireza Miryousefi and Hossein Gharibi, 'View from Iran: World needs rules on cyberattacks', *The Christian Science Monitor*, 14 February 2013, <http://www.csmonitor.com/Commentary/Opinion/2013/0214/View-from-Iran-World-needs-rules-on-cyberattacks-video>.

[156] Governo italiano, *La posizione italiana sui principi fondamentali di Internet*, 17 September 2012, p 5, <http://www.governo.it/backoffice/allegati/69257-8014.pdf>.

[157] UN Doc A/64/129/Add.1, 9 September 2009, p 7.

[158] Dutch Government Response to the AIV/CAVV Report on Cyber Warfare, pp 5–6.

[159] UN Doc A/65/154, 20 July 2010, pp 9–10.

[160] *Conceptual Views on the Activities of the Armed Forces of the Russian Federation in the Information Space*, 9 September 2000, p 6, <http://www.ccdcoe.org/strategies/Russian_Federation_unofficial_translation.pdf> (CCDCOE's unofficial translation).

[161] UN Doc A/65/154, 20 July 2010, p 15. [162] UN Doc A/68/98, 24 June 2013, p 8.

[163] Article 1(2) of Protocol I Additional to the 1949 Geneva Conventions on the Protection of Victims of International Armed Conflicts, text in UNTS, Vol 1125, pp 3 ff.

[164] Antonio Cassese, 'The Martens Clause: Half a Loaf or Simply Pie in the Sky?', *European Journal of International Law* 11 (2000), pp 189–90, 212–13.

[165] *Legality of the Threat or Use of Nuclear Weapons*, Advisory Opinion, 8 July 1996, ICJ Reports 1996 ('*Nuclear Weapons*'), para 78.

the relevant treaties is therefore permitted' and proclaims 'the applicability of the principles mentioned regardless of subsequent developments of types of situation or technology'.[166] The fact that international humanitarian law treaties can extend to weapons developed after their adoption is also confirmed by the inclusion in Protocol I Additional to the Geneva Conventions of Article 36, which states that

[i]n the study, development, acquisition or adoption of a new weapon, means or method of warfare, a High Contracting Party is under an obligation to determine whether its employment would, in some or all circumstances, be prohibited by this Protocol or by any other rule of international law applicable to the High Contracting Party.[167]

In the ICRC's view, then, 'means and methods of warfare which resort to cyber technology are subject to IHL [international humanitarian law] just as any new weapon or delivery system has been so far when used in an armed conflict by or on behalf of a party to such conflict. If a cyber operations [*sic*] is used against an enemy in an armed conflict in order to cause damage, for example by manipulation of an air traffic control system that results in the crash of a civilian aircraft, it can hardly be disputed that such an attack is in fact a method of warfare and is subject to prohibitions under IHL'.[168] At the United Nations, the ICRC recalled 'the obligation of all parties to conflicts to respect the rules of international humanitarian law if they resort to means and methods of cyberwarfare, including the principles of distinction, proportionality and precaution'.[169] It should also be noted that the parties to a conflict may always conclude special agreements between themselves to expand their obligations under international humanitarian law.[170] Agreements may be concluded, for instance, to clarify the application of the *jus in bello* to cyber operations in a particular conflict, or to submit to special protection certain data, software and cyber infrastructure.

The problem with the extension of existing rules and principles to new scenarios such as cyber operations is that they do not take into account their uniqueness and might prove to be too general. As a product of the Westphalian order, for instance, existing rules of international law apply to and imply the existence of territory with geographical borders over which states exercise sovereignty or at least jurisdiction, while cyberspace is an apparently borderless, ever changing man-made domain. As has been observed, however, 'components of cyberspace are not immune from territorial sovereignty nor from the exercise of State jurisdiction'.[171] In fact, it should

---

[166] Yves Sandoz, Christophe Swinarski, and Bruno Zimmermann (eds), *Commentary on the Additional Protocols of 8 June 1977 to the Geneva Conventions of 12 August 1949* (Dordrecht: Nijhoff, 1987), para 55.

[167] On this provision, see also Chapter 4, Section II, p 170 ff.

[168] ICRC, *International Humanitarian Law and the Challenges of Contemporary Armed Conflicts*, pp 36–7, <http://www.icrc.org/eng/assets/files/red-cross-crescent-movement/31st-international-conference/31-int-conference-ihl-challenges-report-11-5-1-2-en.pdf>.

[169] UN Doc A/C.1/66/PV.9, 11 October 2011, p 21.

[170] See Art 3(3) Common to the Geneva Conventions; Art 6 of Geneva Conventions I, II, and III; Art 7 of Geneva Convention IV. The text of the Conventions is in UNTS, Vol 75, pp 31 ff, 85 ff, 135 ff, 287 ff.

[171] Wolff Heintschel von Heinegg, 'Territorial Sovereignty and Neutrality in Cyberspace', *International Law Studies* 89 (2013), p 126. See also Eneken Tikk, 'Ten Rules for Cyber Security',

not be forgotten that cyberspace consists of physical and syntactic (or logical) layers: the former includes the physical infrastructure through which the data travel wired or wireless, including servers, routers, satellites, cables, wires, and the computers, while the latter includes the protocols that allow data to be routed and understood, as well as the software used and the data.[172] Cyber operations can thus be seen as 'the reduction of information to electronic format and the actual movement of that information between physical elements of cyber infrastructure'.[173] The internet itself is nothing else than 'a set of inter-connected computer networks linked to state territory and, thus, is liable to the exercise of sovereign jurisdiction on a territorial basis'.[174] Cyber operations, then, can be 'territorialized' by focusing on the location of the cyber infrastructure used to conduct the operations and on where the effects occur.[175] Therefore, '[i]f a cyber action will result in kinetic or kinetic-like effect (e.g., changing the function of a physical system, or file manipulation that results in a financial loss), the target location is the physical location of the effect'.[176] In its 2013 Report, the GGE confirmed that 'State sovereignty and international norms and principles that flow from sovereignty apply to State conduct of ICT-related activities, and to their jurisdiction over ICT infrastructure within their territory'.[177]

2. The role of customary international law

While treaties must be respected only by those states that have ratified them, customary rules are binding on all subjects of international law (with the exception

Survival 53 (June–July 2011), p 121 ('Information infrastructure located within a state's territory is subject to that state's territorial sovereignty'). See Rules 1–3 of the Tallinn Manual, pp 15–23.

[172] David J Betz and Tim Stevens, 'Analogical Reasoning and Cyber Security', *Security Dialogue* 44 (2013), p 151; Jonathan Zittrain, 'A Mutual Aid Treaty for the Internet', Governance Studies at Brookings, 27 January 2011, p 5, <http://www.brookings.edu/research/papers/2011/01/27-internet-treaty-zittrain>. See also Duncan B Hollis, 'Stewardship Versus Sovereignty? International Law and the Apportionment of Cyberspace', CyberDialogue 2012, March 2012, p 7, <http://www.cyberdialogue.citizenlab.org/wp-content/uploads/2012/2012papers/CyberDialogue2012_hollis.pdf>; Johann-Christoph Woltag, 'Computer Network Operations Below the Level of Armed Force', ESIL Conference Paper no 1/2011, pp 16–17, <http://www.esil-sedi.eu/node/82>.

[173] Nils Melzer, *Cyberwarfare and International Law*, UNIDIR, 2011, p 5, <http://www.isn.ethz.ch/Digital-Library/Publications/Detail/?lng=en&id=134218>.

[174] Teresa Scassa and Robert J Currie, 'New First Principles? Assessing the Internet's Challenges to Jurisdiction', *Georgetown Journal of International Law* 42 (2011), p 1079.

[175] On the exercise of the principles of territorial sovereignty and territorial jurisdiction in cyberspace, see Heintschel von Heinegg, 'Territorial Sovereignty', p 134. China has for instance claimed that 'the free flow of information should be guaranteed under the premises that national sovereignty and security must be safeguarded' and that 'each country has the right to manage its own cyberspace in accordance with its domestic legislation' (UN Doc A/61/161, 18 July 2006, p 4). Venezuela has also stated that 'any violation of information security is contrary to the legitimate right of States to full exercise of their sovereignty' (UN Doc A/59/116/Add.1, 28 December 2004, p 6). The United States is exploring ways to define national borders in cyberspace (Scott D Applegate, 'The Principle of Maneuver in Cyber Operations', in *2012 4th International Conference on Cyber Conflict*, edited by Christian Czosseck, Rain Ottis, and Katharina Ziolkowski (CCDCOE, 2012), p 192).

[176] *Joint Terminology for Cyberspace Operations*, p 14.

[177] UN Doc A/68/98, 24 June 2013, p 8.

of local customs and, possibly, the case of persistent objectors).[178] There is no hierarchy between the two sources: treaties can amend or repeal a custom and vice versa, with prevalence determined by principles like *lex posterior derogat priori* and *lex specialis derogat generali* (subsequent and special laws prevail over previous and general laws). Article 38(1) of the ICJ Statute defines customary international law as 'evidence of a general practice accepted as law'. Customary international law, which is generally non-written, is then created by the convergence of two elements: practice (*usus*, or *diuturnitas*) by a sufficiently representative number of states and other subjects of international law (for instance, international organizations) and 'evidence of a belief that this practice is rendered obligatory by the existence of a rule of law requiring it'[179] or, at least, by social, political or economic exigencies (*opinio juris ac necessitatis*).[180]

The role of customary international law in relation to cyber operations is twofold. First, existing *jus ad bellum* and *jus in bello* customary rules extend to cyber operations amounting to a use of force or acts of hostilities, respectively, in the same way as the relevant treaty provisions do: what has been written in the previous Section, then, applies to customary norms as well. From this point of view, '[t]here is no need for State practice to develop separately as regards every concrete weapon employed in an armed attack'.[181] Secondly, it cannot be excluded that customary international law rules specific to cyber warfare might be in the process of forming and eventually ripen. In this regard, more than ten years ago D'Amato predicted that 'computer network attack will soon be the subject of an outright prohibition under customary international law'.[182] Other commentators, however, have been more sceptical and have argued that no customary international law has yet developed because the phenomenon is still too recent and there is no state practice.[183] The Introduction of the Tallinn Manual adopts a more cautious approach and explains that 'because State cyber practice and publicly available expressions of *opinio juris* are sparse, it is sometimes difficult to definitively conclude that any cyber-specific customary international law norm exists'.[184] In order to verify whether these affirmations are correct, one has first to establish

---

[178] Tullio Treves, *Diritto internazionale* (Milano: Giuffrè, 2005), pp 233–5.
[179] *North Sea Continental Shelf (Germany v Denmark/The Netherlands)*, Judgment, 20 February 1969, ICJ Reports 1969 ('*North Sea Continental Shelf*'), para 77; *Military and Paramilitary Activities in and against Nicaragua (Nicaragua v US)*, Merits, Judgment, 27 June 1986, ICJ Reports 1986 ('*Nicaragua*'), para 183; *Nuclear Weapons*, para 64.
[180] Antonio Cassese, *International Law*, 2nd edn (Oxford: Oxford University Press, 2005), p 156
[181] Yoram Dinstein, 'Cyber War and International Law: Concluding Remarks at the 2012 Naval War College International Law Conference', *International Law Studies* 89 (2013), p 280.
[182] Anthony D'Amato, 'International Law, Cybernetics, and Cyberspace', *International Law Studies* 76 (2002), p 69.
[183] See Michael N Schmitt, 'Computer Network Attack and the Use of Force in International Law: Thoughts on a Normative Framework', *Columbia Journal of Transnational Law* 37 (1998–99), p 921, who concludes that '[a] customary norm may develop over time, but it does not exist at present' as '[n]either practice, nor *opinio juris*, is in evidence'; Shackelford, 'From Nuclear War', p 219.
[184] Tallinn Manual, p 5.

what amounts to state practice.[185] If it is indeed impossible to find cyber operations clearly attributable to states, *usus* as an element of custom also includes '[v]erbal acts, and not only physical acts, of States', such as '[d]iplomatic statements (including protests), policy statements, press releases, official manuals (e.g. on military law), instructions to armed forces, comments by governments on draft treaties, legislation, decisions of national courts and executive authorities, pleadings before international tribunals, statements in international organizations and the resolutions those bodies adopt'.[186] When describing state practice, the 2005 ICRC Study of *Customary International Humanitarian Law* also lists 'military manuals, national legislation, national case-law, instructions to armed and security forces, military communiqués during war, diplomatic protests, opinions of official legal advisers, comments by governments on draft treaties, executive decisions and regulations, pleadings before international tribunals, statements in international organizations and at international conferences and government positions taken with respect to resolutions of international organizations'.[187]

Military manuals, in particular, are an important element of state practice.[188] In the *Tadić* case, the International Criminal Tribunal for the former Yugoslavia (ICTY)'s Appeals Chamber famously found that '[w]hen attempting to ascertain State practice with a view to establishing the existence of a customary rule or a general principle, it is difficult, if not impossible, to pinpoint the actual behaviour of the troops in the field for the purpose of establishing whether they in fact comply with, or disregard, certain standards of behaviour'.[189] This is because 'not only is access to the theatre of military operations normally refused to independent observers (often even to the ICRC) but information on the actual conduct of

---

[185] The UN International Law Commission (ILC) has included the formation and evidence of customary international law in its programme of work. In 2013, a First Report was published by the Special Rapporteur, Sir Michael Wood (UN Doc A/CN.4/663, 17 May 2013).

[186] *Statement of Principles Applicable to the Formation of General Customary International Law*, in International Law Association (ILA), Report of the Sixty-Ninth Conference (London, 2000), p 725. See also Jean-Marie Henckaerts and Louise Doswald-Beck, *Customary International Humanitarian Law* (Cambridge: Cambridge University Press, 2005), Vol I, p xxxii; Ian Brownlie, *Principles of Public International Law*, 7th edn (Oxford: Oxford University Press, 2008), pp 6–7; Tullio Treves, 'Customary International Law', in *Max Planck Encyclopedia of Public International Law* (Oxford: Oxford University Press, 2012), Vol II, p 940; Yoram Dinstein, *The Conduct of Hostilities under the Law of International Armed Conflict* (Cambridge: Cambridge University Press, 2010), p 10; Michael Wood, 'State Practice', *Max Planck Encyclopedia of Public International Law* (2012), Vol IX, p 510. As Gray maintains, interpreting state practice means looking at what states say, not necessarily at what they do (Christine Gray, *International Law and the Use of Force* (Oxford: Oxford University Press, 2008), p 418).

[187] Henckaerts and Doswald-Beck, *Customary International Humanitarian Law*, Vol I, p xxxviii. These documents are at the same time state practice and evidence of *opinio juris*: in fact, '[i]t is...often difficult or even impossible to disentangle the two elements' (ILA, *Statement of Principles*, p 718). See also Robert Kolb and Richard Hyde, *An Introduction to the International Law of Armed Conflicts* (Oxford and Portland: Hart, 2008), p 52.

[188] According to Garraway, '[w]hereas international manuals seek to provide an agreed version of the law, national manuals provide evidence of state practice and *opinio juris* in relation to the states by which they are issued' (Charles Garraway, 'The Use and Abuse of Military Manuals', *Yearbook of International Humanitarian Law* 7 (2004), p 431).

[189] *Prosecutor v Tadić*, Case No IT–94–1, Decision on the Defence Motion for Interlocutory Appeal on Jurisdiction, 2 October 1995, para 99.

hostilities is withheld by the parties to the conflict; what is worse, often recourse is had to misinformation with a view to misleading the enemy as well as public opinion and foreign Governments'.[190] These words are even more fitting in the cyber scenario. The Appeals Chamber concluded that '[i]n appraising the formation of customary rules or general principles one should...be aware that, on account of the inherent nature of this subject-matter, reliance must primarily be placed on such elements as official pronouncements of States, military manuals and judicial decisions'.[191] Even if one must be 'cautious not to infuse them with a normative character that may have been unintended by the promulgating States',[192] then, military manuals 'are directly relevant for what states, or more precisely, the armed forces as a state's organ whose practice is relevant for the purposes here discussed, actually do'.[193] Unfortunately, most military manuals have been adopted before 2000 and therefore do not expressly refer to military cyber operations. To the best of this author's knowledge, the only exceptions are the British *Manual of the Law of Armed Conflict*[194] and the US *Commander's Handbook on the Law of Naval Operations*,[195] which only contain cursory references to cyber operations.

On the other hand, a significant number of states have adopted cyber security strategies and doctrines that often contain express and extensive references to international law: as has been observed, 'legal evolution is likely to occur in significant part through defensive planning doctrine and declaratory policies issued in advance of actual cyber-attack crises'.[196] As 'official pronouncements of States', 'policy statements' and 'instructions to armed and security forces', these documents are not only helpful as an assistance in treaty interpretation, but can also be evidence of state practice and could 'declare, and seek to impose on those who are subject to its guidance, a certain *attitude* to the law, or an *interpretation* of the law, or an operational *intent* that relates to existing law either supportively or in

---

[190] *Tadić*, Decision on the Defence Motion, para 99.
[191] *Tadić*, Decision on the Defence Motion, para 99.
[192] Michael N Schmitt, 'The Law of Targeting', in *Perspectives on the ICRC Study on Customary International Humanitarian Law*, edited by Elizabeth Wilmshurst and Susan Breau (Cambridge: Cambridge University Press, 2007), p 134. See also Garraway, 'The Use and Abuse', p 440. According to Post, the position of military manuals in international law 'largely corresponds to that of national legislation, i.e., as having evidentiary value' (Harry HG Post, 'Some Curiosities in the Sources of the Law of Armed Conflict Conceived in a General International Legal Perspective', in *Diversity in Secondary Rules and the Unity of International Law*, edited by Lambertus ANM Barnhoorn and Karel C Wellen (The Hague, Boston, London: Nijhoff, 1995), p 100).
[193] Michael Bothe, 'Comments', in *International Economic Law and Armed Conflict*, edited by Harry HG Post (Dordrecht: Nijhoff, 1994), p 35. See also Yoram Dinstein, 'The Creation of Customary International Law', *Recueil des cours* 322 (2006) 2006, p 272 ('military manuals—published by the high command as binding instructions to the armed forces—constitute meaningful signposts on the road leading to custom-making'); Post, 'Some Curiosities', p 99 ('[m]ilitary manuals of the most powerful nations may certainly be said to have played (and still do play) an important role in the formative process of the customary law of armed conflict').
[194] UK Ministry of Defence, *The Manual of the Law of Armed Conflict* (Oxford: Oxford University Press, 2004), p 118.
[195] *The Commander's Handbook on the Law of Naval Operations*, July 2007, pp 8–17, <http://www.usnwc.edu/getattachment/a9b8e92d-2c8d-4779-9925-0defea93325c/>.
[196] Matthew C Waxman, 'Self-Defensive Force against Cyber Attacks: Legal, Strategic and Political Dimensions', *International Law Studies* 89 (2013), p 116.

some problematic way'.[197] It is true that they mostly reflect policy, and not legal, considerations, but when they expressly refer to international law one cannot see why they should be denied any value: as Matthew Waxman suggests, 'legal analysis and development cannot be divorced from strategy and politics'.[198]

Finally, *usus* also includes official statements made by states, including those in debates in international fora such as the UN organs.[199] As has already been noted, for instance, the UN General Assembly invited the UN member states to submit their views on information security to the Secretary-General. '[O]pinions of official legal advisers' are also a particularly valuable example of verbal acts: a notable case is the speech on international law in cyberspace pronounced by the then US State Department's Legal Advisor, Harold Koh, at the US CYBERCOM.[200]

State practice, however, must be 'extensive and virtually uniform'.[201] True, documents and statements on the legal aspects of military cyber operations come from a relatively limited number of states, but this is not an insurmountable obstacle to the formation of a custom. As Guzman observes, '[f]or many rules of CIL [customary international law], powerful states dominate the question of state practice. The group may grow still smaller once it is recognized that only states with a stake in the issue must be considered'.[202] The ILA Report on the formation of customary international law points out that the extensive character of state practice is more a qualitative than a quantitative criterion: 'if all major interests ("specially affected States") are represented, it is not essential for a majority of States to have participated (still less a great majority, or all of them)'.[203] Specially affected states are primarily those that had the opportunity to engage in the relevant practice. The ICRC Study on *Customary International Humanitarian Law* argues, for instance, that, in relation to the legality of blinding weapons, the specially affected states include those that are developing such weapons.[204] It is, therefore, at the states that have developed military cyber capabilities that one has to mainly look at in order to establish whether any 'general practice accepted as law' has sedimented.

Furthermore, the fact that cyber operations are still a relatively new phenomenon does not necessarily prevent the formation of customary international law. The ICJ famously found that 'the passage of only a short period of time is not necessarily, or of itself, a bar to the formation of a new rule of customary international law'.[205]

---

[197] Alyson JK Bailes and Anna Wetter, 'Security Strategies', in *Max Planck Encyclopedia of Public International Law* (2012), Vol IX, p 87 (emphasis in original).
[198] Waxman, 'Self-Defensive force', p 110.
[199] Wood, 'State Practice', p 512.
[200] Guymon (ed), *Digest of United States Practice*, pp 593 ff.
[201] The ICJ found that 'an indispensable requirement would be that within the period in question, short though it might be, State practice...should have been both extensive and virtually uniform in the sense of the provision invoked;—and should moreover have occurred in such a way as to show a general recognition that a rule of law or legal obligation is involved' (*North Sea Continental Shelf*, para 74).
[202] Andrew T Guzman, 'Saving Customary International Law', *Michigan Journal of International Law* 27 (2005–06), p 151.
[203] ILA, *Statement of Principles*, p 737.
[204] Henckaerts and Doswald-Beck, *Customary International Humanitarian Law*, Vol I, p xxxviii.
[205] *North Sea Continental Shelf*, para 74.

Therefore, '[s]ome customary rules have sprung up quite quickly: for instance, sovereignty over air space, and the régime of the continental shelf, because a substantial and representative quantity of State practice grew up rather rapidly in response to a new situation'.[206] The idea of fast-developing customs, or *diritto spontaneo*, was elaborated by Roberto Ago almost sixty years ago:[207] the unusual rapidity by which certain customary international law rules have crystallized allegedly occurs in periods of fundamental and unprecedented changes, for instance because of technological advances.[208] In such 'Grotian moments', *opinio juris* becomes more important than *usus*.[209] In international humanitarian law, the subordination of practice to *opinio* in relation to norms based on the laws of humanity or the dictates of public conscience may also be inferred from the above-mentioned Martens Clause.[210] The ICTY, for instance, found that the Clause 'clearly shows that principles of international humanitarian law may emerge through a customary process under the pressure of the demands of humanity or the dictates of public conscience, even when State practice is scant or inconsistent. The other element, in the form of *opinio necessitatis*, crystallising as a result of the imperatives of humanity or public conscience, may turn out to be the decisive element heralding the emergence of a general rule or principle of humanitarian law'.[211] Therefore, international humanitarian law customs may arise even in the absence of extensive and uniform operational state practice, providing that a significant number of specially affected states have expressed their legal views on the matter.[212] A not too dissimilar approach was adopted by the ICJ when it founded the customary nature of certain treaty provisions of international humanitarian law on 'elementary considerations of humanity', without accompanying this view with conclusive evidence of state practice.[213]

---

[206] ILA, *Statement of Principles*, p 731.

[207] Roberto Ago, 'Science juridique et droit international', *Recueil des cours* 90 (1956–II), pp 931 ff.

[208] Michael P Scharf, 'Seizing the "Grotian Moment": Accelerated Formation of Customary International Law in Times of Fundamental Change', *Cornell International Law Journal* 43 (2010), pp 444, 450.

[209] Scharf, 'Seizing', p 468.

[210] Cassese, 'The Martens Clause', p 214. See also Dieter Fleck, 'State Responsibility Consequences of Termination of or Withdrawal from Non-proliferation Treaties', in *Non-proliferation Law as a Special Regime*, edited by Daniel H Joyner and Marco Roscini (Cambridge: Cambridge University Press, 2012), p 259; Theodor Meron, 'The Martens Clause, Principles of Humanity, and Dictates of Public Conscience', *American Journal of International Law* 94 (2000), pp 87–8.

[211] *Prosecutor v Kupreskić*, Case no IT-95-16-T, Trial Chamber Judgment, 14 January 2000, para 527. See Cassese, 'The Martens Clause', p 214; Robert Kolb, 'Selected Problems in the Theory of Customary International Law', *Netherlands International Law Review* 50 (2003), p 124.

[212] Cassese, 'The Martens Clause', p 214; Meron, 'The Martens Clause', p 88. It has been argued that this applies not only to international humanitarian law, but also to the rules on the use of force, 'where the practice is difficult to weigh, as much for what is done as for what is not done' (Kolb, 'Selected Problems', p 129).

[213] See eg *Nicaragua*, para 218 (with regard to Common Art 3 of the 1949 Geneva Conventions on the Protection of Victims of War). See the comments of Giulio Bartolini, 'Armed Forces and the International Court of Justice: The Relevance of International Humanitarian Law and Human Rights Law to the Conduct of Military Operations', in *Armed Forces and International Jurisdictions*, edited by Marco Odello and Francesco Seatzu (Cambridge, Antwerp, Portland: Intersentia, 2013), pp 61–2.

It can be concluded that 'the prevailing position continues to demand fulfilment of the classic two elements of State practice and *opinio juris* [but] there is also a clear tendency not to follow the two elements as strictly as originally envisaged'.[214] In particular, and in spite of some isolated, if influential, contrary views,[215] it is now generally accepted that practice can consist not only of actions, but also of verbal acts, and that the subjective element could be decisive in the formation of customs, especially in the case of prohibitory rules of international humanitarian law.[216] Of course, stating that customary international law specific to cyber operations has already formed exclusively on the basis of cyber security strategies, a few military manuals and a limited number of unattributed cyber attacks would certainly be an exaggeration. At least some uniform operational practice, in addition to verbal acts, seems necessary to avoid natural law setbacks.[217] This, however, does not mean that verbal acts could not indicate *trends* of the direction towards which customary international law is starting to develop in this area, trends that it is useful to identify also from the perspective of a future, if still uncertain, treaty regulating cyber warfare. It is in this light that the present book will examine the above-mentioned documents.

3. The *Tallinn Manual on the International Law Applicable to Cyber Warfare*

If, therefore, existing international law applies in the cyber context, the lawyer's task is to examine the traditional norms, conceived in relation to kinetic scenarios, and identify potential difficulties in their application to different types of cyber operations. It is from this perspective that NATO's CCDCOE invited a group of experts to prepare the *Tallinn Manual on the International Law Applicable to Cyber Warfare*, published in early 2013.[218] The Manual, that aims to identify how the *lex lata* applies to cyber operations above the level of the use of force, includes a set of 95 Rules accompanied by commentaries and does not reflect NATO doctrine or

---

[214] Robert Heinsch, 'Methodology of Law-Making. Customary International Law and New Military Technologies', in *International Humanitarian Law and the Changing Technology of War*, edited by Dan Saxon (Leiden: Brill, 2013), p 36.

[215] See the response of the United States to the ICRC Study, according to which 'the Study places too much emphasis on written materials, such as military manuals and other guidelines published by States, as opposed to actual operational practice by States during armed conflict' (John B Bellinger, III and William J Haynes, II, 'A US Government Response to the International Committee of the Red Cross Study *Customary International Humanitarian Law*', *International Review of the Red Cross* 89 (2007), p 445). The United States, however, does not deny that verbal acts can amount to state practice but only that they can replace operational practice, and actually recognizes that military manuals are 'important indications of State behavior and *opinio juris*' (p 445).

[216] Heinsch, 'Methodology', pp 25–6.   [217] Heinsch, 'Methodology', p 35.

[218] Tallinn Manual, p 5. The CCDCOE has also published a National Cyber Security Framework Manual, which focuses on law enforcement in peacetime (Alexander Klimburg (ed), *National Cyber Security Framework Manual* (CCDCOE, 2012), <http://ccdcoe.org/369.html>).

the official position of any state or organization.[219] It is essentially a scholarly exercise and its rules are of course not binding.[220]

The Manual has been criticized in relation to the composition of the Group of Experts, the methodology employed, its scope, and certain aspects of its contents.[221] The Group of Experts that drafted the Manual comprised international law academics, practitioners, serving or former military officials, technical experts, as well as observers from NATO, the ICRC, and the US CYBERCOM, all participating in their personal capacity. It included, however, only 'military and academic lawyers and technical experts from but a few Western states'.[222] It is indeed a fact that, of the 23 members of the Group of Experts, nine (including the Project's Director) were from the United States, while none was from states that are reportedly heavily involved in cyber operations, both as authors and targets, such as Russia, China, Iran, or Israel.[223] If this can certainly be seen as a limitation, it should not be forgotten that the members participated in the initiative in their individual capacity: even if a Russian expert had been invited, he or she would have not necessarily expressed the views of the Russian government.

As to the methodology employed, only the conclusions on which unanimity among the Group of Experts (but not the observers) was reached were translated into black-letter rules: the most controversial international law aspects of cyber operations were therefore left unresolved, although the divergent positions were noted in the Commentary.[224] Overall, it seems fair to say that the Experts were very cautious to avoid taking any risks when drafting the rules, which are often a mere restatement of existing treaty provisions with the addition of the adjective 'cyber'.[225] In addition to the relevant treaties, the sources used by the Group of Experts include the ICJ jurisprudence as well as the case law of international criminal tribunals, in particular the ICTY, and the works of the ILC. The Manual also heavily relies on the ICRC Commentaries to the Geneva Conventions and Additional Protocols and on more or less successful private codifications, such as the above-mentioned HPCR Manual, the *San Remo Manual on International Law Applicable to Armed Conflicts at Sea*,[226] and the *Manual on the Law of*

---

[219] Tallinn Manual, p 11. Although it took part in the drafting of the Manual, in particular, the ICRC did not endorse all the views expressed therein (Cordula Droege, 'Get Off My Cloud: Cyber Warfare, International Humanitarian Law, and the Protection of Civilians', *International Review of the Red Cross* 94 (2012), p 541).
[220] Tallinn Manual, p 11.
[221] Dieter Fleck, 'Searching for International Rules Applicable to Cyber Warfare—A Critical First Assessment of the New *Tallinn Manual*', *Journal of Conflict and Security Law* 18 (2013), pp 331 ff.
[222] Fleck, 'Searching for International Rules', p 335.
[223] See the critical comments of Rain Liivoja and Tim McCormack, 'Law in the Virtual Battlespace: The Tallinn Manual and the *Jus in Bello*', Melbourne Legal Studies Research Paper no 650, 23 July 2013, pp 4, 12, <http://papers.ssrn.com/sol3/papers.cfm?abstract_id=2297159>.
[224] Tallinn Manual, pp 5–6. See the critical review of Fleck, 'Searching for International Rules', pp 336 ff.
[225] The Manual itself acknowledges that '[a]t times, the text of a Rule closely resembles that of an existing treaty norm' (Tallinn Manual, p 6).
[226] *San Remo Manual on International Law Applicable to Armed Conflicts at Sea, Prepared by a Group of International Lawyers and Naval Experts Convened by the International Institute of Humanitarian*

Non-International Armed Conflict,[227] as well as on the ICRC Study on *Customary International Humanitarian Law*, although in a 'persuasive, but not dispositive' function.[228] With regard to national sources, the Manual's Commentary essentially refers to the military manuals of only four states (Germany, Canada, the United Kingdom, and the United States) on the basis that they are considered 'especially useful', that some members of the Group of Experts participated in their drafting and that they are publicly available.[229] This very narrow selection, however, should have been more extensively justified, and in any case other manuals would have met the identified selection criteria. Finally, the Manual refers to only one cyber security strategy, the 2011 White House's *International Strategy for Cyberspace*, and overlooks the many others that have been adopted, which are often more explicit and more significant.

The Manual only briefly addresses, or does not address at all, important issues.[230] In particular, there is little analysis of how the principle of non-intervention applies to cyber operations.[231] This is particularly troublesome if one considers that the Group of Experts was not able to conclusively establish the threshold for cyber operations to be considered a use of force. There is also little discussion of cyber exploitation operations, even though they could also qualify as acts of hostilities.[232] Furthermore, the Manual does not discuss at length crucial problems such as attribution criteria and evidentiary standards.[233] On the other hand, it is not clear whether Rule 24, an essentially international criminal law provision, was really necessary in a *jus ad bellum/jus in bello* codification.[234]

Only time will tell whether the Tallinn Manual will be as successful as the *San Remo Manual on Armed Conflict at Sea* in influencing state conduct. Although, as the Commentary itself acknowledges, 'any claim that every assertion in the Manual represents an incontrovertible restatement of international law would be an exaggeration',[235] the Manual is, in any case, a good starting point for further analysis and should be commended for advancing the understanding of the international law applicable to cyber warfare. The present book will therefore often refer to it.

Law. Text in Adam Roberts and Richard Guelff, *Documents on the Laws of War* (Oxford: Oxford University Press, 2000), pp 573 ff.

[227] Michael N Schmitt, Charles HB Garraway, and Yoram Dinstein, *The Manual on the Law of Non-International Armed Conflict With Commentary* (Sanremo: International Institute of Humanitarian Law, 2006), <http://www.iihl.org/iihl/Documents/The%20Manual%20on%20the%20Law%20of%20NIAC.pdf>.

[228] Tallinn Manual, p 8.   [229] Tallinn Manual, p 8.

[230] Fleck, 'Searching for International Rules', pp 346 ff.

[231] See Tallinn Manual, pp 44–5. The principle of non-intervention in the cyber context is discussed below, Chapter 2, Section II.1.3.

[232] See Tallinn Manual, pp 192–5. See Chapter 4, Section IV of this book.

[233] The only references to evidence are contained in Rules 7 and 8 (Tallinn Manual, pp 34–6). See Section IV of this Chapter and Section III.6 of Chapter 2.

[234] Tallinn Manual, p 91.   [235] Tallinn Manual, p 5.

IV. Identification and Attribution Problems

Well before the cyber age, in the *Nicaragua* Judgment the ICJ conceded that 'the problem is not...the legal process of imputing the act to a particular State...but the prior process of tracing material proof of the identity of the perpetrator'.[236] These difficulties, however, are even more evident in the cyber context, where identifying who is behind a cyber operation presents significant technical problems. Anonymity is in fact one of the greatest advantages of cyberspace. The internet, in particular, is a decentralized system where the communications protocol divides the sent data into several packets that take different unpredictable pathways to reach their destination before being reassembled.[237] An IP address identifies the origin and the destination of the data: with the cooperation of the Internet Service Provider (ISP) through which the system corresponding to the IP address is connected to the internet, it could be associated with a person, group, or state. The IP address, however, may have been 'spoofed', or the corresponding computer system may only be a 'stepping stone' for an attacker located elsewhere.[238]

Nonetheless, the challenges in the identification of the attackers should not be an excuse not to tackle the international legal aspects of cyber operations. After all, identifying the authors of hostile actions is a problem also in other contexts, for instance international terrorism: as the United States declared, the ambiguities of cyberspace 'simply reflect the challenges in applying the [UN] Charter framework that already exists [*sic*] in many contexts'.[239] It is also not impossible that the author of a cyber operation is eventually identified: traditional intelligence gathering and cyber exploitation, used in support of traceback technical tools, could be helpful instruments in this sense.[240] Further developments in computer technology and internet regulations, such as the introduction of the new internet protocol IPv6, might also make identification easier.[241]

---

[236] *Nicaragua*, para 57.
[237] As has been effectively observed, 'the internet is one big masquerade ball. You can hide behind aliases, you can hide behind proxy servers, and you can surreptitiously enslave other computers to do your dirty work' (Joel Brenner, *America The Vulnerable: Inside the New Threat Matrix of Digital Espionage, Crime, and Warfare* (New York: Penguin Press, 2011), p 32).
[238] Scott J Shackelford and Richard B Andres, 'State Responsibility for Cyber Attacks: Competing Standards for a Growing Problem', *Georgetown Journal of International Law* 42 (2011), p 982. The 1998 'Solar Sunrise' attack that broke into the US DoD's system was for instance carried out by an Israeli teenager and Californian students through a computer based in the United Arab Emirates (Shackelford, 'From Nuclear War', p 204).
[239] UN Doc A/66/152, 15 July 2011, p 18.
[240] Nicholas Tsagourias, 'Cyber Attacks, Self-Defence and the Problem of Attribution', *Journal of Conflict and Security Law* 17 (2012), p 234; Owens, Dam, and Lin, *Technology*, pp 140–1; Advisory Council on International Affairs/Advisory Committee on Issues of Public International Law, *Cyber Warfare*, no 77, AIV/No 22 CAVV, December 2001, p 15, <http://www.aiv-advies.nl/ContentSuite/upload/aiv/doc/webversie__AIV77CAVV_22_ENG.pdf>. See the traceback technology described in Jay P Kesan and Carol M Hayes, 'Mitigative Counterstriking: Self-Defense and Deterrence in Cyberspace', *Harvard Journal of Law and Technology* 25 (2011–12), pp 482 ff. The US DoD is apparently seeking to improve attribution capabilities through behaviour-based algorithms (US DoD, *Cyberspace Policy Report*, p 4).
[241] Dinstein, 'Computer Network Attacks', p 112.

Assuming that the authors of a cyber operation are eventually identified, the problem arises as to whether their conduct may be attributed to a state under the law of state responsibility. If identification is essentially a technical matter, attribution is a legal exercise and is 'the key to understanding the motive of an attack and consequently being able to differentiate between a criminal act and warfare in cyberspace'.[242] The above-mentioned 2013 Report of the GGE confirmed that 'States must meet their international obligations regarding internationally wrongful acts attributable to them' in the cyber context.[243] Although it is not entirely implausible that a special regime of international responsibility will develop as a consequence of the unique features of cyber operations, in the present lack of any indications in that sense such conclusion would certainly be premature.[244] The applicable rules are, therefore, those contained in Chapter II of Part One of the Articles on the Responsibility of States for Internationally Wrongful Acts, adopted by the ILC in 2001 and subsequently endorsed by the UN General Assembly ('ILC Articles'), which substantially reflect customary international law.[245]

Several scenarios can be identified. The first and easiest one is the case of 'uniformed' hackers. According to Article 4(1) of the ILC Articles, '[t]he conduct of any State organ shall be considered an act of that State under international law, whether the organ exercises legislative, executive, judicial or any other functions, whatever position it holds in the organization of the State, and whatever its character as an organ of the central Government or of a territorial unit of the State'. Article 4(2) specifies that '[a]n organ includes any person or entity which has that status in accordance with the internal law of the State'. Although details of states' military cyber capabilities are often classified, it appears that several national armies have established cyber units.[246] To the extent that they are organs of a state, their conduct is attributable to that state. This conclusion would not change if the hackers were civilian, and not military, organs. In the United Kingdom, for instance, the Global Operations and Security Control Centre (GOSCC), whose role is 'to proactively and reactively defend MoD [Ministry of Defence] networks 24/7 against cyber attack to enable agile exploitation of MoD information capabilities across all areas of the Department's operations', is formed not only of members of the military but also by MoD civilian and contractor personnel from industry partners, although only military members can be sent to operational theatres.[247] It also seems that the alleged US cyber operations against Iran were

---

[242] Eleanor Keymer, 'The cyber-war', *Jane's Defence Weekly* (47/39, 29 September 2010), p 22.
[243] UN Doc A/68/98, 24 June 2013, p 8.
[244] Article 55 of the ILC Articles provides that '[t]hese articles do not apply where and to the extent that the conditions for the existence of an internationally wrongful act or the content or implementation of the international responsibility of a State are governed by special rules of international law'.
[245] Read the text of the Articles in *Yearbook of the International Law Commission*, 2001, Vol II, Part Two, pp 26–30. On attribution to international organizations, see the Draft Articles on the Responsibility of International Organizations, adopted by the ILC in 2011, in *Yearbook of the International Law Commission*, 2011, Vol II, Part Two, pp 54 ff.
[246] See Section I of this Chapter p 10.
[247] House of Commons Defence Committee, *Defence and Cyber-Security*, Sixth Report of Session 2012–13, Vol I, 18 December 2012, p 17.

'run by intelligence agencies, though many techniques used to manipulate Iran's computer controllers would be common to a military program'.[248]

A state is responsible not only for the conduct of its *de jure* organs, but also of those individuals that are 'completely dependent' on state authorities and can thus be considered *de facto* organs, even if they do not have that status according to the internal law of the relevant state. The complete dependency test was first referred to by the ICJ in *Nicaragua* and then developed in the *Genocide* Judgment.[249] As has been observed, in this exceptional situation 'the reason why a connection between a state and a de facto organ must be intense is that the mere identity of the actor as a state organ suffices for attribution to occur'.[250]

Furthermore, the hackers could be members of parastatal entities, public, semi-public or privatized corporations empowered by internal law to exercise some degree of governmental authority on behalf of state organs:[251] in this case, their conduct will be attributed to the state 'provided the person or entity is acting in that capacity in the particular instance', as stated in Article 5 of the ILC Articles. The notion of 'governmental authority' is intentionally left undefined in Article 5 as it 'depends on the particular society, its history and traditions'.[252] It seems, however, to be a notion broad enough to include both the defence of 'national' portions of cyberspace and the conduct of intelligence gathering or offensive cyber operations by the individual or the entity. Unlike state agents under Article 8 of the ILC Articles, attribution under Article 5 does not require that the acts be committed under the 'effective control' of state authorities or within those limits, as long as internal law has delegated certain governmental functions to the individual or entity in question.[253] National Computer Emergency Response Teams (CERTs), which provide 'initial emergency response aid and triage services to the victims or potential victims of cyber operations or cyber crimes, usually in a manner that involves coordination between the private sector and government entities',[254] are an example of entities authorized to exercise governmental authority in the cyber context.[255] Another example is the Cyber Unit of the Estonian Defence League, which is 'a force of programmers, computer scientists and software engineers...a volunteer organization that in wartime would function under a unified

---

[248] David E Sanger, 'U.S. Blames China's Military Directly for Cyberattacks', *The New York Times*, 6 May 2013, <http://www.nytimes.com/2013/05/07/world/asia/us-accuses-chinas-military-in-cyberattacks.html?pagewanted=all&_r=1&>.

[249] *Nicaragua*, paras 110, 393; *Application of the Convention on the Prevention and Punishment of the Crime of Genocide (Bosnia and Herzegovina v Serbia and Montenegro)*, Merits, Judgment, 26 February 2007, ICJ Reports 2007 ('*Genocide*'), paras 392–3. It is controversial whether *de facto* organs should be ascribed to Art 4 or Art 8 of the ILC Articles.

[250] Marko Milanović, 'State Responsibility for Acts of Non-state Actors: A Comment on Griebel and Plücken', *Leiden Journal of International Law 22* (2009), p 315.

[251] *Yearbook of the International Law Commission*, 2001, Vol II, Part Two, p 42.

[252] *Yearbook of the International Law Commission*, 2001, Vol II, Part Two, p 43.

[253] *Yearbook of the International Law Commission*, 2001, Vol II, Part Two, p 43.

[254] Tallinn Manual, p 258.

[255] Michael N Schmitt, 'International Law in Cyberspace: The Koh Speech and Tallinn Manual Juxtaposed', *Harvard International Law Journal Online* 54 (2012), p 35, <http://www.harvardilj.org/2012/12/online-articles-online_54_schmitt>.

military command'.[256] The Cyber Unit, which protects Estonia's information infrastructure and supports broader objectives of national defence, cooperates in emergency situations with the Estonian CERT to respond to cyber attacks but does not have contractual obligations or payments from the government.[257]

The conduct of organs and of persons or entities empowered to exercise elements of the governmental authority is attributable to the relevant state even if they exceed their authority or contravene the instructions received, providing they act in their official capacity (Article 7 of the ILC Articles). In case of covert operations like cyber operations, however, '[t]he distinction between *ultra vires* and purely private conduct is particularly problematic'.[258] In such cases, it has been suggested that attribution will require that 'the state organ was acting in its *actual* (rather than *apparent*) official capacity'.[259] It is worth pointing out that if, in the case of an individual who is an organ, attribution to a state is avoided if he was acting in a purely private capacity (ie not as an organ), in the case of entities which are organs their conduct is, in practice, always attributable, even if *ultra vires*, since there is no private capacity. In the case of entities, the examination of whether they were acting in the exercise of the relevant governmental authority may be coterminous with the question of whether they were within the scope of their powers. It should also be recalled that Article 91 of the 1977 Protocol I Additional to the Geneva Conventions on the Protection of Victims of War make clear that a belligerent state 'shall be responsible for *all* acts committed by persons forming part of its armed forces', including those committed in a personal capacity, providing they are unlawful under the *jus in bello*.[260]

The hackers could also be private individuals or corporations instructed by states to conduct specific cyber operations.[261] A well-known example is the Russian Business Network (RBN), a cyber crime firm specializing in phishing, malicious code, botnet command-and-control, DDoS attacks and identity theft, which is suspected of having executed the cyber operations against Georgia on behalf of Russia.[262] The existence of Iranian hackers working for the Revolutionary Guard's

---

[256] Tom Gjelten, 'Volunteer Cyber Army Emerges in Estonia', NPR News, 4 January 2011, <http://www.npr.org/2011/01/04/132634099/in-estonia-volunteer-cyber-army-defends-nation>. See also Shackelford and Andres, 'State Responsibility', p 1009.

[257] Shackelford and Andres, 'State Responsibility', p 1009.

[258] Kimberley N Trapp, *State Responsibility for International Terrorism* (Oxford: Oxford University Press, 2011), p 35.

[259] Trapp, *State Responsibility*, p 35 (emphasis in the original).

[260] Article 91, Additional Protocol I (emphasis added). The provision must be read in conjunction with Art 43 of the Protocol, that defines 'armed forces'. Article 3 of Hague Convention IV has a virtually identical formulation. See Marco Sassòli, 'State Responsibility for Violations of International Humanitarian Law', *International Review of the Red Cross* 84 (2002), pp 405–6.

[261] Jonathan A Ophardt, 'Cyber Warfare and the Crime of Aggression: The Need for Individual Accountability on Tomorrow's Battlefield', *Duke Law and Technology Review* 3 (2010), paras 12–18, <http://www.law.duke.edu/journals/dltr/articles/pdf/2010dltr003.pdf>. Such corporations are allegedly paid by governments to carry out elements of the cyber attacks (Watts, 'Combatant Status', p 411).

[262] Tikk, Kaska, Rünnimeri, Kert, Talihärm, and Vihul, *Cyber Attacks Against Georgia*, p 11; Klimburg, 'Mobilising Cyber Power', pp 49–50; John Markoff, 'Before the Gunfire, Cyberattacks', *The New York Times*, 12 August 2008, <http://www.nytimes.com/2008/08/13/technology/13cyber.html>.

paramilitary Basij group and including 'university instructors and students, as well as clerics' has also been reported.[263] Article 8 of the ILC Articles deals with state agents and provides that '[t]he conduct of a person or group of persons shall be considered an act of a State under international law if the person or group of persons is in fact acting on the instructions of, or under the direction or control of, that State in carrying out the conduct'. In the *Nicaragua* case, the ICJ argued that 'United States participation, even if preponderant or decisive, in the financing, organizing, training, supplying and equipping of the *contras*, the selection of its military or paramilitary targets, and the planning of the whole of its operation, is still insufficient in itself...for the purpose of attributing to the United States the acts committed by the *contras* in the course of their military or paramilitary operations in Nicaragua': what has to be proved is that 'that State had effective control of the military or paramilitary operation in the course of which the alleged violations were committed'.[264] In the *Genocide* case, the ICJ returned to the point and clarified that '[i]t must...be shown that this "effective control" was exercised, or that the State's instructions were given, in respect of each operation in which the alleged violations occurred, not generally in respect of the overall actions taken by the persons or groups of persons having committed the violations'.[265]

According to the ICTY, however, '[t]he degree of control may...vary according to the factual circumstances of each case'.[266] Doubting the consistency of the ICJ's effective control test in *Nicaragua* with the 'logic' of the law of state responsibility,[267] the Tribunal adopted a much less restrictive test to attribute the conduct of militarily organized armed groups to a state. Under the ICTY 'overall' control test, for the actions of such groups to engage state responsibility it is sufficient that the state '*has a role in organising, coordinating or planning the military actions* of the military group, in addition to financing, training and equipping or providing operational support to that group...regardless of any specific instructions by the controlling State concerning the commission of each of those acts'.[268] As has been noted, 'the overall control is not control over the act, but over the actor, an organized and hierarchically structured group, at a general level'.[269] Unlike the 'effective control' test, then, the *Tadić* standard focuses on the 'general influence' that a state exercises over a group, and not on specific activities, but, unlike the complete dependency test, it is much less stringent.[270] In the *Genocide* Judgment,

---

[263] Nasser Karimi, 'Iran's paramilitary launches cyber attacks', *The Associated Press*, 14 March 2011, <http://www.washingtonpost.com/wp-dyn/content/article/2011/03/14/AR2011031401029.html?referrer=emailarticle>.
[264] *Nicaragua*, para 115.   [265] *Genocide*, para 400.
[266] ICTY, *Prosecutor v Tadić*, Case No IT-94-1-A, Appeals Chamber Judgment, 15 July 1999, para 117 (emphasis omitted).
[267] *Tadić*, Appeals Chamber Judgment, paras 116 ff.
[268] *Tadić*, Appeals Chamber Judgment, para 137 (emphasis in the original). The Court added that 'if, as in *Nicaragua*, the controlling State *is not the territorial State* where the armed clashes occur or where at any rate the armed units perform their acts, more extensive and compelling evidence is required to show that the State is genuinely in control of the units or groups not merely by financing and equipping them, but also by generally directing or helping plan their actions' (para 138; emphasis in the original).
[269] Milanović, 'State Responsibility', p 317.   [270] Tsagourias, 'Cyber Attacks', p 238.

the ICJ rejected overall control as an attribution standard by noting that it 'has the major drawback of broadening the scope of State responsibility well beyond the fundamental principle governing the law of international responsibility: a State is responsible only for its own conduct, that is to say the conduct of persons acting, on whatever basis, on its behalf'.[271]

It has been suggested that, due to the inherently clandestine nature of cyber activities and the technical difficulty of identifying the authors, the *Tadić* test should be preferred to the *Nicaragua* test when cyber operations are concerned.[272] This view mixes standard of proof with attribution criteria[273] and cannot be shared: indeed, it is exactly because of the identification problems characterizing cyber activities and the potential for abuse of the right of self-defence that the 'effective control' test is preferable, as it would prevent states from being frivolously or maliciously accused of cyber operations. The above-mentioned view also misses an important point: the ICTY applies the overall control test only to the case of an 'organised and hierarchically structured group, such as a military unit or, in case of war or civil strife, armed bands of irregulars or rebels'.[274] For the case of a 'private individual who is engaged by a State to perform some specific illegal acts in the territory of another State (for instance, ... carrying out acts of sabotage)' and of unorganized, non-military and non-hierarchical groups of individuals (which would arguably include groups such as RBN or 'Anonymous'), the ICTY retains the effective control test, ie the need to prove the issue of specific instructions concerning the commission of that illegal act or the state's public retroactive approval of the actions.[275] With specific regard to cyber operations, then, there is no substantial practical discrepancy between the ICJ and the ICTY approaches: both would probably lead in most cases to the application of the effective control test, as 'organised and hierarchically structured' cyber groups do not seem to exist yet.[276] Clear support for the application of the effective control test to cyber operations can also be found in the speech given by the then US State Department's Legal Advisor, Harold Koh, at the US CYBERCOM, where he claims that states are internationally responsible for cyber acts undertaken through 'proxy actors' when they 'act on the State's instructions or under its direction or control'.[277] Azerbaijan also denounced cyber

---

[271] *Genocide*, para 406.
[272] Shackelford, 'From Nuclear War', p 235; Shackelford and Andres, 'State Responsibility', pp 987–8. See also Ryan, Dion, Tikk, and Ryan, 'International Cyberlaw', p 1187.
[273] See, eg, Shackelford and Andres, 'State Responsibility', p 990.
[274] *Tadić*, Appeals Chamber Judgment, para 120 (emphasis omitted).
[275] *Tadić*, Appeals Chamber Judgment, para 118.
[276] It seems, however, that members of Al-Qaeda have conducted low-intensity cyber operations against the United States (Vijay M Padmanabhan, 'Cyber Warriors and the *Jus in Bello*', *International Law Studies* 89 (2013), p 296). Certain armed groups, such as Hamas and Hezbollah, may have also hired cyber criminals in order to conduct cyber operations (James A Lewis, 'The "Korean" Cyber Attacks and Their Implications for Cyber Conflict' (Center for Strategic and International Studies, 2009), p 8, <http://csis.org/files/publication/091023_Korean_Cyber_Attacks_and_Their_Implications_for_Cyber_Conflict.pdf>).
[277] Guymon (ed), *Digest of United States Practice*, p 596.

also denounced cyber attacks conducted by a group of hackers called the 'Armenian Cyber Army' under the 'direction and control' of Armenia.[278]

Hackers could be neither state organs nor state agents, but their conduct could have been incited by state authorities. In 2001, for example, after a US Navy spy plane collided with a Chinese jet fighter in the South China Sea, websites appeared offering instructions to hackers on how to incapacitate US government computers.[279] It also appears that the Russian government might have encouraged 'patriotic hackers' to conduct the 2007 cyber attacks against Estonia.[280] Russian language blogs, forums, and websites also published instructions on how to overwhelm Georgian government websites as well as a list of vulnerable Georgian websites.[281] There is no express regulation of incitement in the ILC Articles on State Responsibility.[282] Incitement would thus entail state responsibility for the incited actions only to the extent it amounts to direction and control (Article 8).[283] After inciting the actions, however, state authorities may subsequently publicly endorse them: in the *Hostages* case, the ICJ found that, although the initial attack on the US Embassy in Tehran was not attributable to Iran, the subsequent adoption of the action by the Iranian authorities as their own and the decision to perpetuate the occupation transformed the occupation and detention of the hostages into acts of the state.[284] Article 11 of the ILC Articles confirms that '[c]onduct which is not attributable to a State under the preceding articles shall nevertheless be considered an act of that State under international law if and to the extent that the State acknowledges *and* adopts the conduct in question as its own'.[285] It is true that '[a]cknowledgement and adoption of conduct by a State might be express (as for example in the *United States Diplomatic and Consular Staff in Tehran* case), or it might be inferred from the conduct of the State in question',[286] but acknowledgement and adoption of cyber operations by a state are unlikely to occur: as already noted, cyber capabilities are the perfect tool for covert operations and one of their

---

[278] Letter dated 6 September 2012 from the Chargé d'affaires a.i. of the Permanent Mission of Azerbaijan to the United Nations addressed to the Secretary-General, UN Doc A/66/897–S/2012/687, 7 September 2012, p 1.

[279] Noah Weisbord, 'Conceptualizing Aggression', *Duke Journal of Comparative and International Law* 20 (2009), p 20.

[280] Catherine Lotrionte, 'Active Defense for Cyber: A Legal Framework for Covert Countermeasures', in *Inside Cyber Warfare*, edited by Jeffrey Carr, 2nd edn (Sebastopol, CA: O'Reilly, 2012), p 282.

[281] Tikk, Kaska, Rünnimeri, Kert, Talihärm, and Vihul, *Cyber Attacks Against Georgia*, pp 9–10.

[282] Incitement, however, is dealt with in the Commentary to Part One, Chapter IV of the ILC Articles (*Yearbook of the International Law Commission*, 2001, Vol II, Part Two, p 65). When expressly provided, incitement can be an unlawful act per se (see eg Art III of the 1948 Convention on the Prevention and Punishment of the Crimes of Genocide, text in UNTS, Vol 78, pp 277 ff).

[283] See *Yearbook of the International Law Commission*, 2001, Vol II, Part Two, p 65.

[284] *US Diplomatic and Consular Staff in Tehran (US v Iran)*, Judgment, 24 May 1980, ICJ Reports 1980, para 74.

[285] Emphasis added. According to the Commentary, 'acknowledgement and adoption' should be distinguished from 'mere support or endorsement': 'as a general matter, conduct will not be attributable to a State under Article 11 where a State merely acknowledges the factual existence of conduct or expresses its verbal approval of it.... The language of "adoption", on the other hand, carries with it the idea that the conduct is acknowledged by the State as, in effect, its own conduct' (*Yearbook of the International Law Commission*, 2001, Vol II, Part Two, p 53).

[286] *Yearbook of the International Law Commission*, 2001, Vol II, Part Two, p 54.

main advantages is exactly that the author can hide under the invisibility cloak of plausible deniability.

Finally, it could be that the cyber operations originate from computer systems located in a certain state or from the cyber infrastructure of a state without any state involvement whatsoever, as in the case of 'hacktivists' and 'patriotic hackers' willing to support a certain political cause. In such case, the hackers' conduct could not be imputed to the state of origin, which may, however, be held responsible for not taking the necessary and reasonable measures to prevent or stop the operations (for instance, by disabling the internet access of the perpetrators or updating the country's firewall settings). In spite of what some commentators have argued,[287] then, the state's wrongful act would not be the cyber operation, but the breach of its obligation 'not to allow knowingly its territory to be used for acts contrary to the rights of other States'.[288] It appears, for instance, that, even though no evidence was found of state organs directing the attacks, Russia at least tolerated the cyber operations against Estonia and Georgia originating from Russian hacker websites.[289] Russia also did not cooperate with Estonia in tracking down those responsible, and a request for bilateral investigation under the Mutual Legal Assistance Treaty between the two countries was rejected by the Russian Supreme Procuratura.[290] Whether the state victim of a cyber operation amounting to an armed attack may invoke self-defence if the operation is attributable to non-state actors and originates from the territory of a state that is unable or unwilling to prevent or terminate it is a question that will be explored in Chapter 2.[291]

V. The Book's Scope and Purpose

In light of the above, it should be clear that existing primary and secondary rules of international law, including the law of state responsibility, the *jus ad bellum* and the *jus in bello*, do apply to cyber operations. It is, however, more controversial *when* and *how* such rules apply to events that are very different from kinetic

---

[287] See eg David E Graham, 'Cyber Threats and the Law of War', *Journal of National Security Law and Policy* 4 (2010), p 93; Sklerov, 'Solving the Dilemma', p 49; Ryan, Dion, Tikk, and Ryan, 'International Cyberlaw', p 1188.

[288] *Corfu Channel (United Kingdom v Albania)*, Merits, Judgment, 9 April 1949, ICJ Reports 1949, p 22. The obligation is reflected in Rule 5 of the Tallinn Manual, p 26. GA Res 55/63 of 4 December 2000 on the criminal misuse of information technologies recommends that states ensure 'that their laws and practice eliminate safe havens for those who criminally misuse information technologies' (para 1). On due diligence in the cyber context, see Chapter 2, Section III.3.

[289] Tikk, Kaska, Rünnimeri, Kert, Talihärm, and Vihul, *Cyber Attacks Against Georgia*, p 13. Another report claims that Russia refused to intervene with regard to the hacker attacks against Georgia in 2008 (Project Grey Goose, *Russia/Georgia Cyber* War, p 8). It has been suggested that the May 2007 cyber operations against Estonia's computer networks would have not been possible without the blessing of Russian authorities (Joshua Davis, 'Hackers Take Down the Most Wired Country in Europe', *Wired Magazine*, issue 15.09, 21 August 2007, <http://www.wired.com/politics/security/magazine/15-09/ff_estonia>).

[290] Shackelford, 'From Nuclear War', p 208; Klimburg, 'Mobilising Cyber Power', p 50.

[291] See Chapter 2, Section III.3, pp 87–88.

scenarios: the present book explores these difficulties. The next Chapter analyses the *jus ad bellum* issues arising from cyber operations, in particular whether they fall under the prohibition of the threat and use of force contained in Article 2(4) of the United Nations Charter and whether the state victim of a cyber operation may react in self-defence under Article 51 of the Charter. Chapter 3 discusses under what conditions the law of armed conflict is applicable to cyber operations without concurrent kinetic hostilities or in the context of an existing traditional armed conflict, while Chapter 4 analyses the limits that the law on the conduct of hostilities imposes on cyber operations. Finally, Chapter 5 considers the duties of neutral and belligerent states under the law of neutrality in the cyber context.

A few *caveats* on what the present book does *not* do. The book will only focus on military cyber operations: therefore, it does not touch upon questions of domestic or international law related to cyber crime and cyber terrorism. Furthermore, cyber operations above the threshold of the use of force will form the primary object of analysis, although discussion will also be conducted of certain operations falling below that threshold, whenever relevant. The application of the *jus pacis*, such as the law of the sea, aviation law, space law, or international communications law,[292] as well as of international criminal law, to cyber operations is also outside the scope of this book: this is not meant to suggest that these regimes are less relevant to cyber operations than the rules on the use of force or that they cease to apply in armed conflict, but only that they deserve specific in-depth treatment in a separate work.

The book will undertake a documentary analysis of different materials, *in primis* relevant *jus ad bellum* and *jus in bello* treaty provisions and customary international law, as applied by international and national courts. Although they are not, in themselves, sources of law and with all the caution motivated by the fact that they reflect operational and policy considerations, reference will also be made to military manuals, cyber security strategies and doctrines and official statements to the extent that they can assist in interpreting existing law and amount to evidence of state practice and *opinio juris*. As to cyber attacks that have already occurred, their exact details, such as the extent of damage caused or the attribution to specific states, are still uncertain: accordingly, they will be used in this book not as precedents or incontrovertible elements of state practice, but as explanatory real-life examples of different types of cyber operations. The present book is different from the Tallinn Manual in that it does not aim to distillate black-letter rules or to merely restate the law, but rather to suggest solutions and interpretations through which existing rules can be effectively applied to regulate a relatively new and unique phenomenon such as cyber operations. The book also deals with topics neglected by the Manual and suggests solutions for those problems on which the Group of Experts could not find agreement.[293]

---

[292] On the application of those regimes to cyber operations, see, among others, DoD, *An Assessment*, pp 26 ff; Schaap, 'Cyber Warfare Operations', pp 161–70; Shackelford, 'From Nuclear War', pp 223–4, 227–8.
[293] Such topics include, for instance, whether merely disruptive cyber operations amount to a 'use of force', 'armed attack' or 'attack', the nature of the nexus between a cyber operation and an armed

The overall goal is to provide a systematic and coherent analysis of the international law applicable to military cyber operations that will be of use to anyone who wants or needs to understand the basic issues of the rules of international law on the use of force and the law of armed conflict. Indeed, cyber operations give the opportunity to discuss some of the most controversial aspects of contemporary international law, such as self-defence against imminent armed attacks and against attacks by non-state actors, the distinction between the use of force and the law enforcement paradigms, the geographical scope of application of the law of armed conflict, the notions of 'combatancy' and 'direct participation in hostilities', and the legal issues arising from remote and automated warfare. While it is true that, until now, nobody has died in a cyber attack,[294] someone *could* have died: the potentially severe humanitarian consequences of certain cyber operations sufficiently justify an investigation on how international law can deal with them, even if such consequences have luckily not occurred yet.

The law is stated as of 30 September 2013.

---

conflict for the operation to be governed by the law of armed conflict, the attribution and evidentiary standards required for a self-defence reaction against a cyber attack, whether data constitute 'objects', whether a cyber operation qualifying as an act of hostilities but short of 'attack' may initiate an armed conflict.

[294] David P Fidler, '*Inter Arma Silent Leges Redux*? The Law of Armed Conflict and Cyber-Conflict', in *Cyberspace and National Security: Threats, Opportunities and Power in a Virtual World*, edited by Derek S Reveron (Washington, DC: Georgetown University Press, 2012), p 73.

2
Cyber Operations and the *jus ad bellum*

I. Introduction

One of the perspectives from which an international lawyer can study the problem of cyber security is that of the *jus ad bellum*. This Latin expression refers to the legal rules establishing when states may use force in international relations. While early efforts date back at least to the 1919 Covenant of the League of Nations and the 1928 Pact of Paris, the present cornerstone of the *jus ad bellum* matrix is the United Nations Charter, in particular its Article 2(4) and Chapter VII, which largely reflect customary international law. *Jus ad bellum* provisions, however, are also contained in regional, sub-regional and bilateral non-aggression and collective defence treaties, which normally include clauses subordinating them to the Charter.[1]

As the 2013 Report of the UN GGE on Developments in the Field of Information and Telecommunications in the Context of International Security recalls, '[i]nternational law, and in particular the Charter of the United Nations, is applicable and is essential to maintaining peace and stability and promoting an open, secure, peaceful and accessible ICT environment'.[2] This Chapter will therefore discuss how the Charter's provisions on the use of force apply to cyber operations. It will first explore if and when a cyber operation amounts to a use of force and is thus prohibited by Article 2(4) and its counterpart in customary international law. It will then move on to discuss whether the state victim of a cyber operation may invoke the right of self-defence by cyber or kinetic means against it: to this purpose, it will be necessary to establish under what conditions a cyber operation amounts to an 'armed attack' and what the legal requirements for the reaction in self-defence are, along with the specific problems arising in connection with their application in the

---

[1] The African Union (AU), for instance, encourages 'the conclusion and ratification of non-aggression pacts between and among African States' and the harmonization of such agreements (Chapter III, para 13(t) of the Solemn Declaration on a Common African Defence and Security Policy, adopted in Sirte by the Second Extraordinary Session of the AU Assembly (27–28 February 2004), <http://www.africa-union.org/Official_documents/Decisions_Declarations/Sirte/Declaration%20on%20a%20Comm.Af%20Def%20Sec.pdf>). See eg the 2005 AU Non-aggression and Common Defence Pact, the 1978 Economic Community of West African States (ECOWAS) Protocol on Non-aggression, the 1981 ECOWAS Protocol Relating to Mutual Assistance in Defence, the 2000 Economic Community of Central African States (ECCAS) Mutual Assistance Pact, the 2004 Central African Economic and Monetary Union (CEMAC) Non-aggression Pact, and the 2006 Great Lakes Protocol on Non-aggression and Mutual Defence.

[2] UN Doc A/68/98, 24 June 2013, p 8.

cyber context. Remedies against cyber operations below the level of armed attack will then be analysed before turning to the role that the UN Security Council can play in relation to cyber operations.

II. Cyber Operations and the Prohibition of the Threat and Use of Force in International Relations

It is common knowledge that Article 2(4) of the UN Charter provides that

[a]ll Members [of the United Nations] shall refrain in their international relations from the threat or use of force against the territorial integrity or political independence of any state, or in any other manner inconsistent with the Purposes of the United Nations.

This provision is generally considered to reflect customary international law and, at least with regard to its core, also *jus cogens*.[3]

For Article 2(4) and its customary counterpart to apply to cyber operations, three conditions must be met. First, the cyber operation needs to be attributed to a state: private individuals or armed groups do not fall within the scope of the provision, not even when they can inflict damage comparable to that caused by states. Secondly, the cyber operation must amount to either a 'threat' or a 'use of force'. Thirdly, the threat or use of force must be exercised in the conduct of 'international relations'. As to the first condition, the problems concerning identification of the origin and attribution of cyber operations to states examined in Chapter I may well be the main obstacle to the application of Article 2(4) in the cyber context.[4] For the purposes of this Chapter, however, it will be assumed that a cyber operation has been conclusively attributed to a state. The reference to 'international relations' in Article 2(4) entails that the use or threat of force must not only be by a state, but also against another state and that, therefore, states are not prohibited by this provision to threaten or resort to cyber operations against individuals or groups, not even when they amount to a threat or use of force, as long as such operations do not affect another state's territorial integrity or political independence.[5]

---

[3] The customary nature of Art 2(4) has been recognized by the ICJ in *Military and Paramilitary Activities in and against Nicaragua (Nicaragua v US)*, Merits, Judgment, 27 June 1986, ICJ Reports 1986 ('*Nicaragua*'), paras 187–90. See also *Legal consequences of the construction of a wall in the occupied Palestinian territory*, Advisory Opinion, 9 July 2004, ICJ Reports 2005 ('*Legal consequences of the construction of a Wall*'), para 87. The Court, however, famously acknowledged that 'customary international law continues to exist and to apply, separately from international treaty law, even where the two categories of law have an identical content' (*Nicaragua*, para 179). Several authors have argued that the core prohibition contained in Art 2(4), that of aggression, is also a peremptory norm of general international law (Roberto Ago, Addendum to the eighth Report on State Responsibility, *Yearbook of the International Law Commission*, 1980, Vol II, Part One, p 44; Rein Müllerson, '*Jus ad Bellum*: Plus Ça Change (Le Monde) Plus C'Est la Même Chose (Le Droit)?', *Journal of Conflict and Security Law* 7 (2002), p 169; Natalino Ronzitti, *Diritto internazionale dei conflitti armati*, 4th edn (Torino: Giappichelli, 2011), p 33).

[4] Chapter I, Section IV.

[5] This would occur, for instance, if the armed group was located on and operated from the territory of a third state that is either unable or unwilling to prevent their actions: see Section III.3 of this Chapter.

As to if and when a cyber operation amounts to a threat or use of force, Rule 11 of the Tallinn Manual provides, with some circularity, that '[a] cyber operation constitutes a use of force when its scale and effects are comparable to non-cyber operations rising to the level of a use of force'.[6] This formulation, which incorporates the so-called 'kinetic equivalence' doctrine, leaves however open the question of what scale and effects a 'non-cyber operation' must possess in order to qualify as a use of force. This and related questions will be examined in the following Sections.

1. Cyber operations as a 'use of force'

Article 2(4) prohibits both the threat and the use of 'force' without defining what 'force' is. The general criteria for the interpretation of treaties are spelt out in Article 31(1) of the 1969 Vienna Convention on the Law of Treaties.[7] If one applies the contextual and literal criteria in order to establish the meaning of 'force', the results are inconclusive. Indeed, according to *Black's Law Dictionary*, 'force' means '[p]ower, violence, or pressure directed against a person or thing':[8] the ordinary meaning of 'force' is thus broad enough to cover not only armed force but also other types of coercion such as economic and political coercion. As far as the context is concerned, the expression 'force' also appears in the Preamble of the Charter and in Articles 41 and 46 where it is preceded by the adjective 'armed', while in Article 44 it is clear that the reference is to armed force only. This contextual argument has often been used by commentators to maintain that, as elsewhere in the Charter 'force' means armed force, this must hold true for Article 2(4) as well, even in the absence of any specification.[9] The opposite argument could also be made: when the drafters wanted to refer to 'armed force', they said so expressly and, as this was not done in Article 2(4), they may have wanted to refer to a broader notion of force. A teleological interpretation of Article 2(4), however, seems to support a narrow reading of the provision that limits its scope to armed force only: indeed, the overall purpose of the Charter is 'to save succeeding generations from the scourge of war',[10] not to ban all forms of coercion. The *travaux préparatoires* also reveal that the drafters did not intend to extend the prohibition to economic coercion and political pressures:[11] a Brazilian amendment also prohibiting 'the threat or use of

---

[6] Rule 11 of the *Tallinn Manual on the International Law Applicable to Cyber Warfare* (Cambridge: Cambridge University Press, 2013), p 45.

[7] It is true that the UN Charter was adopted before the Vienna Convention and that the Convention does not apply to treaties concluded before its entry into force, but the rules on interpretation contained thereby are generally thought to be a codification of customary international law (Stefan Kadelbach, 'Interpretation of the Charter', in *The Charter of the United Nations—A Commentary*, edited by Bruno Simma, Daniel-Erasmus Khan, Georg Nolte, and Andreas Paulus, 3rd edn, Vol I (Oxford: Oxford University Press, 2012), p 75).

[8] Bryan A Garner, ed, *Black's Law Dictionary*, 9th edn (St Paul, MN: Thomson-West, 2009), p 717.

[9] A. Randelzhofer and O. Dörr, 'Article 2 (4)', in *The Charter of the United Nations*, edited by Simma, Khan, Nolte, and Paulus, Vol I, p 209.

[10] UN Charter, Preamble, UNTS, Vol 16, pp 1 ff.

[11] According to Art 32 of the 1969 Vienna Convention, the preparatory works of a treaty are a supplementary means of interpretation.

economic measures' was rejected at the San Francisco Conference.[12] Subsequent UN documents, such as the 1970 Declaration on Friendly Relations,[13] the 1974 Declaration on the Definition of Aggression,[14] and the 1987 Declaration on the Non-Use of Force[15] support the view that Article 2(4) only refers to armed force, while the principle of non-intervention also extends to other forms of coercion.[16]

The question, then, is if and when a cyber operation reaches the level of a use of *armed* force.[17] Although necessary, a coercive intention is not per se sufficient to identify and distinguish cyber operations as a use of armed force. Indeed, 'armed force' is nothing else than an extreme form of intervention that, like economic and diplomatic coercion, is characterized by the intention of the coercing state to compel the victim state into doing or not doing something through a 'dictatorial interference' in its internal or external affairs.[18] Similarly, if cyber operations amounting to a use of armed force were defined with reference to the authors of the forceful action, ie the armed forces, states would easily avoid the application of the prohibition by outsourcing such actions to intelligence agencies or private contractors. Further confirmation that the author criterion is not decisive derives from the fact that incidents involving the armed forces but not the use of weaponry, such as the violation of airspace or territorial waters by military aircraft or ships, are usually treated as violations of sovereignty, but not as a use of force under Article 2(4).[19]

Whether or not cyber operations fall within the scope of Article 2(4) depends ultimately on which of the three main analytic approaches to understanding the nature of a use of armed force is accepted. The instrument-based approach focuses on the means used to conduct an act, ie weapons, and has been traditionally

---

[12] *Documents of the United Nations Conference on International Organization* (London and New York, 1945: United Nations Information Organizations), Vol VI, 559, pp 720–1.

[13] Declaration on Principles of International Law concerning Friendly Relations and Cooperation among States in accordance with the Charter of the United Nations, GA Res 2625 (XXV), 24 October 1970.

[14] Declaration on the Definition of Aggression, GA Res 3314 (XXIX), 14 December 1974.

[15] Declaration on the Enhancement of the Effectiveness of the Principle of Refraining from the Threat or Use of Force in International Relations, GA Res 42/22, 18 November 1987.

[16] Randelzhofer and Dörr, 'Article 2(4)', pp 208–9; Michael N Schmitt, 'Computer Network Attack and the Use of Force in International Law: Thoughts on a Normative Framework', *Columbia Journal of Transnational Law* 37 (1998–99), pp 906–8; Marco Benatar, 'The Use of Cyber Force: Need for Legal Justification?', *Goettingen Journal of International Law* 1 (2009), pp 384–5. Article 19 of the Charter of the Organization of American States (OAS), on the other hand, also bans 'the use of coercive measures of an economic or political character'.

[17] Threats of force are discussed in Chapter 2, Section II.2.

[18] Hersch Lauterpacht, *International Law and Human Rights* (London: Stevens & Sons Ltd, 1950), p 167. As the ICJ emphasized, '[t]he element of coercion, which defines, and indeed forms the very essence of, prohibited intervention, is particularly obvious in the case of an intervention which uses force' directly or indirectly (*Nicaragua*, para 205). Coercion in inter-state relations has been defined as involving 'the government of one State compelling the government of another State to think or act in a certain way by applying various kinds of pressure, threats, intimidation or the use of force' (Christopher C Joyner, 'Coercion', in *Max Planck Encyclopedia of Public International Law* (2012), Vol II, p 297).

[19] Oliver Dörr, 'Use of Force, Prohibition of', *Max Planck Encyclopedia of Public International Law* (2012), Vol X, p 611.

employed to distinguish armed force from economic and political coercion. This approach has been criticized for being centred on instruments defined by their physical characteristics: as such—it has been claimed—it is ill-suited to be extended to digital codes and would lead to the conclusion that cyber operations can never be a use of force under Article 2(4), even when they result in physical damage.[20] The target-based approach argues that cyber operations reach the threshold of the use of armed force when they are conducted against a national critical infrastructure (NCI), whatever their effects on such infrastructure or the nature of the operation might be.[21] This approach, however, is overinclusive in that it would also qualify as a use of force those cyber operations that only cause inconvenience or merely aim to collect information whenever they target an NCI. Another problem with this view—it has been argued—is that there is no generally accepted definition of 'NCI'.[22]

The approach that has received most support is based on the effects of the action: unlike other forms of coercion, a use of armed force has direct destructive effects on property and persons.[23] Therefore, any cyber operation that causes or is reasonably likely to cause the damaging consequences normally produced by kinetic weapons would be a use of armed force.[24] The effects-based approach has been embraced by the United States. An early day US DoD study noted that 'it seems likely that the international community will be more interested in the consequences of a computer network attack than in its mechanism'.[25] In his speech at CYBERCOM, the State Department's Legal Advisor, Harold Koh, argued that 'if the physical consequences of a cyber attack work the kind of physical damage that dropping a bomb or firing a missile would, that cyber attack should equally be considered a use of force'.[26] This view, that limits the application of Article 2(4) to those cyber

---

[20] Stephanie Gosnell Handler, 'The New Cyber Face of Battle: Developing a Legal Approach to Accommodate Emerging Trends in Warfare', *Stanford Journal of International Law* 48 (2012), pp 226–7; Matthew C Waxman, 'Self-Defensive Force against Cyber Attacks: Legal, Strategic and Political Dimensions', *International Law Studies* 89 (2013), p 111.

[21] Walter G Sharp, Sr, *Cyberspace and the Use of Force* (Falls Church: Aegis Research Corpn, 1999), pp 129–32. See similarly Christopher C Joyner and Catherine Lotrionte, 'Information Warfare as International Coercion: Elements of a Legal Framework', *European Journal of International Law* 12 (2001), p 855, who argue that stealing or compromising sensitive military information could also qualify as an armed attack (and *a fortiori* a use of force) 'even though no immediate loss of life or destruction results'.

[22] On the notion of 'NCI', see Chapter 2, Section II.1.2, p 55 ff.

[23] See eg Russell Buchan, 'Cyber Attacks: Unlawful Uses of Force or Prohibited Interventions?', *Journal of Conflict and Security Law* 17 (2012), p 212; Heather Harrison Dinniss, *Cyber Warfare and the Laws of War* (Cambridge: Cambridge University Press, 2012), p 74.

[24] Dinstein has for instance argued that 'what counts is not the specific type of ordnance, but the end product of its delivery to a selected objective' (Yoram Dinstein, 'Computer Network Attacks and Self-Defense', *International Law Studies* 76 (2002), p 103). Silver also opines that 'physical injury or property damage must arise as a direct and foreseeable consequence of the CNA and must resemble the injury or damage associated with what, at the time, are generally recognized as military weapons' (Daniel B Silver, 'Computer Network Attack as a Use of Force under Article 2(4) of the United Nations Charter', *International Law Studies* 76 (2002), pp 92–3).

[25] US Department of Defense, *An Assessment of International Legal Issues in Information Operations*, May 1999, p 18, <http://www.au.af.mil/au/awc/awcgate/dod-io-legal/dod-io-legal.pdf>.

[26] Harold Koh, 'International Law in Cyberspace', Speech at the USCYBERCOM Inter-Agency Legal Conference, 18 September 2012, in CarrieLyn D Guymon (ed), *Digest of United States Practice in International Law*, 2012, p 595, <http://www.state.gov/documents/organization/211955.pdf>.

operations that cause or are reasonably likely to cause the same effects of kinetic weapons, does not take into account, however, that the dependency of modern societies on computers, computer systems, and networks has made it possible to achieve analogous prejudicial results through other, non-destructive means. Aware of this problem, Michael Schmitt has elaborated a set of eight non-exhaustive factors to consider in order to establish when the scale and effects of cyber operations that produce prejudicial consequences of a non-physical nature sufficiently resemble those of a kinetic use of force: severity, immediacy, directness, invasiveness, measurability of effects, military character, state involvement, and presumptive legality.[27] In his view, armed force can be distinguished from other forms of coercion because it causes more significant physical injury or destruction of property, with greater immediacy and in a more direct way. Armed force also amounts to a greater intrusion in the rights of the victim state, the negative consequences of which are easier to measure than in other forms of coercion. Schmitt's criteria, which the author himself describes as 'not legal' and 'merely factors that can be expected to influence States when making use of force appraisals',[28] are not without problems. Directness, for instance, is not necessarily an inherent characteristic of the use of armed force: the Declaration on the Definition of Aggression, qualifies as an 'act of aggression', ie 'the most serious and dangerous form of the illegal use of force',[29] not only bombings and invasions, but also actions which do not necessarily entail direct destructive effects, such as the violation of a stationing agreement, a naval blockade, and allowing the use of the territory by other states for the purpose of perpetrating aggression.[30] In the *Nicaragua* judgment, the ICJ also qualified the arming and training of armed groups—not directly destructive actions—as a use of force.[31] Another problem with the directness criterion is that it does not sufficiently appreciate the fact that one of cyber operations' main characteristics is that they often produce the intended prejudicial effects indirectly as the consequence of the alteration, deletion, or corruption of data or software or the loss of functionality of infrastructure.[32] As to invasiveness, a very common form of cyber attack, 'flood' attacks, is not intrusive, as the attackers do not gain access to the system: they simply 'clog the entryways to the system, rather than get into it'.[33] What causes disruption is the flood of requests that the botnets or the individual computers send to the target system. Measurement of effects of cyber operations is also notoriously difficult: it has yet to be acknowledged, for instance, whether Stuxnet caused any physical damage and, if so, to what extent. With regard to immediacy, the so-called logic or time bombs, which are Trojan

---

[27] Schmitt, 'Computer Network Attack and the Use of Force', pp 914–15. The criteria are also noted in the Commentary to the Tallinn Manual, pp 48–51.
[28] Michael N Schmitt, 'The "Use of Force" in Cyberspace: A Reply to Dr Ziolkowski', in *2012 4th International Conference on Cyber Conflict* (2012), edited by Christian Czosseck, Rain Ottis, and Katharina Ziolkowski, p 314.
[29] Definition of Aggression, Preamble.   [30] Article 3(c), (e), and (f).
[31] *Nicaragua*, para 228.   [32] Harrison Dinniss, *Cyber Warfare*, pp 65–6.
[33] Fred Schreier, *On Cyberwarfare*, DCAF Horizon 2015 Working Paper no 7, 2012, p 51, <http://www.dcaf.ch/Publications/On-Cyberwarfare>.

horses designed to produce their effects only at a certain time or when certain circumstances occur, can cause damage well after the cyber intrusion has taken place. Finally, the presumptive legality factor is based on what Judge Simma has defined as 'an old, tired view of international law',[34] that according to which '[a]cts that are not forbidden are permitted'.[35]

In the present author's view, a 'use of armed force' should be determined by reference to the instruments used, ie weapons. This is consistent with the ordinary meaning of the expression: according to *Black's Law Dictionary*, 'armed' means '[e]quipped with a weapon' or '[i]nvolving the use of a weapon'.[36] Unlike economic or political coercion, then, armed force entails the coercive use of weapons.[37] Accordingly, a limited use of weapons by a state that is not directed at exercising coercion on another state, as in the case of international abductions, police measures at sea, or interception of trespassing aircraft, does not fall under the scope of Article 2(4), but is a violation of other norms, such as the duty to respect another state's sovereignty.[38] Similarly, economic sanctions that cause starvation among the population are not a use of armed force in spite of their severe humanitarian consequences: sanctions may be enforced with the use of weapons, but are not weapons themselves, as implied in Article 41 of the UN Charter.[39]

Although there is no binding definition of 'weapon' either in *jus ad bellum* or *jus in bello* instruments, *Black's Law Dictionary* defines it as '[a]n instrument used or designed to be used to injure or kill someone'.[40] The ICRC Study on Customary International Humanitarian Law defines weapons as 'means to commit acts of violence against human or material enemy forces', whether or not the violence occurs.[41] Rule 1(ff) of the *HPCR Manual on Air and Missile Warfare* also attaches one main characteristic to a weapon: the capability to cause either injury/death of persons or damage/destruction of objects.[42] Similarly, a leading commentator has defined weapons as including 'any arms...munitions...and other devices, components or mechanisms intended to destroy, disable or injure enemy personnel,

---

[34] *Accordance with international law of the unilateral declaration of independence in respect of Kosovo*, Advisory Opinion, 22 July 2010, ICJ Reports 2010, Declaration of Judge Simma, para 2.

[35] Tallinn Manual, p 51. [36] *Black's Law Dictionary*, p 123.

[37] As has been observed, '[t]he essential feature which characterizes the prohibition of the use of force is...not the intrusion into the sovereign realm of another State, nor is it even the element of coercion as such, but only an intrusion or coercion accompanied by the special features of military weaponry and its actual use' (Dörr, 'Use of Force', p 611).

[38] See Olivier Corten, *The Law Against War. The Prohibition on the Use of Force in Contemporary International Law* (Oxford and Portland: Hart, 2010), p 67.

[39] Article 41 of the UN Charter includes the 'complete or partial interruption of...telegraphic, radio, and other means of communication' in the list of measures 'not involving the use of armed force': this, however, is not necessarily helpful in the qualification of cyber operations, as their effects on computerized societies can be far more drastic than those envisaged by the Charter's drafters in relation to the interruption of communications (Horace B Robertson, Jr, 'Self-Defense against Computer Network Attack under International Law', *International Law Studies* 76 (2002), p 138; Schmitt, 'Computer Network Attack and the Use of Force', p 912).

[40] *Black's Law Dictionary*, p 1730.

[41] Jean-Marie Henckaerts and Louise Doswald-Beck, *Customary International Humanitarian Law* (Cambridge: Cambridge University Press, 2005), Vol I, Rule 6, p 23.

[42] HPCR, *Manual on International Law Applicable to Air and Missile Warfare*, (Cambridge: Cambridge University press, 2013), p 49.

matériel or property'.[43] The minimum common denominator of the above definitions is the violent consequences produced by the instrument.[44] Weapons are therefore identified by their effects, not by the mechanisms through which they produce destruction or damage.[45] If this is correct, armed force in the sense of Article 2(4) could be defined as the form of intervention by a state to exercise coercion on another state that involves the use of instruments (weapons) capable of producing violent consequences. At a closer look, then, the debate between the supporters of the instrument-based and effects-based approaches to establish whether cyber operations are a use of force loses much of its significance, as the two approaches should be combined: it is the instrument used that defines armed force, but the instrument is identified by its (violent) consequences. The focus on instrumentality explains why the ICJ qualified arming and training of armed groups as a use of force: although not directly destructive, those activities are strictly related to weapons, as they aim at enabling someone to use them.

If, then, a use of armed force under Article 2(4) requires weapons, the next question that needs to be answered is whether malware can qualify as such. In its Advisory Opinion on the *Legality of the Threat or Use of Nuclear Weapons*, the ICJ made clear that Articles 2(4), 51, and 42 of the UN Charter 'do not refer to specific weapons. They apply to any use of force, regardless of the weapons employed'.[46] There is then no reason why the weapons covered by those provisions should necessarily have explosive effects or be created for offensive purposes only: the use of certain dual-use non-kinetic weapons, such as biological or chemical agents, against a state would undoubtedly be treated by the victim state as a use of force in the sense of Article 2(4). According to Brownlie, this is so because chemical and biological weapons are commonly referred to as forms of 'warfare' and because they can be used to destroy life and property:[47] both arguments would suit at least some malware as well. In particular, several states have included cyber technologies in their military doctrines, refer to cyberspace as to a domain of warfare and have set up military units with specific cyber expertise.[48] The Russian Foreign Minister

---

[43] Yoram Dinstein, *The Conduct of Hostilities under the Law of International Armed Conflict*, 2nd edn (Cambridge: Cambridge University Press, 2010), p 1.

[44] This is also consistent with the definition of 'attack' under Art 49 of Protocol I Additional to the Geneva Conventions on the Protection of Victims of International Armed Conflicts, UNTS, Vol 1125, pp 3 ff, on which see Chapter 4, Section III.1.1.

[45] Katharina Ziolkowski, 'Computer Network Operations and the Law of Armed Conflict', *Military Law and Law of War Review* 49 (2010), p 69.

[46] *Legality of the Threat or Use of Nuclear Weapons*, Advisory Opinion, 8 July 1996, ICJ Reports 1996 ('*Nuclear Weapons*'), para 39.

[47] Ian Brownlie, *International Law and the Use of Force by States* (Oxford: Clarendon Press, 1963), p 362.

[48] See Chapter 1, Section I. The US Air Force argued as early as 1997 that information is a separate realm for warfare in addition to air, land, sea, and space (US Department of Air Force, *Cornerstones of Information Warfare*, 17 April 1997, <http://www.c4i.org/cornerstones.html>). The 2008 US *National Defense Strategy* refers to 'terrorism, electronic, cyber and other forms of warfare' (US DoD, *National Defense Strategy*, June 2008, p 11, <http://www.defense.gov/news/2008%20national%20 defense%20strategy.pdf>). The 2010 US *National Security Strategy* emphasizes the need to ensure that 'the U.S. military continues to have the necessary capabilities across all domains—land, air, sea,

warned that the destructive effect of information weapons 'may be comparable to that of weapons of mass destruction'.[49] Belarus made the same analogy.[50] Panama noted that '[a]n attack in which new information and telecommunications technologies are employed may cause more damage than, for instance, a conventional bombardment'.[51] Kazakhstan observed that 'information technology advances may be misused as information weapons during armed conflicts'.[52] Cuba has also remarked that '[i]nformation and telecommunication systems can be turned into weapons when they are designed and/or used to damage the infrastructure of a State, and as a result, can put at risk international peace and security'.[53] Spain recalled the '[u]se of the Internet as a weapon, i.e., its use as a means to launch attacks against critical infrastructure information systems or the infrastructure of the Internet itself'.[54] The US Air Force includes cyber capabilities in its legal review of weapons under Article 36 of Additional Protocol I[55] and the US *Joint Vision 2020* expressly refers to the employ of non-kinetic weapons in the area of information operations.[56] Finally, the UK *National Security Strategy* emphasizes that 'activity in cyberspace' is 'a military weapon for use by states and possibly others',[57] and the UK Under-Secretary for Security and Counter-terrorism declared that a cyber attack that takes out a power station would be an act of war.[58]

All the above supports the view that worms, viruses, botnet codes, and other malware are now treated as 'just another weapons system, cheaper and faster than a missile, potentially more covert but also less damaging'.[59] As Dinstein suggests, 'cyber...must be looked upon as a new means of warfare—in other words, a weapon: no less and no more than other weapons'.[60] The Commentary

---

space, and cyber' (*National Security Strategy*, May 2010, p 22, <http://www.whitehouse.gov/sites/default/files/rss_viewer/national_security_strategy.pdf>).

[49] Letter dated 23 September 1998 from the Permanent Representative of the Russian Federation to the United Nations addressed to the Secretary-General, UN Doc A/C.1/53/3, 30 September 1998, p 2, <http://www.un.org/ga/search/view_doc.asp?symbol=A/C.1/53/3&Lang=E>.

[50] UN Doc A/54/213, 10 August 1999, p 3.

[51] UN Doc A/57/166/Add.1, 29 August 2002, p 5.   [52] UN Doc A/64/129, 8 July 2009, p 5.

[53] UN Doc A/66/152/Add.1, 16 September 2011, p 2.

[54] UN Doc A/64/129/Add.1, 9 September 2009, p 10.

[55] Air Force Instruction 51–402, 27 July 2011, <http://www2.gwu.edu/~nsarchiv/NSAEBB/NSAEBB424/docs/Cyber-053.pdf>. On Art 36, see Chapter 4, Section II, p 170 ff.

[56] Joint Chiefs of Staff, *Joint Vision 2020—America's Military: Preparing for Tomorrow*, June 2000, p 23, <http://www.fs.fed.us/fire/doctrine/genesis_and_evolution/source_materials/joint_vision_2020.pdf>.

[57] *A Strong Britain in an Age of Uncertainty: The National Security Strategy*, October 2010, p 29, <http://www.direct.gov.uk/prod_consum_dg/groups/dg_digitalassets/@dg/@en/documents/digitalasset/dg_191639.pdf?CID=PDF&PLA=furl&CRE=nationalsecuritystrategy>.

[58] Jamie Doward, 'Britain fends off flood of foreign cyber-attacks', *The Observer*, 7 March 2010, <http://www.theguardian.com/technology/2010/mar/07/britain-fends-off-cyber-attacks>.

[59] James Lewis, 'To Protect the U.S. Against Cyberwar, Best Defense is a Good Offense', *U.S. News and World Report*, 29 March 2010, <http://www.usnews.com/articles/opinion/2010/03/29/to-protect-the-us-against-cyberwar-best-defense-is-a-good-offense.html>. Schmitt also notes that '[w]ith the advent of CNA, today the computer is no less a weapon than an F-16 armed with precision weapons' (Michael N Schmitt, 'Computer Network Attack: The Normative Software', *Yearbook of International Humanitarian Law* 4 (2001), p 56).

[60] Yoram Dinstein, 'Cyber War and International Law: Concluding Remarks at the 2012 Naval War College International Law Conference', *International Law Studies* 89 (2013), p 280.

to the *HPCR Manual on Air and Missile Warfare* confirms that '[m]eans of warfare include non-kinetic systems, such as those used in EW [electronic warfare] and CNAs. The means would include the computer and computer code used to execute the attack, together with all associated equipment'.[61] It further specifies that death, injury, damage, or destruction 'need not result from physical impact...since the force used does not need to be kinetic. In particular, CNA hardware, software and codes are weapons that can cause such effects through transmission of data streams'.[62] Like missiles, cyber weapons include a delivery system, a navigation system and a payload.[63] The delivery system could go from e-mails to malicious links in websites, hacking, counterfeit hardware and software. System vulnerabilities are the main navigation systems that provide entry points for the payload by enabling unauthorized access to the system. The payload is the component that causes harm: if the code, however sophisticated, is designed solely for the purpose of infiltrating a computer and exfiltrating information, as in the cases of Duqu and Flame, it would not be a 'weapon' in the sense highlighted above, as it is neither intended nor capable of causing damage.[64]

As has been seen in Chapter 1, cyber operations can go from cyber exploitation for information gathering, reconnaissance, and surveillance to cyber attacks and, within the latter category, from operations that merely delete, corrupt, or alter data or software to those causing physical damage to property or persons or malfunction of infrastructure with consequent disruption in the provision of services. This diversity of consequences attached to cyber operations prevents from assessing them as a whole and is the reason why the different typologies will be examined separately in the following pages.

1.1 *Cyber attacks causing, or reasonably likely to cause, physical damage to property, loss of life, or injury to persons*

Cyber operations can produce multiple effects.[65] The primary effects are those on the attacked computer, computer system or network, ie the deletion, corruption, or alteration of data or software, or system disruption through a DDoS attack or other cyber attacks. The secondary effects are those on the infrastructure operated by the attacked system or network (if any), ie its partial or total destruction or incapacitation. Tertiary effects are those on the persons affected

---

[61] HPCR Manual, p 31.
[62] HPCR Manual, p 49.
[63] Schreier, *On Cyberwarfare*, pp 66–7.
[64] Thomas Rid and Peter McBurney, 'Cyber-Weapons', *RUSI Journal* 157, no 1 (February 2012), p 11.
[65] William A Owens, Kenneth W Dam, and Herbert S Lin, *Technology, Policy, Law, and Ethics Regarding U.S. Acquisition and Use of Cyberattack Capabilities* (Washington: The National Academies Press, 2009), p 80. See also Pia Palojärvi, *A Battle in Bits and Bytes: Computer Network Attacks and the Law of Armed Conflict* (Helsinki: Erik Castrén Institute of International Law, 2009), p 32; William H Boothby, 'Methods and Means of Cyber Warfare', *International Law Studies* 89 (2013), p 390.

by the destruction or incapacitation of the attacked system or infrastructure, for instance those that benefit from the electricity produced by a power plant incapacitated by a cyber operation. Physical damage to property, loss of life and injury to persons, then, are never the primary effects of a cyber operation: damage to physical property can only be a secondary effect, while death or injury of persons can be a tertiary effect of a cyber operation. As Waxman notes, 'modern society's heavy reliance on interconnected information systems means that the indirect and secondary effects of cyber-attacks may be much more consequential than the direct and immediate ones'.[66] This is, however, not necessarily a problem for the application of the *jus ad bellum* rules: in the *Nicaragua* judgment, the ICJ expressly recognized that intervention that uses armed force can occur either directly or indirectly.[67]

It is virtually uncontested that a cyber attack that causes or is reasonably likely to cause physical damage to property, loss of life or injury to persons would fall under the prohibition contained in Article 2(4) of the UN Charter. No cyber attack, however, has so far been reported to have resulted in injuries or deaths of persons. If one excludes the explosion of a Soviet gas pipeline in Siberia in June 1982, apparently caused by a logic bomb inserted in the computer-control system by the US CIA,[68] the first known use of malicious software designed to produce material damage to physical property by attacking the SCADA system of a NCI is Stuxnet.[69] Using four unknown vulnerabilities, Stuxnet was allegedly designed to force a change in the centrifuges' rotor speed at the Natanz uranium enrichment plant in Iran, inducing excessive vibrations or distortions that would damage the centrifuges.[70] As a result, the IAEA reported that Iran stopped feeding uranium into thousands of centrifuges at Natanz, a claim denied by the Iranian authorities.[71] The causation of physical damage, however, does not necessarily require acting on the software: it may be sufficient to gain access to the computer system and alter or delete, say, transport or medical data for trains to collide, airplanes to crash, or for patients to receive the wrong medical treatment.

One could ask whether there is a minimum threshold of gravity that the destructive consequences of a cyber operation need to reach in order to be a violation of Article 2(4) and not only of the principle of non-intervention. A former US Department of State's Legal Advisor, for instance, appears to distinguish between injury/death of persons on the one hand and damage to property on

---

[66] Matthew C Waxman, 'Cyber Attacks and the Use of Force: Back to the Future of Article 2(4)', *Yale Journal of International Law 36* (2011), p 445.
[67] *Nicaragua*, para 205. See Harrison Dinniss, *Cyber Warfare*, p 66.
[68] Thomas Rid, *Cyber War Will Not Take Place* (London: Hurst & Co, 2013), p 4.
[69] As has been seen, SCADA systems are computer-controlled industrial control system that monitor and control industrial processes of physical infrastructures. Affecting them means affecting the physical infrastructure they monitor and control.
[70] David Albright, Paul Brannan, and Christina Walrond, 'Did Stuxnet Take Out 1,000 Centrifuges at the Natanz Enrichment Plant?', ISIS Report, 22 December 2010, p 6, <http://isis-online.org/uploads/isis-reports/documents/stuxnet_FEP_22Dec2010.pdf>.
[71] William J Broad, 'Report Suggests Problems with Iran's Nuclear Effort', *The New York Times*, 23 November 2010, <http://www.nytimes.com/2010/11/24/world/middleeast/24nuke.html>.

the other when he argues that it is '[c]yber activities that proximately result in death, injury or *significant* destruction' that would be considered a use of force.[72] Beyond the cyber context, Corten maintains that 'there is a threshold below which the use of force in international relations, while it may be contrary to certain rules of international law, cannot violate article 2(4)':[73] examples are international abductions, extraterritorial enforcement measures, international police operations, hot pursuit and police measures at sea, and the interception and neutralization of aircraft entering a state's airspace without authorization. Similarly, in its 2009 Report, the Independent International Fact-Finding Commission on the Conflict in Georgia found that '[t]he prohibition of the use of force covers all physical force which surpasses a minimum threshold of intensity' and that '[o]nly very small incidents lie below this threshold, for instance the targeted killing of single individuals, forcible abductions of individual persons, or the interception of a single aircraft'.[74] There seems to be some cautious support for this view in the ICJ's 1998 *Fisheries Jurisdiction* judgment: while Spain argued that the forcible measures against the *Estai* amounted to a violation of Article 2(4), the Court found that 'the use of force authorized by the Canadian legislation and regulations falls within the ambit of what is commonly understood as enforcement of conservation and management measures' and that '[b]oarding, inspection, arrest and *minimum* use of force for those purposes are all contained within the concept of enforcement of conservation and management measures according to a "natural and reasonable" interpretation of this concept'.[75]

There is nothing, however, in the wording of Article 2(4) suggesting that uses of force should be distinguished according to their gravity: as Ago notes, Article 2(4) prohibits 'any kind of conduct involving any assault whatsoever on the territorial sovereignty of another State, irrespective of its magnitude, duration or purposes'.[76] Having said that, a literal interpretation of Article 2(4) must not lead to results that are 'manifestly absurd or unreasonable':[77] a cyber operation that causes minimal damage such as the destruction of a single computer or server would clearly not fall within the scope of the provision. Whether or not a 'minimum use of force' is a violation of Article 2(4) ultimately depends on the circumstances of the case: what can be said is that the more invasive and damaging the use of (cyber) weapons, the more the affected state will be inclined to treat it as a use of force.

---

[72] Koh, 'International Law in Cyberspace', p 595 (emphasis added). It should, however, be recalled that, according to the US position reflected in Koh's speech, there is no distinction between 'use of force' and 'armed attack'.

[73] Corten, *The Law Against War*, p 55.

[74] Report of the Independent Fact-Finding Mission on the Conflict in Georgia, September 2009, Vol II, p 242, <http://www.ceiig.ch/Report.html>.

[75] *Fisheries Jurisdiction (Spain v Canada)*, Judgment, 4 December 1998, ICJ Reports 1998, para 84 (emphasis added). See similarly the Report of the Commission of Inquiry (Denmark–United Kingdom) on the *Red Crusader* incident, 23 March 1962, *International Law Reports* 35 (1967), pp 485 ff.

[76] Ago, Addendum, p 41. Similarly, Melzer argues that Art 2(4) prohibits all uses of force, regardless of their magnitude and duration (Nils Melzer, *Cyberwarfare and International Law*, UNIDIR, 2011, p 8, <http://www.isn.ethz.ch/Digital-Library/Publications/Detail/?lng=en&id=134218>).

[77] Article 32 of the Vienna Convention on the Law of Treaties.

Whether data as such can be equated to physical property for the purposes of Article 2(4), so that their deletion, alteration or corruption qualifies as a use of force even without physical damage or incapacitation of infrastructure, is a question that it is very difficult to answer. In August 2012, for instance, a virus destroyed the data of about 30,000 company computers of Saudi Aramco, the world's largest oil producer. The deleted data were replaced with a burning American flag.[78] Schmitt argues that the destruction of or damage to data, on its own, is not enough to amount to an armed attack, with the possible exception of the destruction of data 'designed to be immediately convertible into tangible objects, like banking data'.[79] This conclusion could be extended also to the use of force under Article 2(4). Future practice will probably clarify this point.

1.2 Cyber attacks severely disrupting critical infrastructures

If cyber attacks that cause or are reasonably likely to cause material damage to property or persons can be equated to kinetic attacks, there is disagreement on whether disruptive operations, ie those that render ineffective or unusable infrastructures without physically damaging them,[80] also amount to a violation of Article 2(4) of the UN Charter. The Tallinn Manual's rules largely neglect disruptive cyber operations because of lack of consensus among the Group of Experts with regard to their nature and legality, and focus essentially on those resulting in physical damage. It is this book's contention, however, that disruptive cyber operations also fall under the scope of Article 2(4) if the disruption caused is significant enough to affect state security, or, to use the words of the US Presidential Policy Directive 20, 'national security, public safety, national economic security, the safe and reliable functioning of "critical infrastructure," and the availability of "key resources" '.[81] NCIs, in particular, are those governmental or privately-owned infrastructure the incapacitation or destruction of which might impact national security and/or the welfare of the nation and which are normally accessible only by authorized users.[82] Most of these infrastructures are now operated through SCADA

---

[78] 'Saudi Aramco says cyber attack targeted kingdom's economy', *Al Arabiya News*, 9 December 2012, <http://www.alarabiya.net/articles/2012/12/09/254162.html>. Oil production, however, remained uninterrupted.

[79] Michael N Schmitt, 'Cyber Operations in International Law: The Use of Force, Collective Security, Self-Defense, and Armed Conflicts', in *Proceedings of a Workshop on Deterring Cyberattacks: Informing Strategies and Developing Options for U.S. Policy* (Washington: The National Academies Press, 2010), p 164.

[80] The definition is borrowed from that of 'neutralize' contained in US DoD, *Dictionary of Military and Associated Terms*, Joint Publication 1–02, 8 November 2010 (As Amended Through 15 August 2013), p 195.

[81] US Presidential Policy Directive/PPD–20, October 2012, p 3, <http://www.guardian.co.uk/world/interactive/2013/jun/07/obama-cyber-directive-full-text>.

[82] *Estonia's Cyber Security Strategy*, Ministry of Defence, 2008, p 36. The Estonian Cyber Security Strategy emphasizes that '[f]ailures of or disruptions to critical information systems may impact extensively upon the normal functioning of society with unforeseen and potentially disastrous consequences…Large scale information systems breakdowns may result in considerable physical and financial damage and even human casualties' (pp 10–11). On the distinction between 'critical infrastructure' and 'critical information infrastructure', see Elgin M Brunner and Manuel Suter, *International CIIP*

systems, which monitor and control their processes. There is no general agreement, though, on what infrastructures are 'critical': the UN General Assembly recognized that 'each country will determine its own critical information infrastructures'.[83] The differences between definitions and sectors included reflect different national understanding of what is critical, which is dictated by each country's peculiarities.[84] The 2001 US PATRIOT Act, for instance, defines 'critical infrastructure' as 'systems and assets, whether physical or virtual, so vital to the United States that the incapacity or destruction of such systems and assets would have a debilitating impact on security, national economic security, national public health or safety, or any combination of those matters'.[85] The 2003 US *National Strategy to Secure Cyberspace* describes critical infrastructures as 'the physical and cyber assets of public and private institutions in...agriculture, food, water, public health, emergency services, government, defense industrial base, information and telecommunications, energy, transportation, banking and finance, chemicals and hazardous materials, and postal and shipping'.[86] The 2010 US *Joint Terminology for Cyberspace Operations* defines 'critical infrastructure' as '[s]ystems and assets, whether physical or virtual, so vital that the incapacity or destruction of such may have a debilitating impact on the security, economy, public health or safety, environment, or any combination of these matters, across any Federal, State, regional, territorial, or local jurisdiction'.[87] The 2011 *International Strategy for Cyberspace* emphasizes the need to protect in particular the NCI on energy, transportation, financial systems and the defence industrial base.[88] The UK *Cyber Security Strategy* refers to nine sectors that deliver essential services: energy, food, water, transport, communications, government and public services, emergency services, health, and finance.[89] India's *National Cyber Security Policy* highlights that '[f]ocused cyber attacks

Handbook 2008–2009, Center for Security Studies, ETH Zurich, pp 37–8, <http://www.css.ethz.ch/publications/pdfs/CIIP-HB-08-09.pdf>.

[83] See eg GA Res 58/199 of 23 December 2003. The resolution includes among the critical infrastructures 'those used for, inter alia, the generation, transmission and distribution of energy, air and maritime transport, banking and financial services, e-commerce, water supply, food distribution and public health—and the critical information infrastructures that increasingly interconnect and affect their operations'.

[84] See, for an inventory of critical information infrastructure protection policies, Brunner and Suter, *International CIIP Handbook*.

[85] Public Law 107-56, 26 October 2001, Section 1016(e). The text can be read at <http://fl1.findlaw.com/news.findlaw.com/cnn/docs/terrorism/hr3162.pdf>.

[86] *The National Strategy to Secure Cyberspace*, February 2003, p 1, <http://www.us-cert.gov/reading_room/cyberspace_strategy.pdf>. See also *The National Strategy for the Physical Protection of Critical Infrastructures and Key Assets*, February 2003, p 35, <http://www.dhs.gov/xlibrary/assets/Physical_Strategy.pdf>.

[87] Memorandum for Chiefs of the Military Services, Commanders of the Combatant Commands, Directors of the Joint Staff Directorates, *Joint Terminology for Cyberspace Operations*, November 2010, p 5, <http://info.publicintelligence.net/DoD-JointCyberTerms.pdf>.

[88] White House, *International Strategy for Cyberspace. Prosperity, Security and Openness in a Networked World*, May 2011, p 19, <http://www.whitehouse.gov/sites/default/files/rss_viewer/international_strategy_for_cyberspace.pdf>.

[89] *The UK Cyber Security Strategy, Protecting and promoting the UK in a digital world*, November 2011, p 9, <http://www.gov.uk/government/uploads/system/uploads/attachment_data/file/60961/uk-cyber-security-strategy-final.pdf>.

affecting the organisations in critical sectors such as Defence, Energy, Finance, Space, Telecommunications, Transport, Public Essential Services and Utilities, Law Enforcement and Security would lead to national crisis'.[90] The *Cyber Security Strategy for Germany* describes critical infrastructures as 'organizations or institutions with major importance for the public good, whose failure or damage would lead to sustainable supply bottlenecks, considerable disturbance of public security or other dramatic consequences' and includes the following sectors at federal level: energy, information technology and telecommunication, transport, health, water, food, finance and insurance, state and administration, media and culture.[91] Canada's critical infrastructures include energy and utilities, communications and information technology, finance, health care, food, water, transportation, government and manufacturing.[92] The Australian government defines critical infrastructures as 'those physical facilities, supply chains, information technologies and communication networks which, if destroyed, degraded or rendered unavailable for an extended period, would adversely impact on the social or economic well-being of the nation or affect Australia's ability to ensure national security', in particular in the following sectors: 'banking and finance, communications, emergency services, energy, food chain, health (private), water services, mass gatherings, and transport (aviation, maritime and surface)'.[93] Australia's *Cyber Security Strategy*, however, also points out that systems of national interest 'go beyond traditional notions of critical infrastructure' and include 'systems which, if rendered unavailable or otherwise compromised, could result in significant impacts on Australia's economic prosperity, international competitiveness, public safety, social wellbeing or national defence and security'.[94] In Russia's view, 'vital structures' are those 'State's facilities, systems and institutions, deliberate influence on the information resources of which may have consequences that directly affect national security (transport, energy supply, credit and finance, communications, State administrative bodies, the defence system, law-enforcement agencies, strategic information resources, scientific establishments and scientific and technological developments, installations that pose heightened technological and environmental risks, and bodies for eliminating the consequences of natural disasters or other emergency situations)'.[95] Finally, the Commission of the European Union defines critical infrastructures as 'those physical resources, services, and information technology facilities, networks and

---

[90] *National Cyber Security Policy* (discussion draft), March 2011, p 12, <http://deity.gov.in/sites/upload_files/dit/files/ncsp_060411.pdf>.

[91] Federal Ministry of the Interior, *Cyber Security Strategy for Germany*, February 2011, p 15, <http://www.cio.bund.de/SharedDocs/Publikationen/DE/Strategische-Themen/css_engl_download.pdf?__blob=publicationFile>.

[92] UN Doc A/60/95/Add.1, 21 September 2005, p 4.

[93] Australian Government, *Cyber Security Strategy*, 2009, p 20, <http://www.ag.gov.au/www/agd/rwpattach.nsf/VAP/(4CA02151F94FFB778ADAEC2E6EA8653D)-AG+Cyber+Security+Strategy+-+for+website.pdf/$file/AG+Cyber+Security+Strategy+-+for+website.pdf>.

[94] Australian Government, *Cyber Security Strategy*, 2009, p 12. The Strategy acknowledges that '[t]he identification of systems of national interest is not a static process and...must be informed by an ongoing assessment of risk' (p 12).

[95] UN Doc A/54/213, 10 August 1999, p 10.

infrastructure assets which, if disrupted or destroyed, would have a serious impact on the health, safety, security or economic well-being of Citizens or the effective functioning of governments'.[96]

In spite of the differences in the above definitions, their minimum common denominator is that such infrastructures are vital for national security, including individual, societal, and governmental security.[97] It is also possible to identify certain sectors that are considered critical in most, if not all the above-mentioned documents: banking and finance, government, communications, emergency and rescue services, energy, public health, transportation, food and water supply.[98] Although it is not the only decisive factor as the supporters of the target-based approach suggest, the critical character of the targeted infrastructure is an important element to consider in order to establish when a disruptive cyber operation amounts to a use of force under Article 2(4), in particular as a measure of the severity of its effects. It is especially helpful to *exclude* that the operation is a use of force: if the targeted infrastructure is not critical, it is highly unlikely that the consequent disruption will affect a state's essential functions and its internal public order. Whether or not an incapacitating cyber operation amounts to a use of force, however, necessarily depends not only on the critical character of the targeted infrastructure, but also on other factors such as the seriousness of the disruption, its duration, the sophistication of the means employed, and the reliance of the victim state on information systems.[99]

NCIs include defence infrastructures. Because of their consequences on state security, operations (cyber or not) that severely disrupt such infrastructure are likely to be treated as a use of force.[100] The 1999 DoD *Assessment of International Legal Issues in Information Operations*, for instance, argues that 'corrupting the data in a nation's computerized systems for managing its military fuel, spare parts, transportation, troop mobilization, or medical supplies', therefore seriously interfering 'with its ability to conduct military operations', might be treated as a use of

---

[96] EU Commission, *Green Paper on a European Programme on Critical Infrastructure Protection*, COM(2005) 576 final, 17 November 2005, p 20 <http://eur-lex.europa.eu/LexUriServ/LexUriServ.do?uri=COM:2005:0576:FIN:EN:PDF>. Critical information infrastructures are defined as those information and communication technologies 'that are critical infrastructures for themselves or that are essential for the operation of critical infrastructures (telecommunications, computers/software, Internet, satellites, etc)' (p 19). An indicative list of critical infrastructure sectors includes energy, information and communication technologies, water, food, health, financial, public and legal order and safety, civil administration, transport, chemical and nuclear industry, and space and research (p 24).

[97] See Russian Foreign Ministry and Security Council, Convention on International Information Security (Concept), 2011, Art 2 <http://www.mid.ru/bdomp/ns-osndoc.nsf/1e5f0de28fe77fdcc32575d900298676/7b17ead7244e2064c3257925003bcbcc!OpenDocument>.

[98] Brunner and Suter, *International CIIP Handbook*, p 529.

[99] In his speech at CYBERCOM, the then US Department of State's Legal Advisor stated that, in order to establish whether a cyber operation amounts to a use of force, several factors must be evaluated, including 'the context of the event, the actor perpetrating the action...the target and location, effects and intent, among other possible issues' (Koh, 'International Law in Cyberspace', p 595).

[100] Ziolkowski, 'Computer Network Operations', pp 73–4; Nicholas Tsagourias, 'Cyber Attacks, Self-Defence and the Problem of Attribution', *Journal of Conflict and Security Law* 17 (2012), p 232.

force.[101] If a cyber operation seriously disrupting defence functions is considered a use of force, this conclusion must be *a fortiori* correct for a cyber operation that aims at taking control of networked weapons and weapon systems of another state, such as missiles, satellites, and drones.[102]

As to cyber operations that severely disrupt other NCIs such as emergency and rescue services, energy, public health, transportation, food and water supply, in most cases they will also result in some physical damage to property or persons. In particular, prolonged electricity shortage due to a cyber attack disrupting the national grid is likely to have severe negative repercussions on other NCIs and on virtually all sectors of society. But even when no physical consequences arise, as in cyber attacks against the banking and finance, government, and communications sectors, 'a flexible interpretation [of Article 2(4)] according to the evolution of weaponry and the logic behind this provision does not prevent a broadening of its prohibition in order to incorporate new uses of force'.[103] As the ICJ found, 'where the parties have used generic terms in a treaty... [they] must be presumed, as a general rule, to have intended those terms to have an evolving meaning'.[104] The Court gives the example of 'commerce',[105] but the same reasoning could be applied to 'force'. Indeed, the increasing dependency of states on computer systems and networks to provide critical services for the society as well as the increasing severity and sophistication of cyber attacks should be taken into account when interpreting Article 2(4). As a report suggests, focusing only on physically destructive consequences on individuals and property is reductive in the cyber context, as 'modern society depends on the existence and proper functioning of an extensive infrastructure that itself is increasingly controlled by information technology': therefore, '[a]ctions that significantly interfere with the functionality of that infrastructure can reasonably be regarded as uses of force, whether or not they cause immediate physical damage'.[106] Melzer concurs and remarks how the kinetic equivalence doctrine, that considers as a use of force only those cyber operations that cause material damage comparable to kinetic

---

[101] US DoD, *An Assessment*, p 18.

[102] A computer virus has for instance been reported to have infected US drones' cockpits ('US drones infected by key logging virus', *Aljazeera*, 8 October 2011, <http://www.aljazeera.com/news/americas/2011/10/201110816388104988.html>). The virus apparently originated in China and had the purpose of collecting unmanned aerial vehicle (UAV) data. It was inserted in the military's network through compromised PDF files (Robert Johnson, 'New Evidence Suggests China's Hacking into US Drones Using Adobe Reader and Internet Explorer', *Business Insider*, 22 December 2011, <http://www.businessinsider.com/chinas-hacking-into-us-drones-using-adobe-reader-and-internet-explorer-2011-12>). Chinese hackers are also suspected to have interfered with US satellites used for earth observation (Charles Arthur, 'Chinese hackers suspected of interfering with US satellites', *The Guardian*, 27 October 2001, <http://www.guardian.co.uk/technology/2011/oct/27/chinese-hacking-us-satellites-suspected>).

[103] Antonio Segura-Serrano, 'Internet Regulation and the Role of International Law', *Max Planck Yearbook of United Nations Law* 10 (2006), pp 224–5.

[104] *Dispute Regarding Navigational and Related Rights (Costa Rica v Nicaragua)*, Judgment, 13 July 2009, ICJ Reports 2009, para 66.

[105] *Dispute Regarding Navigational and Related Rights*, para 70.

[106] Owens, Dam, and Lin, *Technology, Policy, Law, and Ethics*, p 254. Similarly, Brown and Poellet argue that '[a]lthough no actual kinetic event may occur, the reliance of modern societies on electricity for health care, communications, and the delivery of essential services makes it clear this would

weapons, is too restrictive.[107] Similarly, Tsagourias emphasizes that 'a cyber attack on critical state infrastructure which paralyses or massively disrupts the apparatus of the State should be equated to an armed attack, even if it does not cause any immediate human injury or material damage'.[108]

If, as has been seen, a use of armed force entails the use of instruments capable of producing violent consequences, it is interesting that Panama has noted that misuse of information and telecommunication systems is a 'new form of violence'.[109] The US Air Force also includes disruption in its definition of weapons to be reviewed under Article 36 of Additional Protocol I, where it states that '[w]eapons are devices designed to kill, injure, disable or temporarily incapacitate people, or destroy, damage *or temporarily incapacitate* property or materiel'.[110] It is also significant that the previous definition adopted by the Air Force did not include any reference to temporary incapacitation and expressly excluded 'electronic-warfare devices' from the definition of weapons.[111] The 2004 *National Military Strategy of the United States of America* refers to 'weapons of mass effect', which 'rely more on disruptive impact than destructive kinetic effects' and gives the example of cyber attacks on US commercial information systems or against transportation networks, which 'may have a greater economic or psychological effect than a relatively small release of a lethal agent'.[112]

Some states have also expressly qualified the incapacitation of certain infrastructures as a use of force. Mali has for instance claimed that '[t]he use of an information weapon could be interpreted as an act of aggression if the victim State has reasons to believe that the attack was carried out by the armed forces of another State and was aimed at disrupting...economic capacity, or violating the State's sovereignty over a particular territory'.[113] A Russian senior military officer has reportedly declared that 'the use of Information Warfare against Russia or its armed forces will categorically not be considered a non-military phase of a conflict whether there were casualties or not'.[114] The US DoD's *Assessment of International Legal Issues in Information Operations* refers to a nation's air traffic control system, its banking and financial system, and public utilities and dams as examples of targets that, if shut down by a coordinated computer network attack, might entitle the victim state to self-defence.[115] The 2011 DoD *Cyberspace Policy Report* maintains that the

---

qualify as a kinetic-like effect and would therefore constitute a military attack if the disruption were for a significant period of time' (Gary Brown and Keira Poellet, 'The Customary International Law of Cyberspace', *Strategic Studies Quarterly* 6(3) (2012), p 137).

[107] Melzer, *Cyberwarfare*, p 14. [108] Tsagourias, 'Cyber Attacks', p 231.
[109] UN Doc A/57/166/Add.1, 29 August 2002, p 5.
[110] Air Force Instruction 51–402 (27 July 2011), p 6 (emphasis added).
[111] Air Force Instruction 51–402 (13 May 1994), p 1.
[112] *The National Military Strategy of the United States of America—A Strategy for Today; A Vision for Tomorrow*, 2004, p 1, <http://www.defense.gov/news/mar2005/d20050318nms.pdf>.
[113] UN Doc A/64/129/Add.1, 9 September 2009, p 8.
[114] Quote from the speech of a senior Russian military officer, cited in Vida M Antolin-Jenkins, 'Defining the Parameters of Cyberwar Operations: Looking for Law in All the Wrong Places?', *Naval Law Review* 51 (2005), p 166.
[115] US DoD, *An Assessment*, p 18.

United States reserves the right to use 'all necessary means' against 'hostile acts', including 'significant cyber attacks', directed not only against the US government or military but also the economy.[116] The US Presidential Policy Directive 20 identifies the following 'significant consequences' of cyber operations that require Presidential approval: '[l]oss of life, significant responsive actions against the United States, significant damage to property, serious adverse U.S. foreign policy consequences, or serious economic impact on the United States'.[117] Finally, when submitting its views on information security to the UN Secretary-General, the United States has argued that 'under some circumstances, a *disruptive* activity in cyberspace could constitute an armed attack' (and therefore a use of force).[118]

The above makes sense if one considers that, because of the reliance of modern societies on computers, computer systems and networks, cyber technologies have enabled states to produce results analogous to those of kinetic weapons but without the need of physical damage. An 'interpretive reorientation'[119] of Article 2(4) that also covers cyber attacks causing serious disruption of critical services without destroying the infrastructures would also reflect the trend in modern warfare to favour incapacitation to destruction.[120] For instance, in the 1999 Operation Allied Force against Yugoslavia, NATO targeted switching stations instead of generation stations to enable fast repair after the conflict and used carbon-graphite filaments in such a way as to cause only temporary disruption of electricity.[121] The Rules of Engagement distributed to the US Military Forces in Operation Iraqi Freedom also provided that attacks at the enemy infrastructure, lines of communication, and economic objects should be aimed at disabling and disrupting them, avoiding destruction whenever possible.[122] Like non-lethal weapons, cyber warfare should be understood in this context.[123]

---

[116] US DoD, *Cyberspace Policy Report. A Report to Congress Pursuant to the National Defense Authorization Act for Fiscal Year 2011, Section 934*, November, 2011, p 4 <http://www.defense.gov/home/features/2011/0411_cyberstrategy/docs/NDAA%20Section%20934%20Report_For%20webpage.pdf>.

[117] US Presidential Policy Directive 20, p 3.

[118] UN Doc A/66/152, 15 July 2011, p 18 (emphasis added).

[119] Waxman, 'Cyber Attacks', p 437.

[120] As Anthony D'Amato has put it, '[t]he cybernetic system of international law is...a purposive system. Its rules cannot be interpreted literally or applied mechanically because each rule is simply an indication of how the system should deal with disruptions that may arise' (Anthony D'Amato, 'International Law, Cybernetics, and Cyberspace', *International Law Studies* 76 (2002), p 62).

[121] Dominik Steiger, 'Civilian Objects', *Max Planck Encyclopedia of Public International Law* (2012), Vol II, p 187.

[122] Human Rights Watch, *Off Target. The Conduct of the War and Civilian Casualties in Iraq* (2003), pp 138–9, <http://www.hrw.org/reports/2003/usa1203/usa1203.pdf>.

[123] NATO has defined non-lethal weapons as 'weapons which are explicitly designed and developed to incapacitate or repel personnel, with a low probability of fatality or permanent injury, or to disable equipment, with minimal undesired damage or impact on the environment' ('NATO Policy on Non-Lethal Weapons', Press Statement, 13 October 1999, <http://www.nato.int/docu/pr/1999/p991013e.htm>). The same definition is contained in the *UK Manual of the Law of Armed Conflict* (UK Ministry of Defence, *The Manual of the Law of Armed Conflict* (Oxford: Oxford University Press, 2004), p 118). The recent ICRC Guide to the Legal Review of New Weapons, Means and Methods of Warfare specifies that Art 36 of Additional Protocol I covers all weapons, be they lethal

An objection that is often raised is that a cyber attack that shuts down economic targets such as the stock exchange and disrupts financial markets amounts to economic coercion and not to armed force.[124] This is not correct, for two reasons. First, economic coercion does not have a specific target, while the cyber attack would be undertaken against an identifiable infrastructure. Secondly, while economic coercion, such as an oil embargo, employs the economy as a *means* of pressure, in a cyber attack that incapacitates the financial market or cripples a state's banking system the economy is rather the *target*, while the means employed is malware. Therefore, if the stock exchange or other financial institutions were to be bombed kinetically and the markets disrupted as a consequence, this would certainly be considered a use of armed force, and not economic coercion, even though the economic consequences of the action would probably outweigh the physical damage to the buildings: one cannot see why the same conclusion should not apply when the stock exchange, instead of being bombed, is shut down for an extended period of time by a virus in its computer system.[125] Such scenario would arguably be seen as having more in common with a surgical kinetic attack than with the 1973 Organization of the Petroleum Exporting Countries (OPEC)'s oil embargo. If, however, the disruption caused is not severe, 'the cyber attack may be less likely to be regarded as a use of force than a kinetic attack with the same (temporary) economic effect, simply because the lack of physical destruction would reduce the scale of the damage caused'.[126]

It cannot be overemphasized that it is only cyber attacks that go beyond mere inconvenience and *significantly* disrupt the functioning of *critical* infrastructure that can potentially fall under the scope of Article 2(4).[127] It is only in these cases that the effects of disruption can be equated to those of destruction caused by traditional armed force. A week-long cyber attack that shuts down the national grid, and thus leaves millions of people without electricity, cripples the financial market and the transport system, and prevents government communications is likely to be treated as a use of force, whether or not physical damage ensues.[128] On the other

---

or non-lethal (ICRC, 'A Guide to the Legal Review of New Weapons, Means and Methods of Warfare: Measures to Implement Article 36 of Additional Protocol I of 1977', *International Review of the Red Cross 88* (2006), p 937). On non-lethal weapons, see generally David P Fidler, 'The Meaning of Moscow: "Non-lethal" Weapons and International Law in the Early 21st Century', *International Review of the Red Cross* 87 (2005), p 525.

[124] See eg Elizabeth Wilmshurst, 'The Chatham House Principles of International Law on the Use of Force in Self-Defence', *International and Comparative Law Quarterly* 55 (2006), p 965.

[125] Brown and Poellet, 'The Customary International Law', p 138; Herbert S Lin, 'Offensive Cyber Operations and the Use of Force', *Journal of National Security Law and Policy* 4 (2010), p 74.

[126] Lin, 'Offensive Cyber Operations', p 74.

[127] Antolin-Jenkins, 'Defining the Parameters', p 172; Ziolkowski, 'Computer Network Operations', pp 74–5.

[128] The scenario is inspired to the 'Cyber ShockWave' simulation staged by the Bipartisan Policy Center (Ellen Nakashima, 'War game reveals U.S. lacks cyber-crisis skills', *The Washington Post*, 17 February 2010, <http://articles.washingtonpost.com/2010-02-17/news/36782381_1_cyber-attack-cyber-coordinator-cyber-shockwave>).

hand, a cyber attack that shuts down a university network is not a use of force, even if it causes prolonged and severe disruption, because the infrastructure it affects is not critical. Similarly, although they targeted critical infrastructures (banking and communications), the 2007 DDoS attacks on Estonia caused no material damage or serious disruption and were thus not a violation of Article 2(4), even if they had been conclusively attributed to a state. Indeed, Estonia reacted by using passive cyber defences, conducting criminal investigations, and requesting judicial cooperation.

Those worried that, by qualifying seriously disruptive cyber operations as a use of force, the risk of inter-state conflicts will increase should be reassured: indeed, a use of force, in itself, is not sufficient to entitle the victim state to react in self-defence, unless it is serious enough to amount to an 'armed attack'.[129] Apart from the stigma attached to it, then, the only consequence of qualifying seriously disruptive cyber operations as a use of force is that they could not be undertaken as countermeasure, which certainly is a welcome result, considering the severe negative impact that they might have on the public order of today's digitalized societies.[130]

1.3 Cyber attacks below the level of the use of force

The fact that non-seriously disruptive cyber attacks or cyber attacks seriously disruptive of non-critical infrastructure are not a violation of Article 2(4) of the UN Charter does not mean that they are lawful: when attributed to a state, they can amount to a violation of the principle of non-intervention in the internal affairs of another state.[131] Intervention is 'the manifestation of a policy of force'.[132] What characterizes intervention and differentiates it from a mere violation of sovereignty is the element of coercion: a state exercises abusive pressure on another state in order to coerce it, through certain means, to do or not to do something 'on matters in which each State is permitted, by the principle of State sovereignty, to decide freely', such as 'the choice of a political, economic, social and cultural system, and the formulation of foreign policy'.[133] What distinguishes intervention from a violation of Article 2(4) is that it can occur 'with or without armed force'.[134]

According to the ICJ, the principle of non-intervention is 'part and parcel of customary international law'.[135] Even though it is not expressly mentioned in the

---

[129] See Chapter 2, Section III.1.
[130] Article 50(1)(a) of the ILC Articles on State Responsibility, *Yearbook of the International Law Commission*, 2001, Vol II, Part Two, p 30. See Chapter 2, Section II.1.3.
[131] See Commentary to Rule 10 of the Tallinn Manual, p 44.
[132] *Corfu Channel (United Kingdom v Albania)*, Merits, Judgment, 9 April 1949, ICJ Reports 1949 ('*Corfu Channel*'), p 35.
[133] *Nicaragua*, para 205. According to the ICJ, '[i]ntervention is wrongful when it uses methods of coercion in regard to such choices, which must remain free ones. The element of coercion...defines, and indeed forms the very essence of, prohibited intervention' (para 205). See Maziar Jamnejad and Michael Wood, 'The Principle of Non-intervention', *Leiden Journal of International Law* 22 (2009), pp 345–81.
[134] *Nicaragua*, para 206.   [135] *Nicaragua*, para 202.

UN Charter, the principle has been incorporated in a plethora of international agreements.[136] Paragraph 1 of the 1965 UN General Assembly Declaration on the Inadmissibility of Intervention in the Domestic Affairs of States and the Protection of their Independence and Sovereignty also condemns 'armed intervention and *all other forms of interference* or attempted threats against the personality of the State or against its political, economic and cultural elements', while paragraph 2 proclaims that '[n]o State may use or encourage the use of economic, political *or any other type of measures* to coerce another State in order to obtain from it the subordination of the exercise of its sovereign rights or to secure from it advantages of any kind': the language is broad enough to include cyber attacks below the level of the use of force.[137] In its third principle, the 1970 Declaration on Friendly Relations condemns all forms of intervention by using analogous language.[138] The 1976 Declaration on Non-interference can also be extended to cyber operations where it condemns 'all forms of overt, subtle and highly sophisticated techniques of coercion, subversion and defamation aimed at disrupting the political, social or economic order of other States or destabilizing the Governments seeking to free their economies from external control or manipulation'.[139] Several of the situations described in the subsequent 1981 General Assembly Declaration on the Inadmissibility of Intervention and Interference in the Internal Affairs of States would also perfectly fit certain cyber operations.[140] In particular, the Declaration recalls '[t]he right of States and peoples to have free access to information and to develop fully, without interference, their system of information and mass media and to use their information media in order to promote their political, social, economic and cultural interests and aspirations, based, inter alia, on the relevant articles of the Universal Declaration of Human Rights and the principles of the new international information order' (paragraph 2(I)(c)). A 1989 bilateral treaty between the United States and the Soviet Union on the prevention of dangerous military activities also prohibits 'interfering with command and control networks in a manner which could cause harm to personnel or damage to equipment of the armed forces of the other Party'.[141] Article I(9) of the treaty, in particular, defines '[i]nterference with command and control networks' as 'actions that hamper, interrupt or limit the operation of the signals and information transmission means and systems providing for the control of personnel and equipment of the armed forces of a Party'.

---

[136] See eg Art 8 of the 1933 Convention on the Rights and Duties of States, Art 19 of the OAS Charter, Art 4 of the AU Constitutive Act, Art 2(2)(e) of the Charter of the Association of Southeast Asian Nations (ASEAN), Art 2 of the Charter of the Shanghai Cooperation Organization (SCO), Art 2(4) of the Charter of the Organisation of the Islamic Conference.

[137] GA Res 2131 (XX), 21 December 1965 (adopted without dissent, with one abstention; emphasis added).

[138] Declaration on Friendly Relations. See also Principle VI of the Helsinki Final Act of the Conference on Security and Co-operation in Europe (CSCE), 1 August 1975, *International Legal Materials* 14 (1975), pp 1294–5.

[139] GA Res 31/91, 14 December 1976, para 4.   [140] GA Res 36/103, 9 December 1981.

[141] Article 2(1)(d) of the Agreement on the Prevention of Dangerous Military Activities, *International Legal Materials* 28 (1989), pp 877 ff. The treaty is still in force.

There is no difficulty, then, to qualify disruptive cyber attacks, be they above or below the level of the use of force, as unlawful interventions when they are used 'to coerce another State in order to obtain from it the subordination of the exercise of its sovereign rights and to secure from it advantages of any kind'.[142] Indeed, in the 20th century the notion of 'intervention' was broadened as a consequence of the increased cooperation among states allowing more subtle ways of interference than the use of force:[143] a further broadening of the notion to include cyber attacks is now necessitated by the interconnectivity of networks and the reliance of modern societies on information systems.

Not only may cyber attacks that cause low-level disruption be unlawful interventions, but also those that deface websites in order to foment civil strife in a state or the sending of thousands of e-mails to voters in order to influence the outcome of political elections in another state.[144] In September 2012, for instance, Azerbaijan denounced cyber attacks conducted by a so-called 'Armenian Cyber Army' under the 'direction and control' of Armenia that were 'aimed at glorifying terrorists and insulting their victims, as well as at advocating, promoting and inciting ethnically and religiously motivated hatred, discrimination and violence'.[145] As has been observed, if a broadcast 'is deliberately false and intended to produce dissent or encourage insurgents, the non-intervention principle is likely to be breached'.[146] Hostile propaganda and defamation are condemned in the above-mentioned 1976 Declaration on Non-interference and 1981 Declaration on Non-intervention.[147]

1.4 Cyber exploitation

As has been seen in Chapter 1, cyber exploitation includes the unauthorized access to other computers, computer systems or networks, in order to exfiltrate information, but without affecting the functionality of the accessed system or corrupting, amending or deleting the data resident therein. They are, therefore, never a use of force under Article 2(4) of the UN Charter. As has been observed, '[t]he primary technical difference between cyber attack and cyberexploitation is in the nature of the payload to be executed—a cyber attack payload is destructive whereas a cyberexploitation payload acquires information nondestructively'.[148]

---

[142] Declaration on Friendly Relations, third principle.
[143] Philip Kunig, 'Intervention, Prohibition of', *Max Planck Encyclopedia of Public International Law* (2012), Vol VI, p 290.
[144] Tallinn Manual, p 45. The UN General Assembly has condemned external interference in electoral processes in several resolutions: see eg GA Res 60/164, 6 December 2005.
[145] Letter dated 6 September 2012 from the Chargé d'affaires a.i. of the Permanent Mission of Azerbaijan to the United Nations addressed to the Secretary-General, UN Doc A/66/897–S/2012/687, 7 September 2012, p 1.
[146] Jamnejad and Wood, 'The Principle', p 374.
[147] Paragraph 2(II)(j) of the Declaration on Non-intervention, for instance, prohibits '[t]he duty of a State to abstain from any defamatory campaign, vilification or hostile propaganda for the purpose of intervening or interfering in the internal affairs of other States'.
[148] Lin, 'Offensive Cyber Operations', p 64.

Some cyber exploitation operations are a contemporary form of military reconnaissance or espionage. It should be recalled that espionage is not prohibited by international law, although it is usually criminalized at domestic level.[149] Cyber exploitation operations, however, may be a violation of the sovereignty of the targeted state when they entail an unauthorized intrusion into cyber infrastructure located in another state (be it governmental or private),[150] although not intervention (and even less a use of force), as they lack the coercive element.[151] It is significant that US guidelines allow the military to transmit computer codes to another state's network for reconnaissance purposes, but the code needs to be passive, ie not include a virus or worm that could be activated to do harm at a later date.[152]

The problem is, of course, that software is often dual-use, ie it could be used both for stealing information or conducting a destructive or disruptive attack. Software codes initially used to collect information might also be reprogrammed and subsequently turned into a destructive agent. The problem of dual-use technology, however, is not specific to cyberspace and is already well-known in the field of disarmament, where it is addressed through verification and control mechanisms.[153]

1.5 Activities related to cyber operations

It has been seen that, according to the ICJ, it is not only the direct use of weapons by a state against another state that amounts to a use of force, but also enabling someone else to use those weapons: the arming and training of armed groups, therefore, is a violation of Article 2(4), even though not an armed attack.[154] If this view is applied in the cyber context, one should conclude that the supply of malware by a state to an armed group acting against another state and the training of such group to conduct cyber attacks are also a use of force.[155] This is at least the opinion of the Group of Experts that drafted the Tallinn Manual.[156]

---

[149] Dinstein, 'Computer Network Attacks', p 101; Richard W Aldrich, 'How Do You Know You Are at War in the Information Age?', *Houston Journal of International Law* 22 (1999–2000), p 252; David P Fidler, 'Tinker, Tailor, Soldier, Duqu: Why cyberespionage is more dangerous than you think', *International Journal of Critical Infrastructure Protection* 5 (2012), p 28.

[150] Wolff Heintschel von Heinegg, 'Territorial Sovereignty and Neutrality in Cyberspace', *International Law Studies* 89 (2013), p 129. In *Nicaragua*, the ICJ found that the US reconnaissance flights breached Nicaragua's sovereignty as a result of their trespass into Nicaraguan airspace (*Nicaragua*, para 91).

[151] Johann-Christoph Woltag, 'Cyber Warfare', *Max Planck Encyclopedia of Public International Law* (2012), Vol II, p 989.

[152] Lolita C Baldor, 'U.S. lays out cyber attack guidelines', *The Associated Press*, 23 June 2011, <http://articles.philly.com/2011-06-23/news/29695033_1_cyber-capabilities-guidelines-strategy>.

[153] On the problems of applying verification mechanisms to cyber weapons, see Louise Arimatsu, 'A Treaty for Governing Cyber-Weapons: Potential Benefits and Practical Limitations', in *2012 4th International Conference*, edited by Czosseck, Ottis, and Ziolkowski, pp 101–2.

[154] *Nicaragua*, para 228.

[155] A recent example is the alleged cyber training of members of the Syrian dissidents by the US State Department (Jay Newton-Small, 'Inside America's Secret Training of Syria's Digital Army', *Time*, 13 June 2012, <http://swampland.time.com/2012/06/13/inside-americas-secret-training-of-syrias-digital-army>).

[156] Tallinn Manual, p 46.

This conclusion, however, is correct only when the supply of malware and the training enable the group to conduct cyber attacks amounting to a use of force, and not other types of cyber operations below that threshold.

Similarly, if one transposes Article 3(f) of the Definition of Aggression in the cyber context, a state that knowingly allows another state to use its cyber infrastructure in order to launch a cyber attack amounting to an act of aggression against a third state would commit an act of aggression itself, and therefore would breach the prohibition of the use of force.[157] It appears, for instance, that North Korea's cyber warfare Unit 121 is at least partially stationed in China due to the limited internet connections in North Korea, although the involvement of the Chinese government is unclear.[158] Article 3(f), however, would only potentially apply to when the cyber operation in question amounts to an act of aggression.

2. Cyber operations and threats of force

The 'threat of the use of force' is nowhere defined in the Charter, nor are the *travaux préparatoires* useful to this purpose, but the present author has suggested elsewhere that a threat of force under Article 2(4) can be defined as an explicit or implicit promise, through statements or actions, of a *future* and *unlawful* use of armed force against one or more states, the realization of which depends on the threatener's will.[159] To be a violation of Article 2(4), then, the threat must not be a measure of self-defence or have been authorized by the Security Council under Chapter VII.[160]

A threat of force can be explicit or be implied in certain conduct. With regard to the former, two situations can be envisaged in the cyber context: a cyber threat of force and the threat of cyber force. The first is the threat of a use of force with *traditional* weapons communicated through *cyber* means. In this scenario, cyber tools are merely used to deliver a more traditional effect. Article 2(4) does not specify the methods through which a threat should be conveyed and thus 'communicating a threat via the Internet would be on the same theoretical footing as communicating a threat by traditional methods'.[161] The threat, delivered through cyber or traditional means, could also envisage a possible use of force by *cyber* means by the

---

[157] Article 3(f) of the Definition of Aggression includes among the acts of aggression '[t]he action of a State in allowing its territory, which it has placed at the disposal of another State, to be used by that other State for perpetrating an act of aggression against a third State'.

[158] Richard A Clarke and Robert K Knake, *Cyber War. The Next Threat to National Security and What To Do About It* (New York: Harpercollins, 2010), pp 27–8.

[159] Marco Roscini, 'Threats of Armed Force and Contemporary International Law', *Netherlands International Law Review* 54 (2007), p 235. Threats of force under Art 2(4) must be distinguished from threats to the peace under Art 39 of the UN Charter. A threat of force can well be qualified as a threat to the peace by the Security Council, since it is likely to escalate into actual use of force, but, the 'threat to the peace' concept is much broader and is not necessarily linked to a use of force or even to a violation of international law. On threats to the peace, see Chapter 2, Section V.

[160] Report of the International Fact-Finding Commission on the Conflict in Georgia, Vol II, p 236. On threats of force as a self-defence measure, see James Green and Francis Grimal, 'The Threat of Force as an Action in Self-Defense under International Law', *Vanderbilt Journal of Transnational Law* 44 (2011), pp 285 ff.

[161] Aldrich, 'How Do You Know', p 237.

threatening state, ie the conduct of the cyber operations and cyber related activities identified above as uses of force. In any case, to fall under the scope of Article 2(4), the threat must be clearly identifiable and must be made known to the target state so to exercise its coercive effect.[162]

As to threats resulting implicitly from certain conduct, 'a demonstration of force for the purpose of exercising political pressure'[163] could well amount to a threat under the terms of Article 2(4). A possible example may be cyber warfare simulations and other military exercises, such as those conducted by China's People's Liberation Army,[164] in the context of a situation of tension between two states. Secret cyber warfare exercises, however, are not a threat of force: in order to be a violation of Article 2(4), the threat must be known to the target so to exercise its unlawful coercion. The factual circumstances of each case in the framework of the relations between the concerned states play a fundamental role in determining whether the target state is entitled to feel threatened and thus whether the conduct qualifies as a threat of force.[165] In its Memorial in the Merits phase before the ICJ, for instance, Nicaragua claimed that the 'continuous US military and naval maneuvers adjacent to Nicaraguan borders, officially acknowledged as a program of "perception management"' amounted to a threat of force under Article 2(4) as they formed part of a 'general and sustained policy of force, publicly expounded, intended to intimidate the lawful Government of Nicaragua into accepting the political demands of the United States Government, and resulting in substantial infringements of the political independence of Nicaragua'.[166] However, the Court concluded that, 'in the circumstances in which they were held', the US military manoeuvres did not amount to a breach of the principle forbidding the recourse to the threat or use of force.[167] Similarly, the Arbitral Tribunal in the *Guyana and Suriname* Award held that '*in the circumstances of the present case*, this Tribunal is of the view that the action mounted by Suriname on 3 June 2000 seemed more akin to a threat of military action rather than a mere law enforcement activity'.[168] The Independent Fact-Finding Mission on the Conflict in Georgia established by the Council of the European Union also found that 'as soon as [militarized acts] are non-routine, suspiciously timed, scaled up, intensified, geographically proximate, staged in the exact mode of a potential military clash, and easily attributable to a foreign-policy message, the hostile intent is considered present and the demonstration of force manifest'.[169]

---

[162] See Schmitt, 'Cyber Operations in International Law', p 153.
[163] *Corfu Channel*, para 35.
[164] Institute for Security Technology Studies, *Cyber Warfare. An Analysis of the Means and Motivations of Selected Nation States*, December 2004, pp 32–3, <http://www.ists.dartmouth.edu/docs/cyberwarfare.pdf>. Network warfare exercises have allegedly been conducted also by the United States, France, South Korea, and Japan, as well as NATO (Li Zhang, 'A Chinese Perspective on Cyber War', *International Review of the Red Cross* 94 (2012), p 805).
[165] Roscini, 'Threats', pp 241–3.
[166] *Nicaragua*, ICJ Pleadings 1991, Vol IV, paras 457 and 447, respectively.
[167] *Nicaragua*, para 118.
[168] *Guyana and Suriname*, Arbitral Award, 17 September 2007, *International Legal Materials* 47 (2008), para 445 (emphasis added).
[169] Report of the Independent Fact-Finding Mission on the Conflict in Georgia, Vol II, p 232.

In its 1996 Advisory Opinion on the *Legality of the Threat or Use of Nuclear Weapons*, the ICJ linked the legality of threats, be they explicit or implied in certain conduct, to the legality of the use of force in the same circumstances.[170] This symmetry between threat and use of force finds support in state practice and in legal literature.[171] The considerations on the legality of cyber operations qualifying as a use of force developed above, therefore, also apply, *mutatis mutandis*, to those amounting to threats of force.

III. Cyber Operations and the Law of Self-Defence

Self-defence against cyber attacks can be exercised by physical, electronic or cyber means. Physical self-defence uses traditional weapons to target the cyber infrastructure of the attacker, such as the servers from which the cyber attacks originate, or other physical targets consistently with the requirements of necessity and proportionality and with international humanitarian law. Electronic reactions to a cyber attack employ 'the use of electromagnetic energy, directed energy, or antiradiation weapons to attack personnel, facilities, or equipment with the intent of degrading, neutralizing, or destroying enemy combat capability'.[172] Cyber defences can be passive or active.[173] While passive defences do not involve coercion or unauthorized intrusion into computer systems and therefore are not a use of force,[174] the latter are responses in-kind to a previous cyber attack and are in fact attacks themselves that may fall within the remit of the *jus ad bellum* to the extent that they amount to a use of force.[175]

With the above distinctions in mind, the following Sections will discuss the application of the law of self-defence in the cyber context. It will be seen that the lessons learned in relation to international terrorism are useful for creating a legal paradigm for self-defence against cyber attacks.[176] Indeed, both terrorism and cyber warfare are often characterized by quasi-state capabilities of non-state actors.[177] Both were initially dealt with in the context of law enforcement and inter-state

---

[170] *Nuclear Weapons*, para 47. This conclusion was reaffirmed in the *Guyana-Suriname* Award, para 439.

[171] Corten, *The Law Against War*, pp 111–24. The symmetrical approach has been adopted also in Rule 12 of the Tallinn Manual, according to which '[a] cyber operation, or threatened cyber operation, constitutes an unlawful threat of force when the threatened action, if carried out, would be an unlawful use of force' (p 52). See, for a different view, Romana Sadurska, 'Threats of Force', *American Journal of International Law* 82 (1988), p 258.

[172] US DoD, *Dictionary of Military and Associated Terms*, p 91.

[173] See Chapter 1, Section II, p 14.

[174] Matthew J Sklerov, 'Solving the Dilemma of State Responses to Cyberattacks: A Justification for the Use of Active Defenses Against States Who Neglect Their Duty to Prevent', *Military Law Review* 201 (2009), p 21.

[175] Jay P Kesan and Carol M Hayes, 'Mitigative Counterstriking: Self-Defense and Deterrence in Cyberspace', *Harvard Journal of Law and Technology* 25 (2011–12), p 477.

[176] Duncan Hollis, 'Why States Need an International Law for Information Operations', *Lewis and Clark Law Review* 11 (2007), pp 1026–8.

[177] See William Banks, 'The Role of Counterterrorism Law in Shaping *ad Bellum* Norms for Cyber Warfare', *International Law Studies* 89 (2013), pp 175 ff.

judicial cooperation but, because of their increased scale and gravity, have come to require, at least in certain instances, a military response. In both cases, identification and attribution of attacks present serious evidentiary problems, attacks are launched without warning, and the lapse of time between the launch of the operation and the impact is often extremely short.

1. Cyber operations as an 'armed attack'

The first sentence of Article 51 of the UN Charter famously provides that

[n]othing in the present Charter shall impair the inherent right of individual or collective self-defense if an armed attack occurs against a Member of the United Nations, until the Security Council has taken measures necessary to maintain international peace and security.

The 2011 US *International Strategy for Cyberspace* notes that '[c]onsistent with the United Nations Charter, states have an inherent right to self-defense that may be triggered by certain aggressive acts in cyberspace'.[178] In a US Senate questionnaire in preparation for a hearing on his nomination to head of the new Cyber Command, Lt Gen Alexander also made clear that, while the right to self-defence 'has not been specifically established by legal precedent to apply to attacks in cyberspace, it is reasonable to assume that returning fire in cyberspace, as long as it complied with law of war principles...would be lawful'.[179] The 2003 US *National Strategy to Secure Cyberspace* states that 'a diplomatic or military response in the case of a state sponsored action' will follow 'large' cyber incidents.[180] Presidential Policy Directive 20 more explicitly notes that '[t]he United States Government shall reserve the right to act in accordance with the United States' inherent right of self-defense as recognized in international law, including through the conduct of DCEO [Defensive Cyber Effects Operations]'.[181] In case of a conflict in the 'information space', Russia claims the right of individual and collective self-defence 'with the implementation of any chosen options and means' in accordance with the norms and principles of international law.[182] Italy affirms that states may protect national critical infrastructures from external attacks consistently with international law.[183]

Under Article 51 of the UN Charter, the state targeted by a cyber operation will be entitled to react forcibly in self-defence only to the extent that such operation

---

[178] *International Strategy for Cyberspace*, p 10.
[179] Lolita C Baldor, 'Military asserts right to return cyber attacks', The Associated Press, 14 April 2010, <http://www.sltrib.com/nationworld/ci_14880980>. A separate classified document discusses whether the United States should first ask the government from which the cyber attack originates to deal with it.
[180] The *National Strategy to Secure Cyberspace*, p 28.
[181] US Presidential Policy Directive 20, p 6. On the notion of DCEO, see Chapter I, Section II, p 16.
[182] *Conceptual Views on the Activities of the Armed Forces of the Russian Federation in the Information Space*, 9 September 2000, p 12, <http://www.ccdcoe.org/strategies/Russian_Federation_unofficial_translation.pdf> (CCDCOE's unofficial translation).
[183] Governo italiano, *La posizione italiana sui principi fondamentali di Internet*, 17 September 2012, p 5, <http://download.repubblica.it/pdf/2012/tecnologia/internet.pdf>.

amounts to an 'armed attack'. This requirement applies not only to a defensive reaction with traditional weapons, but also to one with cyber means to the extent that it amounts to a use of force under Article 2(4). In *Nicaragua*, the ICJ acknowledged that a definition of 'armed attack' does not exist in the UN Charter and is not part of treaty law.[184] If it is apparent that an 'armed attack' implies the use of arms, the ICJ made clear that Article 51 applies to 'any use of force, regardless of the weapons employed'.[185] As seen above in the context of Article 2(4),[186] the fact that cyber operations do not employ kinetic weapons does not necessarily mean they are not 'armed': as Zemanek persuasively notes, 'it is neither the designation of a device, nor its normal use, which make it a weapon but the intent with which it is used and its effect. The use of any device, or number of devices, which results in a considerable loss of life and/or extensive destruction of property must therefore be deemed to fulfill the conditions of an "armed" attack'.[187] This conclusion is supported by the Security Council's reaffirmation of the right to self-defence in relation to the 11 September 2001 attacks on the United States, where the 'weapons' employed were hijacked airplanes.[188]

Of course, in order for a cyber operation to amount to an armed attack, it has to be a use of force first, ie an operation that causes or is reasonably likely to cause extrinsic physical damage to persons or property or severe disruption of critical infrastructures. In spite of a contrary opinion,[189] then, stealing sensitive military information by penetrating into the ministry of defence's computers when 'no immediate loss of life or destruction results' does not qualify as an armed attack. In *Nicaragua*, the ICJ referred to the 1974 General Assembly's Definition of Aggression in order to identify an armed attack.[190] 'Aggression', however, is a notion broader than 'armed attack', with the latter being a sub-category of the former.[191] Of the cases listed in Resolution 3314, Article 3(a), (b), (d), and (g) are generally believed to be not only acts of aggression, but also armed attacks.[192] Article 3(b), in particular, refers to 'the

---

[184] *Nicaragua*, para 176.   [185] *Nuclear Weapons*, para 39.
[186] See Chapter II, Section II.1.
[187] Karl Zemanek, 'Armed attack', *Max Planck Encyclopedia of Public International Law* (2012), Vol I, p 599.
[188] See SC Res 1368 (12 September 2001) and SC Res 1373 (28 September 2001). The UN High Level Panel on Threats, Challenges and Changes states that Resolution 1368 'opened the way for United States-led military action against the Taliban regime in self-defence' (*A More Secure World: Our Shared Responsibility*, Report of the High-level Panel on Threats, Challenges and Change, UN Doc A/59/565, 2 December 2004, para 14).
[189] Joyner and Lotrionte, 'Information Warfare', p 855.
[190] *Nicaragua*, para 195.
[191] See Jean Combacau, 'The Exception of Self-Defence in the United Nations Practice', in *The Current Regulation of the Use of Force*, edited by Antonio Cassese (Dordrecht, Boston, and Lancaster: Nijhoff, 1986), p 22; Pierluigi Lamberti Zanardi, 'Indirect Military Aggression', in *The Current Regulation of the Use of Force*, p 114; Terry D Gill, 'The Law of Armed Attack in the Context of the *Nicaragua* Case', *Hague Yearbook of International Law* 1 (1988), p 36; Albrecht Randelzhofer and Georg Nolte, 'Article 51', in *The Charter of the United Nations*, edited by Simma, Khan, Nolte, and Paulus, Vol II, pp 1407–8; Elena Sciso, 'L'aggressione indiretta nella definizione dell'Assemblea generale delle Nazioni Unite', *Rivista di diritto internazionale* 66 (1983), pp 272–5. Article 6 of the 1947 Inter-American Treaty of Reciprocal Assistance (Rio Treaty), UNTS, Vol 21, pp 77 ff, expressly contemplates a case of 'aggression which is not an armed attack', that gives rise only to the obligation to consult.
[192] Gill, 'The Law of Armed Attack', pp 32–3; Sciso, 'L'aggressione indiretta', p 275.

use of *any weapons* by a State against the territory of another State':[193] the language is broad enough to include cyber weapons. On the other hand, the blockade of ports and coasts is not per se an act of aggression also amounting to an armed attack.[194] Indeed, such measures do not entail an 'attack' if ships stay away from the blockaded ports or are turned away peacefully.[195] Therefore, a cyber operation that simply cuts off a country from the internet, without causing physical damage or severe incapacitation of essential services, would not amount to an armed attack, and therefore the right to self-defence would not ripen.[196] Similarly, '[t]he action of a State in allowing its territory, which it has placed at the disposal of another State, to be used by that other State for perpetrating an act of aggression against a third State' (Article 3(f) of the Definition of Aggression) is generally not believed to amount, on its own, to an armed attack.[197] A state that knowingly allows another state to use its cyber infrastructure in order to launch a cyber operation amounting to an act of aggression, then, would breach the prohibition of the use of force, but would not commit an armed attack itself.

It is well known that the ICJ identified 'the most grave forms of the use of force', ie armed attacks, and less grave forms, and adopted the 'scale and effects' standard in order to distinguish them.[198] As a consequence, a use of force is an 'armed attack' only when its scale and effects are grave enough. The reason for this is clear: the discrepancy between the scope of Article 2(4) and Article 51 aims to avoid that non-serious uses of force unnecessarily lead to an escalation of hostilities through the use of force in self-defence by the victim.[199] The gap between 'use of force' and 'armed attack' is not necessarily wide, as the ICJ held in the *Oil Platforms* case where it did not exclude 'the possibility that the mining of a single military vessel

---

[193] Emphasis added.

[194] In the aftermath of the 2007 DDoS attacks against his country, the Estonian Defence Minister compared cyber blockades to naval blockades on ports (NATO Parliamentary Assembly, *NATO and Cyber Defence*, 173 DSCFC 09 E bis, 2009, para 59, <http://www.nato-pa.int/default.asp?SHORTCUT=1782>). It has been noted that '[t]he effects of naval blockades and information warfare attacks can be similar. Naval blockades prevent the transport of people and products into the target country or area, and may paralyze an economy. In the past, where intercontinental communication was largely by ship, a blockade would keep out information as well. An information warfare attack may also make transport of people and products impossible, paralyzing an economy, and it too may block the spread of information (especially in an "infoblockade")' (Lawrence T Greenberg, Seymour E Goodman, and Kevin J Soo Hoo, *Information Warfare and International Law* (Washington: National Defense University Press, 1998), p 19). Indeed, a state's economy might be more dependent on the internet today than it was on maritime shipping in the past (Owens, Dam, and Lin, *Technology, Policy, Law, and Ethics*, p 260).

[195] Greenberg, Goodman, and Soo Hoo, *Information Warfare*, p 19; Sciso, 'L'aggressione indiretta', p 275.

[196] The Group of Experts that drafted the Tallinn Manual was not able to achieve consensus on the existence, establishment and enforcement of 'cyber blockades' (Tallinn Manual, p 198).

[197] Sciso, 'L'aggressione indiretta', p 275.

[198] *Nicaragua*, paras 191, 195. 'Scale' refers to the amount of armed force employed or its duration, while 'effects' is the damage caused (Tom Ruys, *'Armed Attack' and Article 51 of the UN Charter* (Cambridge: Cambridge University Press, 2010), p 139). The scale and effects standard has also been incorporated in Rule 13 of the Tallinn Manual, p 54.

[199] Natalino Ronzitti, 'The Expanding Law of Self-Defence', *Journal of Conflict and Security Law* 11 (2006), p 351.

might be sufficient to bring into play the "inherent right of self-defence",[200] but it does exist: this is confirmed by Article 2 of the 1974 Definition of Aggression, which requires a use of force to be 'of sufficient gravity' in order to be an act of aggression and therefore, *a fortiori*, an armed attack.[201]

Even though '[i]t is almost impossible to fix the threshold of force employed to define the notion of armed attack',[202] Constantinou has usefully tried to specify the scale and effects standard by arguing that an armed attack is 'an act or the beginning of a series of acts of armed force of considerable magnitude and intensity (ie. scale) which have as their consequences (ie. effects) the infliction of substantial destruction upon important elements of the target State namely, upon its people, economic and security infrastructure, destruction of aspects of its governmental authority, ie. its political independence, as well as damage to or deprivation of its physical element namely, its territory', and the 'use of force which is aimed at a State's main industrial and economic resources and which results in the substantial impairment of its economy'.[203] Note that it is both the scale *and* the effects of the use of force that determine the occurrence of an armed attack: a massive DDoS attack involving millions of botnets that only disrupts a NCI for a limited amount of time is certainly significant with regard to its scale, but its effects are not.[204] Dinstein has suggested some examples of cyber attacks serious enough to amount to armed attacks: '[f]atalities caused by the loss of computer-controlled life-support systems; an extensive power grid outage (electricity blackout) creating considerable deleterious repercussions; a shutdown of computers controlling waterworks and dams, generating thereby floods of inhabited areas; deadly crashes deliberately engineered (e.g., through misinformation fed into aircraft computers)' and 'the wanton instigation of a core-meltdown of a reactor in a nuclear power plant, leading to the release of radioactive materials that can result in countless casualties if the neighbouring areas are densely populated'.[205]

Unlike Dinstein's, Constantinou's definition of the scale and effects standard also includes the effects on the industrial and economic resources of the target state. Indeed, as already noted, it is not only cyber operations causing physical damage

---

[200] *Case Concerning Oil Platforms (Iran v US)*, Merits, Judgment, 6 November 2003, ICJ Reports 2003 ('*Oil Platforms*'), para 72. But see the Partial Award of the Eritrea-Ethiopia Commission (EECC), where it argues that '[l]ocalized border encounters between small infantry units, even those involving loss of life, do not constitute an armed attack for purposes of the Charter' (EECC, *Partial Award*, Jus ad bellum, *Ethiopia's Claims no 1–8*, 19 December 2005, RIAA, Vol XXVI, p 465). Dinstein has suggested that the gap between 'use of force' and 'armed attack', if it exists, has been exaggerated by the ICJ and is in reality very narrow: it is only a use of force that does not result in victims or destruction of property (for instance soldiers shooting across a border killing animals) that falls short of an armed attack (Dinstein, 'Cyber War', p 279).

[201] It should be recalled that armed attack is a *species* in the broader *genus* of aggression.

[202] Ronzitti, 'The Expanding Law', p 351.

[203] Avra Constantinou, *The Right of Self-Defence under Customary International Law and Article 51 of the UN Charter* (Athens and Bruxelles: Ant N Sakkoulas/Bruylant, 2000), pp 63–4 (footnote omitted).

[204] Duncan Blake and Joseph S Imburgia, '"Bloodless Weapons"? The Need to Conduct Legal Reviews of Certain Capabilities and the Implications of Defining Them as "Weapons"', *Air Force Law Review* 66 (2010), p 186.

[205] Dinstein, 'Computer Network Attacks', p 105.

that potentially amount to a use of force, but also those that severely incapacitate critical infrastructures so to affect state security: '[i]t is not their physical destruction as such but their unavailability in the sense of not being able to fulfil the purpose for which they have been set that makes an attack on them an armed attack'.[206] As a consequence, a large-scale cyber attack that shuts down NCIs such as the financial market for a prolonged time and cripples a state's economy or causes the collapse of the national currency would, if the effects are serious enough, potentially amount to an 'armed attack' for the purposes of self-defence.[207] That disruptive cyber operations might be an armed attack justifying the right of self-defence seems to be the *opinio* of many states. It is noteworthy that the United States reserves the right to use 'all necessary means' against 'hostile acts' including 'significant cyber attacks' directed not only against the US government or military but also the economy.[208] The United States has also specified that 'under some circumstances, a *disruptive* activity in cyberspace could constitute an armed attack'.[209] The DoD's *Assessment of International Legal Issues* points out that there may be a right of self-defence 'where significant damage is being done to the attacked system or the data stored in it, when the system is critical to national security or to essential national infrastructures, or when the intruder's conduct or the context of the activity clearly manifests a malicious intent'.[210] The document concludes that 'if a coordinated computer network attack shuts down a nation's air traffic control system *along with its banking and financial systems and public utilities…* it may well be that no one would challenge the victim nation if it concluded that it was a victim of an armed attack, or of an act equivalent to an armed attack'.[211] The Head of the US Strategic Command maintained that the White House retains the option to respond with physical force (including nuclear weapons!) in case of a 'disabling' cyber attack against US computer networks.[212] The 2011 AIV/CAVV Report on Cyber Warfare, the *jus ad bellum* conclusions of which have been endorsed by the Dutch government, states that '[a] serious, organised cyber attack on essential functions of the state could conceivably be qualified as an "armed attack" within the meaning of article 51 of the UN Charter if it could or did lead to serious disruption of the functioning of the state or serious and long-lasting consequences for the stability of the state'.[213] Accordingly, 'a cyber attack that targets the entire financial system or prevents the government from carrying out essential tasks, for example an attack on the entire military communication and command network that makes it impossible to deploy the armed forces, could well be equated with an armed attack'.[214] The White Paper on German

---

[206] Tsagourias, 'Cyber Attacks', p 231.   [207] Melzer, *Cyberwarfare*, p 16.
[208] US DoD, *Cyberspace Policy Report*, p 4.
[209] UN Doc A/66/152, 15 July 2011, p 18 (emphasis added).
[210] US DoD, *An Assessment*, p 20.
[211] US DoD, *An Assessment*, p 18 (emphasis added).
[212] Elaine M Grossman, 'U.S. General Reserves Right to Use Force, Even Nuclear, in Response to Cyber Attack', *Global Security Newswire*, 12 May 2009, <http://gsn.nti.org/gsn/nw_20090512_4977.php>.
[213] AIV/CAVV, *Cyber Warfare*, No 77, AIV/No 22 CAVV, December 2011, p 21, <http://www.aiv-advies.nl/ContentSuite/upload/aiv/doc/webversie__AIV77CAVV_22_ENG.pdf>.
[214] AIV/CAVV, *Cyber Warfare*, p 21.

Cyber Operations and the Law of Self-Defence 75

Security Policy also seems to suggest, if indirectly, that 'military attacks from or on cyberspace' against 'Germany's political and economic structures as well as its critical infrastructure' can be countered 'using military means'.[215] A senior Russian military officer is reported to have said that 'considering the possible catastrophic use of strategic information warfare means by an enemy, whether on economic or state command and control systems, or on the combat potential of the armed forces... Russia retains the right to use nuclear weapons first against the means and forces of information warfare, and then against the aggressor state itself'.[216] Finally, Mali has claimed that '[t]he use of an information weapon could be interpreted as an act of aggression if the victim State has reasons to believe that the attack was carried out by the armed forces of another State and was aimed at disrupting the operation of military facilities, destroying defensive *and economic capacity*, or violating the State's sovereignty over a particular territory'.[217]

To be clear, concluding that cyber attacks that severely disrupt the functioning of critical infrastructures can potentially be an 'armed attack' does not automatically entitle the victim state to use force in self-defence in all cases, as such use must still be necessary and proportionate to the purpose of repelling the attack. Whenever passive cyber defences or cyber attacks below the level of the use of force are reasonably effective means to react, for instance, a use of force in self-defence would be unnecessary and/or disproportionate and thus unlawful, even if the disruptive cyber operation amounted to an armed attack.

What has to be emphasized again is that, contrary to what is argued by a former US Department of State's Legal Advisor,[218] not all cyber operations amounting to a use of force, not even those resulting in physical damage to property or persons, will automatically amount to an 'armed attack': the destruction or disruption must be extensive enough to constitute a 'more serious use of force' giving rise to the right of self-defence. With regard to cyber attacks affecting infrastructure functionality but not resulting in material damage, only coordinated cyber attacks seriously disrupting several or all NCIs of a heavily digitized state for a prolonged time will be likely to meet the high scale and effects threshold of an armed attack. As to cyber operations resulting in material damage, the Commentary of the *Tallinn*

---

[215] German Federal Ministry of Defence, *White Paper 2006 on German Security Policy and the Future of the Bundeswehr*, 2006, p 17, <http://www.isn.ethz.ch/Digital-Library/Articles/Special-Feature/Detail/?lng=en&id=156941>.
[216] Quoted in Antolin-Jenkins, 'Defining the Parameters', p 166.
[217] UN Doc A/64/129/Add.1, 9 September 2009, p 8 (emphasis added).
[218] Koh, 'International Law in Cyberspace', p 597. See also the Dissenting Opinion of Judge Schwebel in *Nicaragua*, pp 347–8 (para 173). At least since the *Nicaragua* case, the United States has criticized the limitation of the right of self-defence to 'armed attack' only (see Abraham D Sofaer, 'International Law and the Use of Force', *American Society of International Law Proceedings* 82 (1988), especially pp 422, 425–6). The two concepts of use of force and armed attack are also merged by the Head of the US Cyber Command, General Alexander: '[i]f the President determines a cyber event does meet the threshold of a use of force/armed attack, he may determine that the activity is of such scope, duration, or intensity that it warrants exercising our right to self-defence and/or the initiation of hostilities as an appropriate response' (Responses to advance questions, Nomination of Lt Gen Keith Alexander for Commander, US Cyber Command, US Sen Armed Serv Committee, 15 April 2010, p 12, <http://armed-services.senate.gov/statemnt/2010/04%20April/Alexander%2004-15-10.pdf>).

Manual concludes that Stuxnet was a use of force and, at least according to some of the experts that drafted the Manual, even an 'armed attack'.[219] This latter view probably goes too far: it is doubtful that, from the information available, Stuxnet had scale and effects significant enough to qualify as an armed attack under Article 51 of the UN Charter.[220] This seems also suggested by the fact that Iran qualified the cyber operation as 'nuclear terrorism' and therefore as 'a grave violation of the principles of the UN Charter and international law', but refrained from explicitly using self-defence language, perhaps implying that the level of damage caused was not considered grave enough to determine the existence of an 'armed attack'.[221] Although such declarations may also be explained on political grounds, they cannot be entirely disregarded from a legal perspective.

It is not clear against whose computers and computer networks the cyber attack should be directed in order to be considered an armed attack on the state. It has been claimed, for instance, that, as Google (an American company) is the most powerful presence on the internet, an attack on it would be an attack on the US critical infrastructure.[222] In a traditional attack, the fact that the target is military or civilian would not make any difference: the state where the target is located would be entitled to self-defence because its territorial integrity has been violated. Hence, Dinstein correctly points out that, if a conventional armed attack against a civilian facility on the territory of the target state would amount to an armed attack even if no member of the armed forces is injured or military property damaged, there is no reason to come to a different conclusion with regard to cyber attacks against civilian systems: '[e]ven if the CNA impinges upon a civilian computer system which has no nexus to the military establishment (like a private hospital installation), a devastating impact would vouchsafe the classification of the act as an armed attack'.[223] Most NCIs are not owned by the government, but by the private sector: the governmental or private character of the infrastructure targeted, however, is also not relevant to the determination of the existence of an armed attack against the state, and neither is the fact that the computer system is run by a company possessing the nationality of a third state or that the computer system operated by the victim state is located outside its borders (for instance, in a military base abroad).[224] When the damage caused to a certain state or its nationals is however not intended (a situation that is particularly likely in the cyber context),[225]

---

[219] Tallinn Manual, pp 45, 58.

[220] Mary Ellen O'Connell, 'Cyber Security without Cyber War', *Journal of Conflict and Security Law* 17 (2012), pp 201–2.

[221] Iranian Foreign Minister's address to the UN Security Council, 28 September 2012, <http://iran-un.org/en/2012/09/28/28-september-2012-2/>. In fact, Iran's President Ahmadinejad downplayed the incident and claimed that it affected only 'a limited number of our centrifuges' (quoted in Albright, Brannan, and Walrond, 'Did Stuxnet', p 1).

[222] Misha Glenny, 'In America's new cyber war Google is on the front line', *The Guardian*, 18 January 2010, <http://www.theguardian.com/commentisfree/2010/jan/18/america-cyberwar-google-china-computer>.

[223] Dinstein, 'Computer Network Attacks', p 106.

[224] Dinstein, 'Computer Network Attacks', pp 106–7.

[225] As Schmitt notes, 'the attacker, because of automatic routing mechanisms, may not be able to control, or even accurately predict, the cyber pathway to the target', which increases the risk of unintended consequences (Schmitt, 'Computer Network Attack: The Normative Software', p 56).

it is doubtful that self-defence may be invoked by the accidental victim, for two reasons. First, as an armed attack is nothing else than a form of aggression, it requires *animus aggressionis*.[226] Indeed, according to the ICJ, an armed attack must be carried out 'with the specific intention of harming'.[227] *Animus aggressionis* means a deliberate intention to cause damage to property, people or systems of a certain state. In the cyber context, this hostile intention can be inferred from 'such factors as persistence, sophistication of methods used, targeting of especially sensitive systems, and actual damage done'.[228] Secondly, if the armed attack by state A against state B also produces unintended harmful consequences on property, persons or systems in state C, a reaction in self-defence by state C would not be necessary, as state A will probably stop the attack on C.[229] The problem is more complicated if state A attacks B posing as state C by spoofing or manipulating transmission data to appear as if they originated from state C. In this case, state C appears to attack state B, which might take actions in self-defence against an unaware state C. But even in this case, the reaction in self-defence may not be necessary if the misunderstanding is cleared up. Furthermore, as the Tallinn Manual suggests, the fact that a cyber operation has been launched from, or has been routed through, the governmental cyber infrastructure of a state is not per se sufficient evidence that the state is responsible for the operation.[230]

2. Anticipatory self-defence against an imminent armed attack by cyber means

Article 51 states that an armed attack must 'occur' in order to trigger the right of self-defence by the victim. In *Nicaragua*, the ICJ did not take a position on the problem of self-defence against attacks that have yet to occur, since 'the issue of the lawfulness of a response to the imminent threat of armed attack' was not raised.[231] Similarly, in the case concerning *Armed activities in the territory of the Congo* the Court expressed no view on this issue, as Uganda eventually claimed that its actions were in response to armed attacks that had already occurred.[232] The Court, however, was aware that the security needs that Uganda aimed to protect were 'essentially preventative'[233] and held that 'Article 51 of the Charter may justify a use of force in self-defence only within the strict confines there laid down. It does not allow the use of force by a State to protect perceived security interests beyond these parameters'.[234]

---

[226] Ruys, 'Armed Attack', p 177; Melzer, *Cyberwarfare*, p 16.
[227] *Oil Platforms*, para 64. It is not clear, however, whether the Court wanted to emphasize a general requirement for self-defence or it only intended to limit the requirement to that specific case (Christine Gray, *International Law and the Use of Force*, 3rd edn (Oxford: Oxford University Press, 2008), p 146).
[228] US DoD, *An Assessment*, p 21.
[229] Schmitt, 'Cyber Operations in International Law', p 165.
[230] Rules 7 and 8, Tallinn Manual, pp 34, 36.
[231] *Nicaragua*, para 194.
[232] *Armed Activities on the Territory of the Congo (DRC v Uganda)*, Judgment, 19 December 2005, ICJ Reports 2005 ('*DRC v Uganda*'), para 143.
[233] *DRC v Uganda*, para 143.   [234] *DRC v Uganda*, para 148.

Dinstein employs the notion of 'interceptive' self-defence to indicate a 'reaction to an event that has already begun to happen even if it has not yet fully developed in its consequences'[235] and maintains that, in such case, self-defence may be invoked under Article 51 because an armed attack 'is already in progress, even if it is still incipient'.[236] Others refer to the imminence of the armed attack in order to distinguish anticipatory and pre-emptive reactions. The latter, which refers to inchoate attacks that might (or might not) materialize at some undefined point in the future, is generally considered inconsistent with international law: not only does it run against the letter of Article 51 of the UN Charter but it is also not supported by extensive and uniform state practice and *opinio juris*.[237] Pre-emptive self-defence is also at odds with the requirements of necessity and proportionality that any self-defence reaction must comply with: the further in time and the more uncertain the attack, the less necessary the defensive armed reaction is, and the more difficult the calculation of its proportionality. On the other hand, a right of anticipatory self-defence against an *imminent* armed attack is consistent not only with customary international law, but also with Article 51.[238] It is true that, under a literal reading of this provision, the armed attack must 'occur', but, according to Article 32 of the 1969 Vienna Convention on the Law of Treaties, the application of the Article 31 interpretive criteria should not lead to an interpretation which is 'manifestly absurd or unreasonable'. The rationale of self-defence is to enable the victim to avert an armed attack: if the danger is 'instant, overwhelming, leaving no choice of means, and no moment for deliberation',[239] if, in other words, it is necessary to react in that very moment because otherwise it would be too late to effectively repel the attack, it would be unreasonable to expect that states will await the occurrence of the attack, or of its effects, before reacting.[240]

---

[235] Yoram Dinstein, *War, Aggression and Self-Defence*, 5th edn (Cambridge: Cambridge University Press, 2011), p 203.
[236] Dinstein, *War, Aggression and Self-Defence*, pp 204–5.
[237] Antonio Cassese, *International Law* (Oxford: Oxford University Press, 2005), p 361; Gray, *International Law*, pp 213–16. The doctrine of pre-emptive self-defence was elaborated in the 2002 US *National Security Strategy* (reaffirmed in 2006), that tried to expand the definition of 'imminence' of armed attack well beyond the *Caroline* requirements to cover cases where 'uncertainty remains as to the time and place of the enemy's attack' (*The National Security Strategy of the United States of America*, 20 September 2002, p 15, <http://www.state.gov/documents/organization/63562.pdf>; the 2006 version is available at <http://www.comw.org/qdr/fulltext/nss2006.pdf>, p 23). The 2010 US *National Security Strategy*, adopted under the Obama Administration, does not mention pre-emptive or anticipatory self-defence and limits itself to invoke the right to use of force when 'necessary' (*National Security Strategy* (2010), p 22). In a speech delivered on 16 September 2011, however, John O Brennan, Assistant to the US President for Homeland Security and Counterterrorism, maintained that 'the traditional conception of what constitutes an "imminent" attack should be broadened in light of the modern-day capabilities, techniques, and technological innovations of terrorist organizations' (CarrieLyn D Guymon (ed), *Digest of United States Practice in International Law*, 2011, p 561, <http://www.state.gov/documents/organization/194113.pdf>).
[238] *A More Secure World*, para 188; *In Larger Freedom: Towards Development, Security and Human Rights for All*, Report of the UN Secretary-General, UN Doc A/59/2005, 21 March 2005, p 33.
[239] Letter from Daniel Webster to Henry S Fox (24 April 1841), *British and Foreign State Papers*, Vol 29, pp 1137–8.
[240] Elizabeth Wilmshurst, 'The Chatham House Principles', p 968. Classic examples of imminent attacks are an advancing army or ships on the horizon or a large-scale mobilization of troops by an

Anticipatory self-defence against imminent cyber armed attacks has been incorporated in Rule 15 of the Tallinn Manual.[241] Indeed, the speed of data transmission in cyberspace seems to fit very well with the 'instantaneous' and 'no moment for deliberation' elements of the *Caroline* doctrine.[242] All in all, the fundamental question is: 'when is the last opportunity to take action to thwart or blunt the attacks?'[243] The evaluation must be made in good faith on the basis of the information available at the time. Even the installation of vulnerabilities in another state's computer systems that might be used in a subsequent cyber attack does not per se justify anticipatory self-defence: as noted in the Commentary to Rule 15 of the Tallinn Manual, what counts is not that the initiator acquires the capability of conducting an armed attack, but that it has actually decided to carry out the attack and that the target state has no other choice than to act immediately to respond effectively to the (imminent) attack.[244] In any case, it is doubtful that a reaction in self-defence would be necessary when the vulnerability, once discovered, can be neutralized through the use of passive cyber defences or active defences below the level of the use of force.

Anticipatory self-defence against a cyber attack that preludes an imminent kinetic armed attack, as in the case of Israel's Operation Orchard against a Syrian nuclear facility,[245] or anticipatory self-defence by cyber means against an imminent kinetic armed attack would not create problems significantly different from those already arising in a traditional scenario.[246] It is safe to argue that, if, on the eve of the 1967 Six Days War, Israel had reacted to the massing of troops at its border by its Arab neighbours and to the blockade of the Strait of Tiran not by bombing the Egyptian air force on the ground before the aircraft could take off and deliver the attack on the Jewish state, but by incapacitating Egypt's air force radars and command and control systems with a massive cyber attack, the legality of such attack would have probably not been doubted. In the absence of an associated kinetic attack, however, anticipatory self-defence by cyber or kinetic means against an imminent standalone *cyber* armed attack will be extremely difficult to invoke in practice: in the absence of visible indications, convincingly establishing the origin, nature, and imminence of the cyber attack and the necessity and proportionality

---

unfriendly neighbouring state on its frontiers (William H Taft, IV, *The Legal Basis for Preemption*, 18 November 2002, <http://www.cfr.org/publication.php?id=5250>).

[241] Tallinn Manual, p 63.

[242] Graham H Todd, 'Armed Attack in Cyberspace: Deterring Asymmetric Warfare with an Asymmetric Definition', *Air Force Law Review* 64 (2009), p 99.

[243] Banks, 'The Role', p 183. Russia has claimed, for instance, that '[h]ostile use of information to attack anti-aircraft, anti-missile and other defence communication and control systems leaves a State defenceless before a potential aggressor and deprives it of the possibility of exercising its legitimate right of self-defence' (UN Doc A/56/164/Add.1, 3 October 2001, p 3).

[244] Tallinn Manual, p 65.

[245] Of course, if one considers the attacks on the Syrian nuclear reactor as part of an ongoing armed conflict between the two states concerned, self-defence issues become irrelevant (Dinstein, *War, Aggression and Self-Defence*, pp 199–200).

[246] Terry D Gill and Paul AL Ducheine, 'Anticipatory Self-Defense in the Cyber Context', *International Law Studies* 89 (2013), p 465.

of the reaction may prove to be an impossible task.[247] Indeed, as will be seen, states claiming a right of anticipatory self-defence will have to provide, as a minimum, 'clear and convincing' evidence of the imminent attack.[248] In this situation, then, the choice is between keeping the imminence requirement in its literal temporal meaning so that the intervening state's margin of appreciation is limited and abuses are exorcised but to the cost of restricting the defensive options of the victim state, or opt for more flexible reasonability standards that take into account the peculiar features of cyber operations and the nature and magnitude of the threat.[249] As has been argued, 'one orientation views imminence as a fixed point while the other views it elastically to account for context'.[250] Only future state practice will clarify where the law stands on these issues. While states that pursue aggressive policies or, vice versa, states that are the frequent target of cyber attacks, like China, the Russian Federation, and the United States, are likely to favour the more flexible approach to imminence, states that do not play an active role in the cyber arena and fear possible abuses by more powerful states will probably go for the stricter temporal notion of imminence.

3. Self-defence against cyber attacks by non-state actors

A characteristic that cyber operations and international terrorism have in common is the prominent role played by non-state actors. Indeed, like terrorist attacks, it appears that the majority of cyber operations against states are conducted by individuals and groups. Kinetic attacks by non-state actors traditionally fell within the scope of criminal law and law enforcement. The transnational character and the quasi-state capabilities of armed and terrorist groups have led to the application, at least in certain cases, of the *jus ad bellum* rules. In 2004, the UN Report *A More Secure World* noted however that '[t]he norms governing the use of force by non-State actors have not kept pace with those pertaining to States'.[251] In particular, the UN Charter does not expressly refer to non-state actors in the context of the rules on the use of force. Two alternative approaches can be adopted in order to find a legal basis for a right of self-defence against (cyber or kinetic) armed attacks by non-state actors.[252] The first maintains that there exists a primary rule of international law providing for a right of states to use force in case of an armed attack, whoever the author. Such primary rule might be Article 51 of the UN Charter and/or a customary international law provision. The other approach focuses instead on secondary rules: it maintains that the primary rule only permits self-defence against armed attacks attributable to states, but at the same time relaxes the rules on attribution and attributes the conduct of non-state actors to

---

[247] Gill and Ducheine, 'Anticipatory Self-Defense', p 466. Similar remarks are made with regard to nuclear warfare by Matthew C Waxman, 'Regulating Resort to Force: Form and Substance of the UN Charter Regime', *European Journal of International Law* 24 (2013), p 160.
[248] See Chapter 2, Section III.6.   [249] Waxman, 'Regulating Resort to Force', pp 160–1.
[250] Waxman, 'Regulating Resort to Force', p 162.   [251] *A More Secure World*, para 159.
[252] Derek Jinks, 'State Responsibility for the Acts of Private Armed Groups', *Chicago Journal of International Law* 4 (2003), p 83.

states in a broader spectrum of situations than those envisaged in the ILC Articles on State Responsibility.[253]

The latter approach has been frequently adopted in order to justify armed reactions against states sponsoring terrorism.[254] According to this view, armed attacks by non-state actors may be attributed to the state from where they originate if this state is unable or unwilling to prevent or terminate the attacks. Only when the territorial state is unaware of the terrorist actions conducted from its territory does it avoid attribution.[255] In the cyber context, this approach seems to have been adopted by the Head of the US Cyber Command, General Keith Alexander, where he states that '[e]very government is responsible for actions originating in its own country'.[256] When submitting its views on information security to the UN Secretary-General, Russia also argued that 'States and other subjects of international law must bear international liability for activities in information space which they carry out or which are carried out from territory under their jurisdiction'.[257] In literature, Garnett and Clarke claim that 'in a situation where there have been repeated instances of hostile computer activity emanating from a State's territory directed against another State, it seems reasonable to presume that the host State had knowledge of such attacks and so should incur responsibility'.[258] Similarly, Todd suggests that, because of the 'unique circumstances of cyberspace', 'additional responsibility under international law [exists] when a state knowingly allows a person or entity to use a cyberspace weapon against the people or property of another state'.[259] Finally, Sharp opines that '[t]he complete refusal or unwillingness of a state ... to cooperate in the suppression or prevention of an acknowledged nonstate-sponsored hostile, transnational activity in CyberSpace that originates in its sovereign territory constitutes state-sponsorship of a use of force *ipso facto*'.[260]

As has been seen in Chapter 1, however, the ILC Articles on State Responsibility provide that a state is responsible for the conduct of individuals or groups that are not organs only when they are 'in fact acting on the instructions of, or under the direction or control of, that State in carrying out the conduct' (Article 8).[261] It is true that international law has long provided for a duty of states to prevent acts by

---

[253] See eg Tsagourias, 'Cyber Attacks', p 242.
[254] Christian J Tams, 'The Use of Force Against Terrorists', *European Journal of International Law* 20 (2009), p 385. See also Jörn Griebel and Milan Plücken, 'New Developments Regarding the Rules of Attribution? The International Court of Justice's Decision in *Bosnia v. Serbia*', *Leiden Journal of International Law* 21 (2008), pp 619–20; Andreas Zimmermann, 'The Second Lebanon War: *Jus ad bellum, jus in bello* and the Issue of Proportionality', *Max Planck Yearbook of United Nations Law* 11 (2007), p 120; Brent Michael, 'Responding to Attacks by Non-State Actors: The Attribution Requirement of Self-Defence', *Australian International Law Journal* 16 (2009), p 155; Schmitt, 'Cyber Operations in International Law', p 171.
[255] Tams, 'The Use of Force', pp 385–6.
[256] Responses to advance questions, Nomination of Lt Gen Keith Alexander, p 25.
[257] UN Doc A/58/373, 17 September 2003, p 11.
[258] Richard Garnett and Paul Clarke, 'Cyberterrorism: A New Challenge for International Law', in *Enforcing International Law Norms Against Terrorism*, edited by Andrea Bianchi (Oxford and Portland: Hart, 2004), p 479.
[259] Todd, 'Armed Attack', p 87.   [260] Sharp, Sr, *Cyberspace*, p 9.
[261] See Chapter I, Section IV, p 36 ff.

individuals under their jurisdiction that are harmful to other states and to punish those that engage in such activities.[262] This obligation was for instance affirmed by the ICJ in the *Corfu Channel*[263] and *Tehran hostages* cases.[264] India has made this point with regard to cyberspace: '[b]y creating a networked society and being a part of [a] global networked economy, it is necessary for nation states to realise that they not only have a requirement to protect their own ICT infrastructure but at the same time have a responsibility to ensure that their ICT is not abused, either covertly or overtly, by others to target or attack the ICT infrastructure of another nation state'.[265] When listing the 'established' principles of self-defence in relation to cyberspace, the United States also noted that 'States are required to take all necessary measures to ensure that their territories are not used by other States or non-State actors for purposes of armed activities, including planning, threatening, perpetrating or providing material support for armed attacks against other States and their interests'.[266] Russia's proposed Convention on Information Security requires the states parties to 'take all necessary steps to prevent any destructive information action originating from their own territory or using the information infrastructure under their jurisdiction, as well as cooperate to locate the source of computer attacks carried out with the use of their territory, to repel these attacks and to eliminate their consequences'.[267] The 2013 *Report of the Group of Governmental Experts* created by the UN General Assembly also concluded that states 'should seek to ensure that their territories are not used by non-State actors for unlawful use of ICTs',[268] while the five principles for cyber peace elaborated by the International Telecommunications Union (ITU)'s Secretary-General state, inter alia, that '[e]very country will commit itself not to harbour terrorists/criminals in its own territories'.[269] The Tallinn Manual has incorporated these views in Rule 5, according to which '[a] State shall not knowingly allow the cyber infrastructure located in its territory or under its exclusive governmental control to be used for acts that adversely and unlawfully affect other States'.[270]

---

[262] Hersch Lauterpacht, 'Revolutionary Activities by Private Persons Against Foreign States', *American Journal of International Law* 22 (1928), p 126; Hans Kelsen, *Principles of International Law* (New York: Reinhart & Co, 1952), pp 120–1.
[263] *Corfu Channel*, p 22.
[264] *US Diplomatic and Consular Staff in Tehran (US v Iran)*, Judgment, 24 May 1980, ICJ Reports 1980, para 68. The point has also been reiterated in several General Assembly resolutions, including the 1965 Declaration on the Inadmissibility of Intervention in the Domestic Affairs of States, and the 1970 Declaration on Friendly Relations, which, according to the ICJ, reflect customary international law (*DRC v Uganda*, para 162). See also SC Resolutions 1189 (13 August 1998) and 1373 (28 September 2001) on international terrorism.
[265] Quoted in Sean Kanuck, 'Sovereign Discourse on Cyber Conflict Under International Law', *Texas Law Review* 88 (2010), p 1591.
[266] UN Doc A/66/152, 15 July 2011, p 19.
[267] Draft Convention on International Information Security (Concept), 2011, Art 6(2).
[268] UN Doc A/68/98, 24 June 2013, p 8.
[269] Hamadoun I Touré, 'The International Response to Cyberwar', in *The Quest for Cyber Peace*, edited by Hamadoun I Touré et al, ITU, January 2011, p 103, <http://www.itu.int/dms_pub/itu-s/opb/gen/S-GEN-WFS.01-1-2011-PDF-E.pdf>.
[270] Tallinn Manual, p 26.

This duty to prevent harmful activities originating from one's territory, however, is not a rule of attribution, but a primary rule and it is far from clear whether military action is warranted against the state that does not comply with it: this is exactly what needs to be demonstrated. The first approach, that based on primary rules, is therefore preferable. As Milanović states, '[i]t is the *jus ad bellum* which should address the use of force by non-state actors, not the law of state responsibility'.[271] Indeed, not only is the suggested relaxation of the attribution rules not consistent with the current law of state responsibility, it is also not supported by state practice: even though states have been criticized for not being able or willing to prevent attacks originating from their territories, the intervening states have usually not accused them of being responsible for such attacks.[272]

In spite of what the ICJ controversially held in its Advisory Opinion on the *legality of the Palestinian Wall*,[273] the primary rule providing for a right of self-defence against non-state actors might be Article 51 of the UN Charter itself, which is silent on this issue: it could be argued that, as Article 2(4) limits its scope to inter-state relations and Article 51 does not, the latter is also applicable, by *argumentum a contrario*, to armed attacks by non-state actors. This conclusion would also be consistent with the purpose of Article 51, which is to allow a state to defend itself against armed attacks in case the Security Council fails to exercise its responsibility in the maintenance of international peace and security: from this teleological perspective, it is irrelevant that the armed attack originates from a state or a non-state actor.[274] As Gill points out, '[i]t is not so much *who* is carrying out the use of force, but *what* the scale and effects of such an operation are—which is important in determining whether the operation constitutes an "armed attack"'.[275] The reason why Article 51 does not expressly contemplate non-state actors is that the Charter's drafters did not envisage that they would be able to conduct attacks with consequences comparable to those of states.[276]

Alternatively (or in addition), the primary rule providing for a right of self-defence against non-state actors could be found in customary international law. The conservative position adopted by the ICJ in the *Palestinian Wall* Advisory Opinion was confined to the interpretation of Article 51. In *DRC v Uganda*, the Court was more cautious and, although it held that, as the attacks carried out by rebel groups against Uganda were non-attributable to the DRC, 'the legal and factual

---

[271] Marko Milanović, 'State Responsibility for Acts of Non-State Actors: A Comment on Griebel and Plucken', *Leiden Journal of International Law* 22 (2009), p 323. See similarly Ruys, *'Armed Attack'*, p 493; Jinks, 'State Responsibility', p 95.

[272] Raphaël Van Steenberghe, 'Self-Defence in Response to Attacks by Non-state Actors in the Light of Recent State Practice: A Step Forward?', *Leiden Journal of International Law* 23 (2010), pp 194–5.

[273] *Legal consequences of the construction of a Wall*, para 139. According to Judge Higgins, 'nothing in the text of Article 51...stipulates that self-defence is available only when an armed attack is made by a State' (Separate Opinion of Judge Higgins, para 33). See also Judge Kooijmans's Separate Opinion, para 35 and Judge Buergenthal's Declaration, para 6.

[274] Zimmermann, 'The Second Lebanon War', p 117.

[275] Gill, 'The Law of Armed Attack', p 50 (emphasis in the original).

[276] Van Steenberghe, 'Self-Defence', p 198.

circumstances for the exercise of a right of self-defence by Uganda against the DRC were not present', it concluded that 'the Court has no need to respond to the contentions of the Parties as to whether and under what conditions contemporary international law provides for a right of self-defence against large-scale attacks by irregular forces'.[277] Recent practice shows growing support for the right of self-defence against armed attacks by non-state actors.[278] The Security Council reaffirmed the inherent right of individual and collective self-defence of the United States with regard to the 11 September 2001 attacks by Al-Qaeda.[279] The NATO and Rio Treaty machineries for collective self-defence were also activated in reaction to those attacks.[280] In the 2006 operations against Hezbollah militias based in southern Lebanon, most states recognized that Israel had the right of self-defence against such attacks, although some questioned the proportionality of the reaction.[281] Similarly, in February 2008 the international community did not condemn Turkey's military operation in northern Iraq in order to destroy the Kurdish Workers' Party (PKK) bases, even if the armed group's actions were not supported by or imputable to Iraq.[282] In 2002, Russia also conducted armed operations against Chechen rebels in Georgia claiming that the latter was unable or unwilling to prevent the attacks.[283] Kenya invoked Article 51 of the UN Charter to justify its military operations against Al-Shabaab militants in Somalia.[284] In the cyber context, the United States has affirmed that '[t]he right of self-defence against an imminent or actual armed attack applies whether the attacker is a State actor or a non-State actor'.[285] The Netherlands also argued that both states and non-state

---

[277] *DRC v Uganda*, para 147. Judge Kooijmans and Judge Simma, however, argued that armed attacks carried out by non-State actors which are of sufficient scale and effects entitle to self-defence whether or not they may be attributed to the territorial State (Separate Opinion of Judge Kooijmans, paras 29–32; Separate Opinion of Judge Simma, para 12). A commentator has maintained that the ICJ decisions in the *Nicaragua* and *DRC v Uganda* cases 'should not be understood as ruling out the legitimate use of defensive force against non-State actors unless the armed attacks of such non-State actors are attributable to a State. Instead, the Court's decisions should be understood as requiring that armed attacks be attributable to a State if the State *itself* is to be the subject of defensive uses of force' (Kimberley N Trapp, 'Back to Basics: Necessity, Proportionality, and the Right of Self-Defence Against Non-State Actors', *International and Comparative Law Quarterly* 56 (2007), p 145; emphasis in the original). This distinction appears, however, to be artificial.

[278] It could be noted than even some less recent practice seems to support this view: after all, the *Caroline* incident (the *locus classicus* of the law of self-defence) involved a reaction against non-State actors (see the facts in Robert Y Jennings, 'The *Caroline* and *McLeod* Cases', *American Journal of International Law* 32 (1938), pp 82–8, 92).

[279] See the already mentioned SC Resolutions 1368 (2001) and 1373 (2001).

[280] Ronzitti, 'The Expanding Law', p 348.   [281] Trapp, 'Back to Basics', pp 154–5.

[282] Raphaël van Steenberghe, 'Le Pacte de non-agression et de défense commune de l'Union Africaine: entre unilatéralisme et responsabilité collective', *Revue générale de droit international public* 113 (2009), p 139.

[283] Annex to letter dated 11 September 2002 from the Permanent Representative of the Russian Federation to the United Nations addressed to the Secretary-General, UN Doc S/2002/1012, 12 September 2002.

[284] The Somali representative at the United Nations, however, suggested that such incursions are a violation of Somalia's sovereignty (Vidan Hadzi-Vidanović, 'Kenya invades Somalia invoking the right to self-defence', *EJIL: Talk!*, 18 October 2011, <http://www.ejiltalk.org/kenya-invades-somalia-invoking-the-right-of-self-defence>).

[285] UN Doc A/66/152, 15 July 2011, p 18.

actors can carry out an armed attack, including by cyber means, that entitles the victim state to self-defence.[286]

It is also worth noting that the right of self-defence against non-state actors has been expressly mentioned in certain treaties. According to the 2006 Great Lakes Protocol on Non-Aggression and Mutual Defence, an armed attack can be committed not only by a state, but by non-state actors as well: Article 5(4) provides that 'Member States shall counter acts of aggression committed against any one [sic] of them by armed groups, taking into account the provisions of Articles 6 [on self-defence] and 8', while Article 8(2) expressly refers to 'armed groups…carrying out armed attacks'.[287] The AU Non-aggression and Common Defence Pact also refers to aggression by non-state actors, triggering the right of collective self-defence.[288]

An interesting question put forward but not answered in the Commentary to the Tallinn Manual is 'whether a single individual mounting an operation that meets the scales and effects' of an armed attack could trigger the right of self-defence of the victim state.[289] Indeed, the equalizing effect of cyber technologies may enable even single individuals to cause significant damage on states. If so, one cannot see why the cyber operation should not qualify as an armed attack: the scale and effects standard should be referred to the consequences of the attack and its magnitude, and not to the number or level of organization of those that perpetrate it. It can be doubted, however, that, in such situation, it would be necessary to react using armed force in self-defence instead of resorting to law enforcement mechanisms.

If, as argued above, there is a primary rule that allows states to exercise the right of self-defence against armed attacks by non-state actors, the exercise of such right will of course be submitted to the requirements of necessity and proportionality.[290] The inability or unwillingness of the territorial state to prevent the attacks originating from its territory is what makes the reaction in self-defence in the territory of that state necessary.[291] Again, this argument is not based on the secondary rules on attribution, but rather on the primary rule of self-defence, of which the requirement of necessity is an important element.[292] The intervening state should

---

[286] Dutch Government Response to the AIV/CAVV Report on Cyber Warfare, p 5, <http://www.rijksoverheid.nl/bestanden/documenten-en-publicaties/rapporten/2012/04/26/cavv-advies-nr-22-bijlage-regeringsreactie-en/cavv-advies-22-bijlage-regeringsreactie-en.pdf>.

[287] The Protocol can be read at <http://www.lse.ac.uk/collections/law/projects/greatlakes/1.%20Peace%20and%20Security/1c.%20Protocols/Protocol%20-%20Non-Aggression.pdf>. On the Great Lakes Protocol, see Marco Roscini, 'Neighbourhood Watch? The Great Lakes Pact and *ius ad bellum*', *Zeitschrift für ausländisches öffentliches Recht und Völkerrecht* 69 (2009), pp 931 ff.

[288] Articles 1(c) and 4(b). Text at <http://www.africa-union.org/root/au/Documents/Treaties/text/Non%20Aggression%20Common%20Defence%20Pact.pdf>.

[289] Tallinn Manual, p 59.   [290] On these requirements, see Chapter 2, Section III.4.

[291] Claus Kress, 'Some Reflections on the International Legal Framework Governing Transnational Armed Conflicts', *Journal of Conflict and Security Law* 15 (2010), p 250. A commentator has proposed an enhanced 'unwilling or unable' test, where certain factors should be taken into account in order to determine whether force may be used on the territory of the 'unwilling or unable' host state: prioritization of consent or cooperation with the territorial state, nature of the threat posed by the non-state actor, request to address the threat and time to respond, reasonable assessment of territorial state control and capacity, proposed means to suppress the threat, prior interactions with the territorial state (Ashley S Deeks, '"Unwilling or Unable": Toward a Normative Framework for Extraterritorial Self-Defense', *Virginia Journal of International Law* 52 (2011–12), pp 519 ff).

[292] See similarly Van Steenberghe, 'Self-Defence', p 200.

first try to secure the cooperation of the territorial state and request that it put an end to the attack originating from its territory or, alternatively, that it allow the victim state to do so, unless such request appears futile or immediate action is required.[293] The state from which the cyber attack originates could for instance disable the internet access of the perpetrators or update the country's firewall settings to prevent hackers from accessing the computers under attack. It could also investigate and arrest those responsible for the attack.[294] The amount of time for the response of the territorial state should be assessed in good faith in relation to the imminence and gravity of the threat.[295] If the harbouring state is unwilling or unable to cooperate, for instance because it does not have the required financial or technical resources,[296] then it would become necessary for the victim state itself to react. The requirement of necessity also entails that the defensive reaction must be directed exclusively against the non-state actors when it is only them that are responsible for the armed attack, and not also the harbouring state.[297]

The application of the unable or unwilling standard (as a primary, not secondary, rule) in the cyber context finds support in legal literature[298] as well as in state practice. The US DoD, for instance, has argued that 'the international law of self-defense would not generally justify acts of "active defense" across international boundaries unless the provocation could be attributed to an agent of the nation concerned, *or* until the sanctuary nation has been put on notice and given the opportunity to put a stop to such private conduct in its territory and has failed to do so, or the circumstances demonstrate that such a request would be futile'.[299] In a recent document, the DoD specified that it

adheres to well-established processes for determining whether a third country is aware of malicious cyber activity originating from within its borders. In doing so, DoD works closely with its interagency and international partners to determine:

...

- The ability and willingness of the third country to respond effectively to the malicious cyber activity; and
- The appropriate course of action for the U.S. Government to address potential issues of third-party sovereignty depending upon the particular circumstances.[300]

---

[293] Deeks, '"Unwilling or Unable"', pp 521–5.
[294] The risk with the application of the unable and/or unwilling standard in the cyber context, however, is that it may be used as a pretext by governments to increase control and surveillance on the internet for censorship purposes.
[295] Deeks, '"Unwilling or Unable"', p 525.
[296] Kimberley N Trapp, *State Responsibility for International Terrorism* (Oxford: Oxford University Press, 2011), p 71. The host state might also not be able to do much because the cyber operation is not illegal under domestic law (Owens, Dam, and Lin, *Technology, Policy, Law, and Ethics*, p 275). GA Res 55/63 of 4 December 2000, para 1(a), recommends that states ensure 'that their laws and practice eliminate safe havens for those who criminally misuse information technologies'.
[297] Ruys, 'Armed Attack', p 496.
[298] See eg Ashley Deeks, 'The Geography of Cyber Conflict: Through a Glass Darkly', *International Law Studies* 89 (2013), pp 1 ff; Hollis, 'Why States Need an International Law', p 1050.
[299] US DoD, *An Assessment*, pp 22–3 (emphasis added).
[300] US DoD, *Cyberspace Policy Report*, p 8.

The unable/unwilling standard is one of due diligence.[301] This is particularly evident in the cyber context, where strict liability would be an unacceptable high burden on states, considering the difficulty of preventing cyber intrusions and the ease with which computers can be remotely controlled and identities spoofed.[302] For example, the United States has a very large cyber infrastructure and may not always be able to detect cyber operations that use servers hosted there but that are conducted from other states.

The US *International Strategy for Cyberspace* includes 'cybersecurity due diligence' as an 'emerging norm' essential in cyberspace and defines it as the states' 'responsibility to protect information infrastructures and secure national systems from damage or misuse'.[303] The UN General Assembly has recommended that:

(a) States should ensure that their laws and practice eliminate safe havens for those who criminally misuse information technologies;
(b) Law enforcement cooperation in the investigation and prosecution of international cases of criminal misuse of information technologies should be coordinated among all concerned States;
(c) Information should be exchanged between States regarding the problems that they face in combating the criminal misuse of information technologies;
(d) Law enforcement personnel should be trained and equipped to address the criminal misuse of information technologies;
(e) Legal systems should protect the confidentiality, integrity and availability of data and computer systems from unauthorized impairment and ensure that criminal abuse is penalized;
(f) Legal systems should permit the preservation of and quick access to electronic data pertaining to particular criminal investigations;
(g) Mutual assistance regimes should ensure the timely investigation of the criminal misuse of information technologies and the timely gathering and exchange of evidence in such cases;
(h) The general public should be made aware of the need to prevent and combat the criminal misuse of information technologies;
(i) To the extent practicable, information technologies should be designed to help to prevent and detect criminal misuse, trace criminals and collect evidence;

---

[301] In the *Corfu Channel* case, the ICJ held that 'it cannot be concluded from the mere fact of the control exercised by a State over its territory and waters that that State necessarily knew, or ought to have known, of any unlawful act perpetrated therein, nor yet that it necessarily knew, or should have known, the authors. This fact, by itself and apart from other circumstances, neither involves *prima facie* responsibility nor shifts the burden of proof' (*Corfu Channel*, p 18). In *Nicaragua*, the Court found that Nicaragua had not breached its duty to stop the arms traffic to the opposition in El Salvador, as 'an activity of that nature, if on a limited scale, may very well be pursued unbeknown to the territorial government' (*Nicaragua*, para 158). In the *Genocide* case, the ICJ also found that the obligation to prevent genocide is an obligation of conduct, not of result (*Application of the Convention on the Prevention and Punishment of the Crime of Genocide (Bosnia and Herzegovina v Serbia and Montenegro)*, Merits, Judgment, 26 February 2007, ICJ Reports 2007 ('*Genocide*'), para 430): *a fortiori* this must hold true for the duty to prevent harmful actions by non-state actors operating from one's territory.
[302] See, *contra*, Eric Talbot Jensen, 'Computer Attacks on Critical National Infrastructure: A Use of Force Invoking the Right of Self-Defense', *Stanford Journal of International Law* 38 (2002), pp 236–7.
[303] *International Strategy for Cyberspace*, p 10.

(j) The fight against the criminal misuse of information technologies requires the development of solutions taking into account both the protection of individual freedoms and privacy and the preservation of the capacity of Governments to fight such criminal misuse.[304]

These measures give flesh to due diligence in the cyber context. The Report of the Council of Europe on *International and multi-stakeholder co-operation on cross-border Internet* has also clarified that due diligence amounts to 'reasonable efforts by a state to inform itself of factual and legal components that relate to transboundary disruptions or interferences with the Internet infrastructure and to take appropriate measures in a timely fashion to address them. Such measures would include firstly, formulating policies designed to prevent and respond to disruptions or interferences, or to minimise risk or consequences thereof and secondly implementing these policies.'[305] Due diligence would also require adopting adequate domestic legislation on cyber crime, conducting investigations and prosecutions of those responsible, and cooperating with the states targeted by the cyber attack in their investigations.[306]

4. Necessity, proportionality, and immediacy of the reaction in self-defence

The reaction in self-defence against cyber operations amounting to armed attacks, as any reaction in self-defence against states or non-state actors, must meet the requirements of necessity, proportionality, and immediacy.[307] Even though Article 51 does not refer to them, in the *Nuclear Weapons* Advisory Opinion the ICJ reiterated that '[t]he submission of the exercise of the right of self-defence to the conditions of necessity and proportionality is a rule of customary international law' and '[t]his dual condition applies equally to Article 51 of the Charter, whatever the means of force employed'.[308] In the cyber context, the United States has reaffirmed that a use of force in self-defence against a cyber attack 'must be limited to what is necessary to address an imminent or actual armed attack and must be proportionate to the threat that is faced'.[309] When presenting the 2011 Strategy for Cyberspace, the US Deputy Defense Secretary also stated that '[t]he United States reserves the right...to respond to

---

[304] GA Res 55/63, 4 December 2000, para 1.
[305] Council of Europe, *International and multi-stakeholder co-operation on cross-border Internet*, Interim report of the Ad-hoc Advisory Group on Cross-border Internet to the Steering Committee on the Media and New Communication Services incorporating analysis of proposals for international and multi-stakeholder co-operation on cross-border Internet, H/Inf (2010) 10, 2010, para 73, <http://www.coe.int/t/dghl/standardsetting/media/MC-S-CI/MC-S-CI%20Interim%20Report.pdf>.
[306] Christopher E Lentz, 'A State's Duty to Prevent and Respond to Cyberterrorist Acts', *Chicago Journal of International Law* 10 (2010), pp 820–2.
[307] Dinstein, *War, Aggression and Self-Defence*, pp 230–4. See Rules 14 and 15 of the Tallinn Manual, pp 61, 63.
[308] *Nuclear Weapons*, para 41.   [309] UN Doc A/66/152, 15 July 2011, p 19.

serious cyber attacks with a proportional and justified military response at the time and place of our choosing'.[310] The UK Foreign Secretary included the 'need for governments to act proportionately in cyberspace and in accordance with national and international law' in his seven principles for the international use of cyberspace.[311] The Netherlands has also stated that the use of force in response to an 'armed cyber attack' has to comply with the requirements of necessity and proportionality.[312] Finally, Canada's *Cyber Security Strategy* emphasizes that '[t]he severity of the cyber attack determines the appropriate level of response and/or mitigation measures'.[313]

Necessity requires that the forcible reaction be a means of last resort and the only effective way to repel the armed attack:[314] as a minimum, it entails an obligation to identify the author, verify that the cyber attack is not an accident, that it was really aimed at the state invoking self-defence, and that the matter cannot be settled by less intrusive means (eg, by preventing the hackers from accessing the networks and computers under attack through the use of passive cyber defences or by conducting a counter cyber operation not amounting to a use of force). The 2012 Presidential Policy Directive 20, for instance, requires to first try law enforcement or network defence techniques before resorting to active defences against a cyber attack on the United States and provides that, if used, such defences must employ 'the least intrusive methods feasible to mitigate a threat'.[315] In particular, the adoption of law enforcement measures could be explored in cooperation with the state from which non-state actors' cyber attacks originate.[316] As Article 51 expressly states, a reaction in self-defence is also unnecessary if the Security Council is taking measures to maintain or restore international peace and security. On the

---

[310] Remarks on the Department of Defense Cyber Strategy, As Delivered by Deputy Secretary of Defense William J Lynn, III, 14 July 2011, <http://www.defense.gov/speeches/speech.aspx?speechid=1593>.

[311] William Hague, Speech at the Munich Security Conference: Security and Freedom in the Cyber Age—Seeking the Rules of the Road, 11 February 2011, cited in Daniel J Ryan, Maeve Dion, Eneken Tikk, and Julie JCH Ryan, 'International Cyberlaw: A Normative Approach', *Georgetown Journal of International Law* 42 (2011), p 1172.

[312] Dutch Government Response to the AIV/CAVV Report on Cyber Warfare, p 5.

[313] *Canada's Cyber Security Strategy: For a Stronger and More Prosperous Canada* (2010), p 3, <http://www.publicsafety.gc.ca/cnt/rsrcs/pblctns/cbr-scrt-strtgy/cbr-scrt-strtgy-eng.pdf>.

[314] As Ago puts it, '[t]he reason for stressing that action taken in self-defence must be *necessary* is that the State attacked (or threatened with imminent attack, if one admits preventive self-defence) must not, in the particular circumstances, have had any means of halting the attack other than recourse to armed force.... Self-defence will be valid as a circumstance precluding wrongfulness of the conduct of the State only if that State was unable to achieve the desired result by different conduct involving either no use of armed force at all or merely its use on a lesser scale' (Ago, Addendum, p 69; emphasis in the original).

[315] US Presidential Policy Directive 20, p 8. See also the 2011 US *International Strategy for Cyberspace* ('we will exhaust all options before military force whenever we can; will carefully weigh the costs and risks of action against the costs of inaction; and will act in a way that reflects our values and strengthens our legitimacy, seeking broad international support whenever possible' (*International Strategy for Cyberspace*, p 14)); AIV/CAVV, *Cyber Warfare*, p 22.

[316] Gill and Ducheine, 'Anticipatory Self-Defense', p 449; Laurie R Blank, 'International Law and Cyber Threats from Non-State Actors', *International Law Studies* 89 (2013), p 418.

other hand, the fact that subsequently the self-defence measure turns out to be ineffective against the armed attack does not mean that it was not necessary. The conclusion would perhaps be different only if it was clear *ex ante* that the reaction had no chance to repel the armed attack.[317]

Proportionality is strictly linked to necessity and is the other side of the same coin.[318] The quantum of armed force used in the defensive reaction could be balanced either against the scale and effects of the armed attack or against the purpose of repelling the armed attack. The latter view is preferable, as 'more force may be necessary, or less force may be sufficient, to repel the attack or defeat one that is imminent' than that employed in the armed attack.[319] If intended in this sense, proportionality tends to merge with necessity. Proportionality does not mean in kind and allows both a kinetic and a cyber response to a cyber attack, as well as a cyber response to a kinetic attack.[320] In fact, a response in-kind against a cyber attack may not always be possible or effective, either because the victim state does not have the technology to hack back or because the aggressor is a low-technology state, or a non-state actor, with no digital infrastructure to hit.[321]

The problem with calculating proportionality in the cyber context resides in the speed and covert nature of cyber attacks: it might be difficult to readily establish their magnitude and consequences.[322] Financial institutions and companies might also be reluctant to provide information on the damage suffered because of business confidentiality.[323] In relation to the cyber reaction, proportionality could also be difficult to calculate in advance because of the interconnectivity of information systems: as with biological weapons, malware sent through cyberspace might

---

[317] The ICJ supported this view in the *Oil Platforms* case, where it found that, if the target of the self-defence reaction has no military function, its destruction cannot be considered effective and therefore necessary to repel the armed attack (*Oil Platforms*, para 74). Similarly, in *DRC v Uganda*, the Court pointed out that 'the taking of airports and towns many hundreds of kilometres from Uganda's border would not seem proportionate to the series of transborder attacks it claimed had given rise to the right of self-defence, nor to be necessary to that end' (*DRC v Uganda*, para 147).

[318] Ago, Addendum, p 69. The ICJ expressed the view that proportionality does not per se exclude the use of a weapon in self-defence 'in all circumstances' (*Nuclear Weapons*, para 42).

[319] Tallinn Manual, p 63. In favour of balancing the force used in self-defence with the purpose of repelling the armed attack, see Ago, Addendum, p 69; Randelzhofer and Nolte, 'Article 51', p 1426; Harrison Dinniss, *Cyber Warfare*, p 103.

[320] See UK House of Commons Defence Committee, *Defence and Cyber-Security*, Sixth Report of Session 2012–13, Vol I, 18 December 2012, p 24. This 'flexible response', where 'harmful action within the cyber domain can be met with parallel response in another domain' (Peter W Singer and Noah Schachtman, 'The Wrong War: The Insistence on Applying Cold War Metaphors to Cybersecurity is Misplaced and Counterproductive', Brookings Institution, 15 August 2011, <http://www.brookings.edu/research/articles/2011/08/15-cybersecurity-singer-shachtman> (citing the US DoD's Cyber Strategy)) was well known to Cold War strategists, who applied it to nuclear warfare.

[321] Greenberg, Goodman, and Soo Hoo, *Information Warfare*, p 32; Blank, 'International Law', p 419.

[322] Matthew Hoisington, 'Cyberwarfare and the Use of Force Giving Rise to the Right of Self-Defense', *Boston College International and Comparative Law Review* 32 (2009), p 452.

[323] Eneken Tikk, Kadri Kaska, Kristel Rünnimeri, Mari Kert, Anna-Maria Talihärm, and Liis Vihul, *Cyber Attacks Against Georgia: Legal Lessons Identified* (CCDCOE, November 2008), p 17, <http://www.carlisle.army.mil/DIME/documents/Georgia%201%200.pdf>; Waxman, 'Cyber Attacks', p 444.

spread uncontrollably. In any case, a disproportionate reaction would not per se turn a self-defence measure into an unlawful reprisal, but 'only renders the State responsible of an act of excess (or abuse) of self-defence'.[324] All in all, meeting the proportionality criterion is essentially a technical issue: customized cyber reactions in self-defence are possible if the software is written with this purpose in mind. The code could, for instance, be designed in a way as to be activated only by the presence of certain characteristics. This requires a high degree of information on the targeted systems, which can be obtained through traditional intelligence collection and/or cyber exploitation.[325] Stuxnet is a good example of such customized cyber operation. Unlike most malware, Stuxnet did little harm to computers and networks that did not meet specific configuration requirements. While the worm was promiscuous, it made itself inert if the specific Siemens software used at Iran's Natanz uranium enrichment plant was not found on infected computers, and contained safeguards to prevent each infected computer from spreading the worm to more than three others. The worm was also programmed to erase itself on 24 June 2012.[326]

The requirement of immediacy of the reaction, which should not be confused with the imminence of the armed attack in relation to anticipatory self-defence, reflects the fact that the ultimate purpose of self-defence is not to punish the attacker, but rather to repel an armed attack.[327] Immediacy does not mean 'instantaneous' and must be applied flexibly: what it entails is that 'there must not be an undue time-lag between the armed attack and the exercise of self-defence in response'.[328] As has been observed, '[a] state does not...forfeit its right of self-defence because it is incapable of instantly responding or is uncertain of who is responsible for the attack or from where the attack originated'.[329] Some flexibility in assessing the immediacy of the reaction is especially required in the case of cyber attacks: if a state's military computer systems and networks have been incapacitated by the attack, for instance, it might take some time for it to be able to react in self-defence either by cyber or kinetic means. The gathering of sufficient evidence that allows to confidently point the finger against a certain state or non-state actor can also be a time-consuming task in the cyber context.[330]

---

[324] Ronzitti, 'The Expanding Law', p 355.
[325] Owens, Dam, and Lin, *Technology, Policy, Law, and Ethics*, p 123.
[326] Jeremy Richmond, 'Evolving Battlefields: Does Stuxnet Demonstrate a Need for Modifications to the Law of Armed Conflict?', *Fordham International Law Journal* 35 (2011–12), p 856.
[327] Although not listed by the ICJ as one of the requirements for the exercise of self-defence, customary international law upholds its existence (Dinstein, *War, Aggression and Self-Defence*, pp 230–1). However, not everyone agrees that immediacy is a requirement of the self-defence reaction: see for instance Thomas M Franck, 'Terrorism and the Right of Self-Defense', *American Journal of International Law* 95 (2001), p 840. According to Melzer, immediacy is not a third requirement of the exercise of self-defence, but is rather a temporal aspect of necessity (Melzer, *Cyberwarfare*, p 17). Immediacy has been incorporated in Rule 15 of the Tallinn Manual, p 63.
[328] Dinstein, *War, Aggression and Self-Defence*, p 233.
[329] Gill and Ducheine, 'Anticipatory Self-Defense', p 451.
[330] See Chapter 2, Section III.6.

5. Collective self-defence in reaction to a cyber armed attack: the cases of NATO and the European Union

Article 51 contemplates not only individual but also collective self-defence: in this case, the state using defensive force reacts against an armed attack that targeted another state. Collective self-defence against cyber armed attacks has been incorporated in Rule 16 of the Tallinn Manual[331] and has often been emphasized in US documents. The 2011 White House's *International Strategy for Cyberspace*, for instance, notes that '[a]ll states possess an inherent right to self-defense, and we recognize that certain hostile acts conducted through cyberspace could compel actions under the commitments we have with our military treaty partners. We reserve the right to use all necessary means—diplomatic, informational, military, and economic—as appropriate and consistent with applicable international law, in order to defend our Nation, our allies, our partners, and our interests'.[332] In 2011, Australia and the United States took the occasion of the 60th anniversary of the 1951 Security Treaty between Australia, New Zealand, and the United States (ANZUS Treaty) to jointly declare that 'in the event of a cyber attack that threatens the territorial integrity, political independence or security of either of our nations, Australia and the United States would consult together and determine appropriate options to address the threat'.[333] The Australian Defence Minister explained that a 'substantial' cyber attack against either country would determine a response similar to that that followed the attacks on the United States on 11 September 2001.[334]

Collective self-defence is submitted to the same conditions as individual self-defence, ie the occurrence of an armed attack, and the necessity, proportionality, and immediacy of the reaction, whose application in the cyber context has already been examined in the previous Sections. In addition, as clarified by the ICJ in the *Nicaragua* judgment, collective self-defence also requires that the victim state declares itself to be the victim of an armed attack and requests assistance to repel it.[335]

One way of exercising collective self-defence is through a military alliance established to that purpose.[336] The most significant collective self-defence international organization today is NATO. The Organization adopted a cyber defence policy in 2008, which was revised in June 2011 together with the adoption of an associated Action Plan for its implementation. NATO has also created a Cyber Defense Management Authority, a Computer Incidence Response Capability and the CCDCOE. The Organization has conducted cyber defence exercises with

---

[331] Tallinn Manual, p 67.
[332] *International Strategy for Cyberspace*, p 14. See also US *Department of Defense Strategy for Operating in Cyberspace*, July 2011, pp 9–10; Lithuania's *Programme for the Development of Electronic Information Security (Cyber-Security) for 2011–20*, 29 June 2011, p 3, <http://www.ird.lt/doc/teises_aktai_en/EIS(KS)PP_796_2011-06-29_EN_PATAIS.pdf>.
[333] Joint Statement on Cyberspace, 15 September 2011, <http://www.foreignminister.gov.au/releases/2011/kr_mr_110916a.html>.
[334] Simon Mann, 'Cyber war added to ANZUS pact', *The Sydney Morning Herald*, 16 September 2011, <http://www.smh.com.au/national/cyber-war-added-to-anzus-pact-20110915-1kbuv.html>.
[335] *Nicaragua*, para 199.   [336] Dinstein, *War, Aggression and Self-Defence*, pp 286–9.

the participation of teams from member states and signed memoranda of understanding (MoU) in relation to cyber security with some member states, including Estonia, Slovakia, Turkey, the United Kingdom, and the United States.[337]

The main question in the NATO context is whether cyber operations against member states should fall under Article 4 of the NATO Treaty, which provides for an obligation to consult 'whenever, in the opinion of any of them, the territorial integrity, political independence or security of any of the Parties is threatened', or Article 5.[338] Article 5(1) provides that

> [t]he Parties agree that an armed attack against one or more of them in Europe or North America shall be considered an attack against them all and consequently they agree that, if such an armed attack occurs, each of them, in exercise of the right of individual or collective self-defence recognised by Article 51 of the Charter of the United Nations, will assist the Party or Parties so attacked by taking forthwith, individually and in concert with the other Parties, such action as it deems necessary, including the use of armed force,to restore and maintain the security of the North Atlantic area.[339]

It is to be recalled that there is no automatism in the collective reaction to an armed attack and that NATO states have always considered that the decision of what support to lend was ultimately theirs.[340] The question is whether the *casus fœderis* in the form of an 'armed attack' against one or more of the members also includes cyber attacks.

Although recognizing that '[t]he next significant attack on the Alliance may well come down a fibre optic cable',[341] the position of NATO and its member states on the applicability of the duty of assistance in collective self-defence under Article 5 of the North Atlantic Treaty in case of a cyber attack is unclear. The Strategic Concept adopted in 1999 stated that, in addition to traditional armed attacks, 'Alliance security interests can be affected by other risks of a wider nature, including ... the disruption of the flow of vital resources'.[342] In January 2008, NATO adopted a Policy on Cyber Defence that was endorsed by the heads of state and government at the Bucharest Summit in April of the same year.[343] Paragraph 47 of the

---

[337] Touré, 'The International Response to Cyberwar', p 93.

[338] Article 5 has been invoked only once in NATO's history in response to the 11 September 2011 attacks against the United States. Article 4 has been formally used in February 2003, when Turkey requested consultations on the effects of the impending Operation Iraqi Freedom on its security (Ulf Häussler, 'Cyber Security and Defence from the Perspective of Articles 4 and 5 of the NATO Treaty', in *International Cyber Security Legal and Policy Proceedings*, edited by Eneken Tikk and Anna-Maria Talihärm (Tallinn: CCDCOE, 2010), p 103, <http://www.ccdcoe.org/245.html>).

[339] See also Art 3(1) of the Rio Treaty. Unusually, the 2005 AU Non-aggression and Common Defence Pact employs the broader notion of 'aggression' instead of 'armed attack' (Art 4(b)). The OAS has adopted a *Cyber Security Strategy* in 2004, which however focuses on cyber crime and cyber terrorism, and not on the military uses of cyberspace by states (text at <http://www.oas.org/cyber/documents/AG-RES.%202004%20Cyber%20Security%20Strategy%20(complete).pdf>).

[340] Häussler, 'Cyber Security', p 107.

[341] *NATO 2020: Assured Security; Dynamic Engagement*, Analysis and Recommendations of the Group of Experts on a New Strategic Concept for NATO, 17 May 2010, p 45, <http://www.nato.int/nato_static/assets/pdf/pdf_2010_05/20100517_100517_expertsreport.pdf>.

[342] *The Alliance Strategic Concept*, 24 April 1999, para 24, <http://www.nato.int/cps/en/natolive/official_texts_27433.htm>.

[343] The exact details of the Policy remain classified.

Summit's Final Declaration emphasizes 'the need for NATO and nations to protect key information systems in accordance with their respective responsibilities; share best practices; and provide a capability to assist Allied nations, upon request, to counter a cyber attack'.[344] The 2010 Report of the Group of Experts on the New Strategic Concept for NATO maintains this ambiguity where it refers to 'less conventional threats to the Alliance', such as 'cyber assaults', 'which may or may not reach the level of an Article 5 attack'.[345] The document further points out that large-scale cyber attacks against NATO's command and control systems or energy grids 'could readily warrant consultations under Article 4 and could possibly lead to collective defence measures under Article 5'[346] and that 'whether an unconventional danger—such as a cyber attack ... triggers the collective defence mechanisms of Article 5 ... will have to be determined by the NAC [North Atlantic Council] based on the nature, source, scope, and other aspects of the particular security challenge'.[347] NATO's *New Strategic Concept*, adopted in November 2010, limits itself to provide that the Organization will further develop its ability 'to prevent, detect, defend against and recover from cyber-attacks, including by using the NATO planning process to enhance and coordinate national cyber-defence capabilities, bringing all NATO bodies under centralized cyber protection, and better integrating NATO cyber awareness, warning and response with member nations, without further specifications on how far the reaction can go and on what legal basis.[348] It is, however, interesting that the *New Strategic Concept* avoids adding 'armed' to the 'attack' that gives rise to collective defence and refers to 'emerging security challenges [that] threaten the fundamental security of individual Allies or the Alliance as a whole'.[349] The revised Policy on Cyber Defence, adopted by the Defence Ministers in June 2011, states that NATO will maintain its strategic ambiguity and flexibility on possible responses to different types of crises involving a cyber component, subject to the political decisions of the North Atlantic Council.[350] On 4 June 2013, NATO defence ministers held the first-ever meeting dedicated to cyber defence and agreed that the Alliance's cyber-defence capabilities should be fully operational by autumn 2013. The defence ministers also agreed that discussions on how NATO can support and assist Allies that request assistance in case of a cyber attack should continue at their next meeting in October 2013.[351]

It is worth noting that a NATO member state, Estonia, was the target of a DDoS attack in 2007. Even though Article 4 was not formally invoked, it seems that consultations took place after the attack,[352] while the Estonian Defence

---

[344] Read the text of the Final Declaration at <http://www.summitbucharest.ro/en/doc_201.html>.
[345] *NATO 2020*, p 9.   [346] *NATO 2020*, p 45.   [347] *NATO 2020*, p 20.
[348] *Active Engagement, Modern Defence. Strategic Concept for the Defence and Security of the Members of the North Atlantic Treaty Organisation*, 19 November 2010, para 19, <http://www.nato.int/cps/en/natolive/official_texts_68580.htm>.
[349] *Active Engagement, Modern Defence*, para 4(a).
[350] NATO, *Defending the Networks—The NATO Policy on Cyber Defence* (2011), p 2, <http://www.nato.int/nato_static/assets/pdf/pdf_2011_09/20111004_110914-policy-cyberdefence.pdf>.
[351] 'NATO and cyber defence', <http://www.nato.int/cps/en/natolive/topics_78170.htm?>.
[352] Häussler, 'Cyber Security', p 120.

Minister was said to consider the invocation of Article 5.[353] During the crisis, however, the Minister claimed that 'NATO does not define cyber-attacks as a clear military action. This means that the provisions of Article V...will not automatically be extended' and that 'this matter needs to be resolved in the near future'.[354] On 23 April 2010, Estonia concluded a MoU with NATO to facilitate exchange of information and to create a mechanism of assistance in case of cyber attacks.[355] Although the MoU is not for public release, in response to a question from this author an Estonian official from the Ministry of Defence replied that the MoU sets up a framework of support, information exchange and consultations in case of cyber attacks against Estonia and does not consider cyber attacks as armed attacks against NATO.[356]

NATO member states' position is also ambiguous. The UK *National Security Strategy (2009 Update)* refers to the fact that 'some allies of the UK, to which we have an obligation *under Article V* of the NATO Charter, could be threatened from other states, through military *or other means*'.[357] The UK Minister for the Armed Forces declared that Article 5 of the NATO Treaty is 'potentially' applicable to a cyber attack, while the Assistant Chief of Defence Staff, General Shaw, contended that a cyber attack would be an 'armed attack' for the purposes of Article 5 of the NATO Treaty if its effects are 'severe'.[358] The Dutch government has affirmed that '[i]n the digital domain, as elsewhere, it is not always easy to establish when article 5 would come into operation. That is always a question that must be tackled at political level'.[359] With regard to the 2007 cyber attack on Estonia, a German official is reported to have stated that, while such attack did not activate Article 5 of the North Atlantic Treaty, this could change in the future if the attacks become more sophisticated.[360] A Report by an Italian Parliamentary Committee also suggests that the notion of 'collective security' in the context of Article 5 of the NATO Treaty should be broadened so to include cyber attacks.[361]

---

[353] Scott J Shackelford, 'From Nuclear War to Net War: Analogizing Cyber Attacks in International Law', *Berkeley Journal of International Law* 27 (2009), p 194.

[354] Ian Traynor, 'Russia Accused of Unleashing Cyberwar to Disable Estonia', *The Guardian*, 17 May 2007, <http://www.guardian.co.uk/world/2007/may/17/topstories3.russia>.

[355] 'NATO and Estonia conclude agreement on cyber defence', 23 April 2010, <http://www.nato.int/cps/en/natolive/news_62894.htm>.

[356] E-mail on file with the author.

[357] UK Cabinet Office, *Security for the Next Generation, The National Security Strategy of the United Kingdom: Update 2009*, p 41 (emphasis added). The Strategy points out that security threats 'go beyond the traditional military domains of land, sea and air, to include weapons of mass destruction, and the increasing importance of cyberspace' (p 7).

[358] UK House of Commons Defence Committee, *Defence and Cyber-Security*, p 25.

[359] Dutch Government Response to the AIV/CAVV Report on Cyber Warfare, p 7.

[360] Quoted in James A Lewis, 'The "Korean" Cyber Attacks and Their Implications for Cyber Conflict', Center for Strategic and International Studies, October 2009, p 3, <http://csis.org/publication/korean-cyber-attacks-and-their-implications-cyber-conflict>.

[361] COPASIR, *Relazione sulle possibili implicazioni e minacce per la sicurezza nazionale derivanti dall'utilizzo dello spazio cibernetico*, Doc XXXIV, no 4, 7 July 2010, p 53, <http://www.parlamento.it/documenti/repository/commissioni/bicamerali/COMITATO%20SICUREZZA/Doc_XXXIV_n_4.pdf>.

In order to solve the conundrum, one should resort to the rules on treaty interpretation. It is submitted that the expression 'armed attack' in collective self-defence treaties should be interpreted consistently with Article 51 of the UN Charter, for at least two reasons. First, Article 31(3)(c) of the Vienna Convention on the Law of Treaties requires that a treaty is interpreted also taking into account 'any relevant rules of international law applicable in the relations between the parties': Article 51 is the matrix of collective self-defence treaties such as the NATO Treaty, as Article 5 of the NATO Treaty expressly recognizes. Secondly, the NATO Treaty contains a subordination clause (Article 7) with respect to the UN Charter.[362] Even lacking one, the primacy of the Charter would be guaranteed by Article 103 of the Charter, according to which '[i]n the event of a conflict between the obligations of the Members of the United Nations under the present Charter and their obligations under any other international agreement, their obligations under the present Charter shall prevail'. The above considerations lead to conclude that 'armed attack' in Article 5 of the NATO Treaty has the same scope as 'armed attack' in Article 51 of the UN Charter. Cyber operations will amount to an armed attack under the former at the same conditions of the latter.[363]

It should be noted that Article 5 of the NATO Treaty specifies that, in order to activate the collective self-defence mechanism, the armed attack against one of the states parties must occur in Europe or North America. Article 6 further notes that an armed attack on one or more of the parties includes an armed attack

- on the territory of any of the Parties in Europe or North America, on the Algerian Departments of France, on the territory of or on the Islands under the jurisdiction of any of the Parties in the North Atlantic area north of the Tropic of Cancer;
- on the forces, vessels, or aircraft of any of the Parties, when in or over these territories or any other area in Europe in which occupation forces of any of the Parties were stationed on the date when the Treaty entered into force or the Mediterranean Sea or the North Atlantic area north of the Tropic of Cancer.[364]

The fact that cyber attacks take place in and through cyberspace, and not in the specific geographical locations indicated in Articles 5 and 6, does not preclude that such attacks can potentially fall under the scope of those provisions. Indeed, as has been seen,[365] it is at where the cyber operations originate and where their consequences occur that one has to look at in order to 'territorialize' them: therefore, a cyber operation that causes loss of life or injuries to persons, physical damage to property, or severe disruption of the functioning of critical infrastructure located in Europe or North America would meet the geographical requirements of Articles 5 and 6. In any case, it is now well-established that 'coercive actions by NATO may

---

[362] See Art 30(2) of the Vienna Convention on the Law of Treaties.
[363] See Chapter 2, Section III.1. In particular, the cyber attack will have to be a use of force that is serious enough to reach the 'scale and effects' threshold of an armed attack. Häussler, for instance, applies the doctrine of kinetic equivalence in the NATO context and suggests that 'Article 5 of the North Atlantic Treaty may be triggered in this context if the scale and gravity of the overall effect of that first strike corresponds with the kinetic equivalent' (Häussler, 'Cyber Security', p 118).
[364] Footnotes omitted.   [365] See Chapter 1, Section III.1, pp 23–4.

take place even outside the geographic boundaries of the European and Atlantic region if functionally linked to the need to respond to threats that could endanger stability and security in the area'.[366]

The European Union has also become increasingly concerned with defence against cyber attacks. The *Cyber Security Strategy* of the European Union states that '[a] particularly serious cyber incident or attack could constitute sufficient ground for a Member State to invoke the EU Solidarity Clause (Article 222 of the Treaty on the Functioning of the European Union)',[367] while no mention is made of Article 42(7) of the Treaty on the European Union, which contains the 'obligation of aid and assistance by all the means in their power, in accordance with Article 51 of the United Nations Charter' in case of 'armed aggression'. Article 222 provides that the European Union 'shall mobilise all the instruments at its disposal, including the military resources made available by the Member States' to counter certain threats, in particular terrorist, man-made or natural disasters. The Solidarity Clause is broad enough to justify a coordinated military response in case of such new threats, which, according to the *EU Cybersecurity Strategy*, now also include 'particularly serious cyber incidents or attacks'.

It is submitted that, while Article 222 may be invoked in the event of any 'serious cyber incident or attack', Article 42(7) is still potentially applicable to those that amount to 'armed aggression' under the same conditions and for the same reasons highlighted above with regard to NATO. Indeed, Article 42(7) both recalls Article 51 of the UN Charter and subordinates its own operation to the 'commitments under the North Atlantic Treaty Organisation, which, for those States which are members of it, remains the foundation of their collective defence and the forum for its implementation'. It is, therefore, entirely possible that both the Solidarity and the Mutual Defence Clauses could be triggered by a state-sponsored cyber attack against a EU member state that causes material damage to property, loss of life or injury to persons, or severe disruption of the functioning of critical infrastructures,[368] although a reaction under Article 42(7), unlike one based on Article 222, would be purely intergovernmental and would not require coordination at the EU level.

6. The standard of proof required for the exercise of self-defence against cyber attacks

The standard of proof is 'the *quantum* of evidence necessary to substantiate the factual claims made by the parties'.[369] Evidence is required to prove both the objective (be it an act or an omission) and subjective elements of an internationally wrongful

---

[366] Enzo Cannizzaro, 'NATO's New Strategic Concept and the Evolving Legal Regulation of the Use of Force', *The International Spectator* 36, no 1 (2001), p 70.
[367] European Commission, *Cybersecurity Strategy of the European Union: An Open, Safe and Secure Cyberspace*, 7 February 2013, p 19, <http://ec.europa.eu/information_society/newsroom/cf//document.cfm?doc_id=1667>.
[368] On the EU notion of 'critical infrastructure', see Chapter 2, Section II.1.2., p 55 ff.
[369] James A Green, 'Fluctuating Evidentiary Standards for Self-Defence in the International Court of Justice', *International and Comparative Law Quarterly* 58 (2009), p 165.

act. In the cyber context, the state invoking self-defence against cyber attacks will therefore have to demonstrate: (a) that the cyber attack actually occurred and that its scale and effects reached the threshold of an 'armed attack'; and (b) that it was attributable to a certain state or non state-actor. It is well-known that, while in civil law systems there are no specific standards of proof that judges have to apply as they can evaluate the evidence produced according to their personal convictions, in common law there is a rigid classification of standards, including (from the most stringent to the least) beyond reasonable doubt (ie indisputable evidence, normally used in criminal proceedings), clear and convincing (or compelling) evidence (ie more than probable but short of indisputable) and preponderance of evidence or balance of probabilities (more likely than not, probable).[370] Green also adds a fourth standard, *prima facie* evidence, that merely requires indicative proof of the contention.[371] International law does not prescribe a general standard of proof for all internationally wrongful acts, and international courts and tribunals have determined their own standards in each case, not always in a consistent manner.[372] As the ICJ held in the *Nicaragua* Judgment, 'within the limits of its Statute and Rules... [the Court] has freedom in estimating the value of the various elements of evidence'.[373] In fact, the Court has often avoided clearly indicating the standards of proof it has employed in the cases before it.[374]

If there is no uniform standard of proof in international law, there may be special standards for discrete areas or specific claims. As has been observed, 'the context or "type" of dispute in question should be relevant to the evidentiary standard employed in establishing relevant facts'.[375] Indeed, when the charge is the same, it seems logical that the evidentiary standard should also be the same.[376] There are indications that claims related to *jus ad bellum* violations, in particular in relation to the invocation of self-defence against an armed attack, require 'clear and convincing evidence'.[377] When justifying its 2001 armed operation against Afghanistan, the US Permanent Representative to the United Nations referred to the fact that the US government had 'clear and compelling information that the Al-Qaeda organization, which is supported by the Taliban regime in Afghanistan, had a central role in the [11 September 2001] attacks'.[378] The same language was

---

[370] Mary Ellen O'Connell, 'Rules of Evidence for the Use of Force in International Law's New Era', *America Society of International Law Proceedings* 100 (2006), p 45; Marko Milanović, 'State Responsibility for Genocide', *European Journal of International Law* 17 (2006), p 594; Green, 'Fluctuating Evidentiary Standards', p 167.

[371] Green, 'Fluctuating Evidentiary Standards', p 166.

[372] Green, 'Fluctuating Evidentiary Standards', p 165.   [373] *Nicaragua*, para 60.

[374] This has been criticized by judges from common law countries (see eg *Oil Platforms*, Separate Opinion of Judge Buergenthal, paras 41–6; Separate Opinion of Judge Higgins, paras 30–9).

[375] Green, 'Fluctuating Evidentiary Standards', p 167.

[376] Green, 'Fluctuating Evidentiary Standards', pp 170–1. The author suggests that one consistent standard should apply to all cases of self-defence (whatever magnitude the consequences of the violation of the prohibition of the use of force might have) and both to the objective and subjective elements of the internationally wrongful act (p 169).

[377] Mary Ellen O'Connell, 'Evidence of Terror', *Journal of Conflict and Security Law* 7 (2002), pp 22 ff.

[378] Letter dated 7 October 2001 from the Permanent Representative of the United States of America to the United Nations Addressed to the President of the Security Council, UN Doc S/2001/946, 7 October 2001.

used by NATO's Secretary-General.³⁷⁹ In the context of the proposed intervention to react against the alleged use of chemical weapons by the Syrian government, the US President stated that attacking another country without a UN mandate and without 'clear evidence that can be presented' would raise questions of international law.³⁸⁰ In the *Corfu Channel* Judgment, the ICJ referred to 'conclusive evidence', 'a degree of certainty' and inferences of fact that 'leave *no room* for reasonable doubt' in relation to the minelaying.³⁸¹ Although it did not identify a specific standard, in the *Nicaragua* Judgment, the Court referred to 'convincing evidence' of the facts on which a claim is based and to the lack of 'clear evidence' of the degree of control exercised by the United States over the *contras*.³⁸² In the *Oil Platforms* case, the ICJ rejected evidence with regard to Iran's responsibility for minelaying that was 'highly suggestive, but not conclusive' and argued that 'evidence indicative of Iranian responsibility for the attack on the *Sea Isle City*' was insufficient.³⁸³ In *DRC v Uganda*, the ICJ referred again to facts 'convincingly established by the evidence', 'convincing evidence', and 'evidence weighty and convincing'.³⁸⁴ Confusingly, however, in other parts of the Judgment the Court seemed to employ a *prima facie* or preponderance of evidence standard, in particular when it had to determine whether the conduct of armed groups against the DRC was attributable to Uganda.³⁸⁵ The EECC also found that there was 'clear' evidence that events in the vicinity of Badme were minor incidents and did not reach the magnitude of an armed attack.³⁸⁶

The above seems to indicate that at least clear and convincing evidence is required for claims related to the use of force.³⁸⁷ Indeed, as self-defence is an exception to the prohibition of the use of force, the standard of proof should be high enough to limit its invocability to exceptional circumstances and thus avoid abuses.³⁸⁸ As Judge Higgins suggested in her Separate Opinion attached to the *Oil Platforms*

---

³⁷⁹ Statement at NATO Headquarters, 2 October 2001, <http://www.nato.int/docu/speech/2001/s011002a.htm>.

³⁸⁰ Julian Borger, 'West reviews legal options for possible Syria intervention without UN mandate', *The Guardian*, 26 August 2013, <http://www.theguardian.com/world/2013/aug/26/united-nations-mandate-airstrikes-syria>. Indeed, the Report of the UN Secretary-General's Investigation found 'clear and convincing evidence' of the use of chemical weapons in the armed conflict (United Nations Mission to Investigate Allegations of the Use of Chemical Weapons in the Syrian Arab Republic, *Report on the Alleged Use of Chemical Weapons in the Ghouta Area of Damascus on 21 August 2013*, 13 September 2013, p 5, <http://www.un.org/disarmament/content/slideshow/Secretary_General_Report_of_CW_Investigation.pdf>).

³⁸¹ *Corfu Channel*, pp 17, 18 (emphasis in the original). See Green, 'Fluctuating Evidentiary Standards', p 177.

³⁸² Green, 'Fluctuating Evidentiary Standards', p 172. See *Nicaragua*, paras 29, 109.

³⁸³ *Oil Platforms*, paras 71, 61. See Green, 'Fluctuating Evidentiary Standards', pp 172–3.

³⁸⁴ *DRC v Uganda*, paras 72, 91, 136.

³⁸⁵ Green, 'Fluctuating Evidentiary Standards', pp 175–6.

³⁸⁶ EECC, *Partial Award*, Jus ad bellum, para 12. See O'Connell, 'Rules of Evidence', p 45.

³⁸⁷ Waxman cautions, however, against 'setting a single, specific proof standard' and argues that level of threat, intention and capability should be taken into account when determining such standard in each case (Matthew Waxman, 'The Use of Force Against States That *Might* Have Weapons of Mass Destruction', *Michigan Journal of International Law* 31 (2009–10), p 61).

³⁸⁸ Mary Ellen O'Connell, 'Lawful Self-Defense to Terrorism', *University of Pittsburgh Law Review* 63 (2002), p 898.

Judgment, 'the graver the charge the more confidence there must be in the evidence relied on'.[389] In the *Genocide* case, the ICJ confirmed that 'claims against a State involving charges of exceptional gravity must be proved by evidence that is fully conclusive.... The same standard applies to the proof of attribution for such acts' (therefore both to the objective and subjective element of the internationally wrongful act).[390] It is not entirely clear, however, whether the Court linked gravity to the importance of the norm allegedly breached or the magnitude of the violation. It would seem more correct to refer the gravity to the former, as, if the evidentiary standard depended on the latter, 'some States could have a perverse incentive to sponsor more devastating attacks so as to raise the necessary burden of proof and potentially defeat accountability.'[391] Either way, the prohibition of the use of force is certainly a crucial provision and, as an 'armed attack' is a serious form of the use of force,[392] the magnitude of the violation is inherent in that notion.[393]

If clear and convincing evidence is required in relation to claims of self-defence, the question arises whether there is a special standard when the alleged armed attack is conducted by cyber means. In spite of its crucial importance, the Tallinn Manual does not discuss in depth the standard of proof required in the cyber context: the only cursory references to evidence are contained in Rules 7 and 8.[394] The Project Grey Goose Report on the 2008 cyber operations against Georgia relies only on circumstantial evidence, if of high level, to suggest that the Russian government was responsible for the operations.[395] In contrast, the CCDCOE *Report on Georgia* concludes that 'there is no conclusive proof of who is behind the DDOS attacks, even though finger pointing at Russia is prevalent by the media'.[396] In a Senate questionnaire in preparation for a hearing on his nomination to head of the new US Cyber Command, General Alexander argued that 'some level of mitigating action' may be taken against cyber attacks 'even when we are not certain who is responsible'.[397] Similarly, in its reply to the UN Secretary-General on issues related to information security, the United States claimed that 'the identity and motivation of the perpetrator(s) can only be inferred from the target, effects and other circumstantial evidence surrounding an incident': indeed, 'high-confidence attribution of identity to perpetrators cannot be achieved in a timely manner, if ever, and success often depends on a high degree of transnational cooperation'.[398] The US DoD's *Assessment of International Legal Issues in Information Operations* also argues that '[s]tate sponsorship might be persuasively established by such factors

---

[389] *Oil Platforms*, Separate Opinion of Judge Higgins, para 33.
[390] *Genocide*, para 209. See also *Corfu Channel*, p 17.
[391] Scott J Shackelford and Richard B Andres, 'State Responsibility for Cyber Attacks: Competing Standards for a Growing Problem', *Georgetown Journal of International Law* 42 (2011), p 990.
[392] *Nicaragua*, para 191.
[393] Green, 'Fluctuating Evidentiary Standards', p 170.
[394] Tallinn Manual, pp 34–6.
[395] *Russia/Georgia Cyber War—Findings and Analysis*, Project Grey Goose: Phase I Report, 17 October 2008, <http://www.scribd.com/doc/6967393/Project-Grey-Goose-Phase-I-Report>.
[396] Tikk, Kaska, Rünnimeri, Kert, Talihärm, and Vihul, *Cyber Attacks*, p 12.
[397] Responses to advance questions, Nomination of Lt Gen Keith Alexander, p 12.
[398] UN Doc A/66/152, 15 July 2011, pp 16–17.

as signals or human intelligence, the location of the offending computer within a state-controlled facility, or public statements by officials'.[399] In other unspecified situations, the study suggests that relevant circumstances might be even looser and include 'the state of relationships between the two countries, the prior involvement of the suspect state in computer network attacks, the nature of the systems attacked, the nature and sophistication of the methods and equipment used, the effects of past attacks, and the damage which seems likely from future attacks'.[400] The above views seem to suggest a standard of proof lower than clear and convincing evidence on the basis that identification and attribution are more problematic in a digital environment than in the analogue world.[401]

It is difficult, however, to see why the standard of proof should be lower simply because it is more difficult to reach it. Such standard exists not to penalize the claimant, but to protect the defendant against false attribution, which, thanks to tricks like IP spoofing, onion routing and the use of botnets, is a particularly serious risk in the cyber context. The above mentioned views are also far from being unanimous, even within the US departments: the Air Force Doctrine for Cyberspace Operations, for instance, states that attribution of cyber operations should be established with 'sufficient confidence and verifiability'.[402] A report prepared by Italy's Parliamentary Committee on the Security of the Republic goes further and requires to demonstrate 'unequivocally' ('in modo inequivocabile') that an armed attack by cyber means originated from a state and was instructed by governmental structures.[403] The document also suggests that state attribution needs 'irrefutable digital "evidence"' ('"prove" informatiche inconfutabili'), which—the Report concedes—is a condition very difficult to meet.[404] Germany also highlighted the need for 'reliable attribution' of malicious cyber activities in order to avoid '"false flag" attacks', misunderstandings and miscalculations.[405] Finally, the AIV/CAVV Report, that has been endorsed by the Dutch government, requires 'reliable intelligence... before a military response can be made to a cyber attack' and 'sufficient certainty' about the identity of the author of the attack.[406] In its response to the Report, the Dutch government argued that self-defence may be

---

[399] US DoD, *An Assessment*, p 21.   [400] US DoD, *An Assessment*, p 21.

[401] See eg David E Graham, 'Cyber Threats and the Law of War', *Journal of National Security Law and Policy* 4 (2010), p 93 (the author seems, however, to confuse attribution criteria and standard of proof). Others have claimed that '[t]*he fact that a harmful cyber incident is conducted via the information infrastructure subject to a nation's control is* prima facie *evidence that the nation knows of the use and is responsible for the cyber incident*' (Ryan, Dion, Tikk, and Ryan, 'International Cyberlaw', p 1185; emphasis in the original).

[402] *Cyberspace Operations*, Air Force Doctrine Document 3–12, 15 July 2010, p 10, <http://www2.gwu.edu/~nsarchiv/NSAEBB/NSAEBB424/docs/Cyber-060.pdf>.

[403] *Relazione sulle possibili implicazioni*, p 26.

[404] *Relazione sulle possibili implicazioni*, p 26.

[405] Permanent Mission of the Federal Republic of Germany to the United Nations, Note Verbale No 516/2012, 5 November 2012, p 1, <http://www.un.org/disarmament/topics/informationsecurity/docs/Germany_Verbal_Note_516_UNODA.pdf>. Laurie Blank also observes that 'the victim State must tread carefully and seek as much clarity regarding the source of the attack as possible to avoid launching a self-defense response in the wrong direction' (Blank, 'International Law', p 417).

[406] AIV/CAVV, *Cyber Warfare*, p 22.

exercised against cyber attacks 'only if the origin of the attack and the identity of those responsible are sufficiently certain'.[407]

All in all, clear and convincing evidence seems the appropriate standard not only for claims of self-defence against traditional armed attacks, but also for those against cyber operations: a *prima facie* or preponderant standard of proof might lead to specious claims and false attribution, while a beyond reasonable doubt standard would be unrealistic: 'the degree of burden of proof...adduced ought not to be so stringent as to render the proof unduly exacting'.[408] A clear and convincing standard, on the other hand, 'obliges a state to act reasonably, that is, in a fashion consistent with the normal state practice in same or similar circumstances. Reasonable states neither respond precipitously on the basis of sketchy indications of who has attacked them nor sit back passively until they have gathered unassailable evidence'.[409] The Tallinn Manual, therefore, correctly maintains that neither the fact that a cyber operation originates from a state's governmental cyber infrastructure nor that it has been routed through the cyber infrastructure located in a state are sufficient evidence for attributing the operation to those states.[410]

Those that criticize a clear and convincing evidence standard for cyber operations rely on the fact that, due to the speed at which such operations may occur and produce their consequences, the requirement of a high level of evidence may in fact render impossible for the victim state to safely exercise its right of self-defence. Such concerns, however, are ill-founded. Indeed, if the cyber attack was a standalone one that instantaneously produced its damaging effects, a reaction in self-defence would probably not be necessary. If, on the other hand, the cyber attack is continuing or is formed by a series of smaller-scale cyber attacks, the likelihood that clear and convincing evidence can be collected would considerably increase.[411]

The standard of proof should be distinguished from the burden of proof, which (when narrowly intended) only identifies the litigant that has the *onus* of meeting that standard.[412] It is normally the party that relies upon a certain fact that is required to prove it (*onus probandi incumbit actori*).[413] Some authors have suggested that, in the cyber context, there should be a reversal of the burden of proof from the state that invokes self-defence to the state where the cyber attack originates.[414] As the ICJ found in the *Corfu Channel* case, however, the fact that

---

[407] Dutch Government Response to the AIV/CAVV Report on Cyber Warfare, p 5.
[408] *Certain Norwegian Loans (France v Norway)*, Merits, Judgment, 6 July 1957, ICJ Reports 1957, Separate Opinion of Judge Sir Hersch Lauterpacht, p 39.
[409] Michael N Schmitt, 'Cyber Operations and the *Jus ad Bellum* Revisited', *Villanova Law Review* 56 (2011–2012), p 595.
[410] Tallinn Manual, pp 34–6. For a view opposite to that of the Tallinn Manual, see Eneken Tikk, 'Ten Rules for Cyber Security', *Survival* 53(3) (2011), p 122, according to whom '[t]he fact that a cyber attack has been launched from an information system located in a state's territory is evidence that the act is attributable to that state'.
[411] A similar reasoning is made, in relation to the identification of the state responsible for the cyber attack, by Dinstein, 'Cyber War', p 282.
[412] Green, 'Fluctuating Evidentiary Standards', p 165.
[413] *Military and Paramilitary Activities in and against Nicaragua (Nicaragua v US)*, Jurisdiction, Judgment, 26 November 1984, ICJ Reports 1984, para 101.
[414] Clarke and Knake, *Cyber War*, p 249.

evidence is located exclusively on the territory of one party does not result in a reversal of the burden of proof.[415] The Court, also conceded that

the fact of this exclusive territorial control exercised by a State within its frontiers has a bearing upon the methods of proof available to establish the knowledge of that State as to such events. By reason of this exclusive control, the other State, the victim of a breach of international law, is often unable to furnish direct proof of facts giving rise to responsibility. Such a State should be allowed a more liberal recourse to inferences of fact and circumstantial evidence. This indirect evidence is admitted in all systems of law, and its use is recognized by international decisions. It must be regarded as of special weight when it is based on a series of facts linked together and leading logically to a single conclusion.[416]

Similarly, the fact that evidence is contained in classified documents, as is often the case in the cyber context, does not result in a reversal of the burden of proof: in both the *Genocide* and *Corfu Channel* cases, the ICJ did not demand the production of classified documents by the respondent states, attracting however the criticism of the minority judges.[417]

7. The duty to report the self-defence measures to the UN Security Council

Article 51 of the UN Charter requires states adopting measures in individual and collective self-defence to report them 'immediately' to the UN Security Council. Such an obligation might be difficult to comply with in the case of a cyber attack in self-defence: it has been seen that, because of their inherent features and the current architecture of cyberspace, cyber operations are the perfect tool for covert actions. Does this mean that a cyber attack in self-defence would be unlawful if it is not immediately reported to the Security Council?

The question arose in more general terms in the *Nicaragua* case, as the US paramilitary operations in and against Nicaragua were conducted covertly through the CIA and other agencies. The Court found that the duty to report does not reflect customary international law (at least at the time of the judgment),[418] but that it could be 'one of the factors indicating whether the State in question was itself convinced that it was acting in self-defence'.[419] Non-compliance, however, does not affect the legality of an otherwise lawful exercise of self-defence: as clearly highlighted by Judge Schwebel in his Dissenting Opinion in the same case, it would be 'bizarre' if it would follow from the duty to report that aggressors are free to act covertly while those who defend themselves are not.[420] Indeed, '[a] State cannot be

---

[415] *Corfu Channel*, p 18. [416] *Corfu Channel*, p 18.
[417] Tsagourias, 'Cyber Attacks', p 235. See the Dissenting Opinion of Vice-President Al-Khasawneh appended to the *Genocide* Judgment, para 35.
[418] *Nicaragua*, para 200. But see Gray, *International Law*, pp 101–2, who argues that states have in fact been complying with the duty to report after the *Nicaragua* judgment.
[419] *Nicaragua*, para 200.
[420] *Nicaragua*, Dissenting Opinion of Judge Schwebel, para 222. See similarly Ronzitti, 'The Expanding Law', p 356, who argues that non-compliance with the duty to report is an 'excusable

deprived, and cannot deprive itself, of its inherent right of individual or collective self-defence because of its failure to report measures taken in the exercise of that right to the Security Council'.[421]

It can therefore be concluded that the covert character of defensive cyber operations does not per se render them unlawful under Article 51 of the UN Charter, providing that all other requirements for the exercise of self-defence are met. The inclusion of Rule 17 in the Tallinn Manual, according to which '[m]easures involving cyber operations undertaken by States in the exercise of the right of self-defence pursuant to Article 51 of the United Nations Charter shall be immediately reported to the United Nations Security Council', appears, therefore, of limited practical relevance.

IV. Remedies Against Cyber Operations Short of Armed Attack

So far no cyber operation crossing the threshold of an armed attack in the sense highlighted above has occurred. The current scenario is that of 'campaigns of diffuse, low-intensity attacks'.[422] Paraphrasing Christine Gray, one could well say that low-intensity cyber attacks are the most common form of cyber force between states.[423] The advantages are numerous: they are less expensive, easier to conduct, there is less risk of a full-scale response by the victim (or of any response at all), and 'may be effective to retard a target's economic, social and technological development'.[424]

In his famous essay on Article 2(4) of the UN Charter, Thomas Franck wrote that '[m]odern warfare...tends...to proceed along two radically different lines, one too small and the other too large to be encompassed effectively by Article 51'.[425] The cyber attacks conducted so far appear to be at the lowest end of this scale, which seems to lead to the conclusion that they escape the application of Article 51 of the UN Charter. Indeed, in *Nicaragua*, the ICJ famously found that uses of force short of armed attacks only entitle the victim state to react by adopting non-forcible proportionate countermeasures.[426] The EECC upheld this conclusion.[427] In his Separate Opinion in the *Oil Platforms* case, however, Judge Simma took a different view and proposed

a distinction between (full-scale) self-defence within the meaning of Article 51 against an 'armed attack' within the meaning of the same Charter provision on the one hand and, on

---

violation'; Pierluigi Lamberti Zanardi, *La legittima difesa nel diritto internazionale* (Milano: Giuffrè, 1972), pp 275–6.

[421] *Nicaragua*, Dissenting Opinion of Judge Schwebel, para 230.
[422] Sean Watts, 'Low-Intensity Computer Network Attack and Self-Defense', *International Law Studies* 87 (2011), p 60.
[423] Gray, *International Law*, p 177 (referring to frontier incidents).
[424] Watts, 'Low-Intensity Computer Network Attack', pp 72–3.
[425] Thomas M Franck, 'Who Killed Article 2(4)? Or: Changing Norms Governing the Use of Force by States', *American Journal of International Law* 64 (1970), p 812.
[426] *Nicaragua*, para 249. [427] EECC, *Partial Award*, Jus ad Bellum, para 12.

the other, the case of hostile action, for instance against individual ships, below the level of Article 51, justifying proportionate defensive measures on the part of the victim, equally short of the quality and quantity of action in self-defence expressly reserved in the United Nations Charter.[428]

Judge Simma suggested that the reference in paragraph 249 of the *Nicaragua* judgment to 'proportionate counter-measures' not involving the use of force should be interpreted as 'defensive military action "short of" full-scale self-defence'.[429] What is crucial in his view is the proportionality of the reaction.[430] However, as Michael Schmitt notes, the principle of proportionality already applies to self-defence reactions, so all Simma's view does is lower the threshold for the use of force in self-defence from 'armed attack' to 'use of force', which is in clear contradiction with the letter of the Charter.[431] It is also not consistent with Article 50(1)(a) of the ILC Articles on State Responsibility, according to which countermeasures may not affect the prohibition of the threat and use of force as contained in the UN Charter.

The non-judicial remedies against cyber operations not amounting to an armed attack are then acts of retorsion and non-forcible countermeasures, as well as resort to the UN Security Council.[432] If retorsion, ie unfriendly acts not involving any breach of international law, may be undertaken at any time, countermeasures are 'measures that would otherwise be contrary to the international obligations of an injured State *vis-à-vis* the responsible State, if they were not taken by the former in response to an internationally wrongful act by the latter in order to procure cessation and reparation'.[433] The 'injured state'[434] adopting cyber countermeasures also has to comply with those requirements provided in Part Three, Chapter II of the ILC Articles on State Responsibility that reflect customary international law. In particular, the injured state must first call upon the responsible state to discontinue

---

[428] *Oil Platforms*, Separate Opinion of Judge Simma, para 12.
[429] *Oil Platforms*, Separate Opinion of Judge Simma, para 12.
[430] See also Rosalyn Higgins, *Problems and Process: International Law and How We Use it* (Oxford: Clarendon, 1994), p 251 ('Is the question of *level* of violence by regular forces not really an issue of *proportionality*, rather than a question of determining what is "an armed attack"?'; emphasis in the original).
[431] Schmitt, 'Cyber Operations in International Law', p 160. As has been already mentioned, the United States has maintained that there is no difference between the scope of 'use of force' and 'armed attack' (Sofaer, 'International Law', pp 422, 425–6).
[432] On the role played by the UN Security Council, see Section V of this Chapter. Another possible remedy is the resort to an international court: see Marco Roscini, 'World Wide Warfare—*Jus ad bellum* and the Use of Cyber Force', *Max Planck Yearbook of United Nations Law* 14 (2010), pp 111–13.
[433] ILC, Draft articles on Responsibility of States for Internationally Wrongful Acts, with commentaries ('ILC Commentary'), *Yearbook of the International Law Commission*, 2001, Vol II, Part Two, p 128.
[434] According to Art 42 of the ILC Articles on State Responsibility, '[a] State is entitled as an injured State to invoke the responsibility of another State if the obligation breached is owed to: (*a*) that State individually; or (*b*) a group of States including that State, or the international community as a whole, and the breach of the obligation: (i) specially affects that State; or (ii) is of such a character as radically to change the position of all the other States to which the obligation is owed with respect to the further performance of the obligation'.

the internationally wrongful act and/or provide reparation[435] and, apart from the case of 'urgent countermeasures',[436] must notify it of the decision to take countermeasures and offer to negotiate.[437] An obligation to notify cyber countermeasures, however, is probably unrealistic, as it deprives the operations of one of their main advantages, ie their anonymity. Furthermore, if the injured state notifies its intention to adopt cyber countermeasures, the wrongdoing state may immunize itself by reinforcing its active and passive cyber defences. Having said that, '[t]he injured State need not specify the content or timing of the measures'.[438] Article 52(1)(b) of the ILC Articles, therefore, still leaves some room for covert operations, including cyber ones.

The purpose of the countermeasure must also be to ensure compliance with international law and the measure must be 'as far as possible' reversible.[439] From this perspective, cyber operations aiming to incapacitate infrastructure without destroying it can be a particularly useful instrument. Finally, the countermeasure must be proportionate, ie 'commensurate with the injury suffered, taking into account the gravity of the internationally wrongful act and the rights in question'.[440] As the ILC Commentary states, '[c]ountermeasures are more likely to satisfy the requirements of necessity and proportionality if they are taken in relation to the same or a closely related obligation'.[441] A reaction in-kind to an unlawful cyber operation, however, might not be possible because the injured state does not possess the technology to conduct it, or because the wrongdoing state does not have networked infrastructure. Furthermore, in the cyber context even a reaction in-kind does not guarantee proportionate results: it could be, for instance, that the state against which the cyber countermeasure is adopted is much more digitally reliant than the state against which the initial wrongful cyber operation was undertaken, which could magnify the damaging effects of the cyber countermeasure. In any case, no cyber operation amounting to a use of force, ie those causing physical damage to property, death/injury to persons or severe disruption of the functioning of critical infrastructures, may be carried out in countermeasure.[442]

If the general rule, then, is that only cyber operations amounting to an armed attack trigger the application of Article 51 while others may be responded exclusively by adopting acts of retorsion and non-forcible countermeasures and by

---

[435] Article 52(1)(a) of the ILC Articles on State Responsibility.
[436] Article 52(2) of the ILC Articles on State Responsibility.
[437] Article 52(1)(b) of the ILC Articles on State Responsibility.
[438] Yuji Iwasawa and Naoki Iwatsuki, 'Procedural Conditions', in *The Law of International Responsibility*, edited by James Crawford, Alain Pellet, and Simon Olleson (Oxford: Oxford University Press, 2010), p 1152.
[439] Article 49 of the ILC Articles on State Responsibility.
[440] Article 51 of the ILC Articles on State Responsibility. See *Case concerning the Gabčíkovo-Nagymaros Project (Hungary v Slovakia)*, Judgment, 25 September 1997, ICJ Reports 1997, para 85.
[441] ILC Commentary, p 129.
[442] Other obligations that may not be affected by countermeasures are obligations for the protection of fundamental human rights, obligations of a humanitarian character prohibiting reprisals, obligations arising from peremptory norms of general international law, obligations under any dispute settlement procedure applicable between the wrongdoing and the responsible states, and obligations related to the inviolability of diplomatic or consular agents, premises, archives, and documents (Art 50 of the ILC Articles on State Responsibility).

referring the situation to the UN Security Council, it is submitted that, in at least two cases, a cyber operation short of armed attack may also determine an armed reaction by the victim. The first case is that of a low-intensity cyber attack that prepares an imminent kinetic or cyber armed attack:[443] even when the cyber operation does not reach the threshold of an 'armed attack' itself, the victim state might still be in a position to invoke anticipatory self-defence if the armed attack which the low-intensity cyber operation preludes to is imminent.[444] As already mentioned, for instance, right before the 2008 Russian invasion, several Georgian governmental websites had already been the target of brief but debilitating cyber attacks that continued throughout the conflict: the shutting down of crucial websites severed communication from the Georgian government in the initial phase of the conflict.[445] It also appears that the 2007 bombing by Israel of a nuclear facility in Syria was preceded by a cyber attack that neutralized ground radars and anti-aircraft batteries.[446] In these scenarios, the characterization of the cyber operation is irrelevant: what counts is that it is an indicator of an imminent armed attack, and it is that (and not the low-intensity cyber attack) which gives rise to the right of anticipatory self-defence. As argued above, the legality of anticipatory self-defence can be accepted only within the strict limits of the *Caroline* doctrine: the low-intensity cyber attack must be a clear indication of the imminence of the subsequent armed attack and it must be necessary to react anticipatorily because otherwise it would be too late. It is therefore questionable that the right of self-defence under Article 51 may be triggered by 'computer network activities' amounting to an 'imminent threat' of an armed attack, as claimed by the former US Department of State's Legal Advisor Harold Koh:[447] an imminent threat of an armed attack is not an imminent armed attack, and it is not even a threat. According to Schmitt, three factors must be taken into account when establishing the right to respond in anticipatory self-defence against a cyber attack that does not amount per se to an armed attack under Article 51: '1) The CNA is part of an overall operation culminating in armed attack; 2) The CNA is an irrevocable step in an imminent (near-term) and probably unavoidable attack; and 3) The defender is reacting in advance of the attack itself during the last possible window of opportunity available to effectively counter the attack'.[448] Schmitt's analysis appears to be a correct application of the *Caroline* criteria to the cyber context. On the other hand, claiming that a self-defence reaction is allowed whenever there is a penetration into 'those sensitive

---

[443] Anticipatory self-defence has already been discussed in relation to an imminent cyber operation amounting to an armed attack (see Chapter 2, Section III.2). What is discussed here is the different case of anticipatory self-defence against an ongoing cyber operation not amounting to an armed attack which preludes to an imminent (cyber or kinetic) armed attack.

[444] Robertson, Jr, 'Self-Defense', p 139.

[445] Tikk, Kaska, Rünnimeri, Kert, Talihärm, and Vihul, *Cyber Attacks*, pp 4–5, 15; Gosnell Handler, 'The New Cyber Face', p 224.

[446] Misha Glenny, 'Cyber armies are gearing up in the cold war of the web', *The Guardian*, 25 June 2009, <http://www.guardian.co.uk/commentisfree/2009/jun/25/cybercrime-nato-cold-war>; Gosnell Handler, 'The New Cyber Face', p 223.

[447] Koh, 'International Law in Cyberspace', p 595.

[448] Schmitt, 'Computer Network Attack and the Use of Force', p 933.

systems that are critical to a state's vital national interests'[449] appears more an application of the doctrine of pre-emptive, rather than anticipatory, self-defence: the mere introduction of a non-destructive code in another state's computer in order to collect intelligence, sophisticated as it might be as in the case of Flame, or even of a destructive one that has not been activated, does not in itself prelude to an armed attack and does not therefore justify a reaction in self-defence.

The second case is that of a series of debilitating 'pin-prick' cyber attacks originating from the same author that, individually considered, do not entitle to self-defence but that, cumulatively, reach the 'scale and effects' threshold of armed attacks. It has been suggested, for instance, that the 2007 DDoS attacks on Estonia were 'a coordinated set of cyberattacks that collectively rose to the level of an armed attack'.[450] The doctrine of the accumulation of events in the context of self-defence has been invoked with respect to repeated small-scale attacks by armed groups and is increasingly acquiring legal currency.[451] Israel is well known for invoking it in order to justify armed reactions against terrorist attacks.[452] The 2006 military operation in Lebanon, for instance, originated from the abduction and killing of some Israeli soldiers: while the operation was condemned for lack of proportionality, the acquiescence of the international community to Israel's invocation of the right of self-defence may be seen as tacit acceptance of the doctrine of the accumulation of events in light of the broader campaign of Hezbollah against Israel.[453] More recently, it seems that Kenya relied on the accumulation of events doctrine when it invoked nine incidents from 2009 to 2011 involving Al-Shabaab incursions into Kenyan territory from Somalia. Kenya's position was criticized, as there does not seem to be any distinctive pattern resulting from the incursions and it is also questionable whether the incursions, even taken as a whole, reached a sufficient threshold of gravity: there was no significant material damage or loss of life as a result, even if instability in the border region and economic losses as a consequence of the attacks on tourism did ensue.[454] The criticism focused, however, on whether the conditions to invoke the accumulation of events doctrine existed, not on the doctrine itself.

Even though the ICJ has so far avoided taking a clear position on this issue, it has suggested that an accumulation of events cannot be ruled out for the purposes

---

[449] Sharp, Sr, *Cyberspace*, p 129. See similarly Joyner and Lotrionte, 'Information Warfare', p 855.

[450] Sklerov, 'Solving the Dilemma', p 76.

[451] Randelzhofer and Nolte, 'Article 51', p 1409; Tams, 'The Use of Force', p 388; Theresa Reinold, 'State Weakness, Irregular Warfare, and the Right to Self-Defense Post-9/11', *American Journal of International Law* 105 (2011), p 271; Ruys, *'Armed Attack'*, p 172; Van Steenberghe, 'Self-Defence', p 203. The doctrine of accumulation of events also affects the proportionality and immediacy of the reaction, which need to be assessed against the series of events as a whole, and not against each event individually considered, therefore broadening the scope and duration of the permissible defensive reaction.

[452] Norman M Feder, 'Reading the U.N. Charter Connotatively: Toward a New Definition of Armed Attack', *New York University Journal of International Law and Politics* 19 (1987), p 415.

[453] Reinold, 'State Weakness', p 266; Enzo Cannizzaro, 'Contextualizing Proportionality: *jus ad bellum* and *jus in bello* in the Lebanese War', *International Review of the Red Cross* 88 (2006), pp 782–3.

[454] Hadzi-Vidanović, 'Kenya invades Somalia'.

of self-defence. In the *Nicaragua* judgment, the Court implied that Nicaragua's incursions into Honduras and Costa Rica could amount, 'singly or collectively', to an armed attack.[455] More explicit support for this doctrine can be found in Judge Schwebel's Dissenting Opinion, where he argues that Nicaragua's subversive activity 'is cumulatively tantamount to armed attack upon El Salvador'.[456] In *Cameroon v Nigeria*, the Court declined to pronounce on the accumulation of events claim on the basis that 'neither of the Parties sufficiently proves the facts which it alleges, or their imputability to the other Party'.[457] In the *Oil Platforms* case, the ICJ did not expressly reject the doctrine of the accumulation of events, although the Court found it not applicable to the case before it.[458] Similarly, in *DRC v Uganda*, the Court left all options open.[459]

It is submitted that an accumulation of events can be taken into account for the purposes of establishing whether there is an armed attack only to the extent that the individual events are part of a 'composite' armed attack consisting of numerous low-intensity attacks. Composite wrongful acts are the object of Article 15 of the ILC Articles on State Responsibility, which describes them as 'a series of actions or omissions defined in aggregate as wrongful' (Article 15(1)). A composite act is different from a series of repeated acts, that occurs when 'there are distinct acts which succeed each other and are breaches of the same nature'.[460] The attacks must thus be linked in time, cause and source to be 'accumulated'.[461] The existence of such a pattern can only be ascertained on a case-by-case basis in the light of the relevant circumstances. The Commentary to Article 15 points out that '[t]he number of actions or omissions which must occur to constitute a breach of the obligation is also determined by the formulation and purpose of the primary rule':[462] in self-defence, it is the scale and effects threshold of an armed attack that needs to be reached. This was not the case of the DDoS attacks on Estonia, that, even cumulatively, only caused limited disruption.

Focusing on the composite character of the armed attack in order to delimit the application of the accumulation of events doctrine would prevent the risk of the 'open-ended licence to use force' envisaged by Tams.[463] Indeed, in case of a composite act, 'the breach extends over the entire period starting with the first of the actions or omissions of the series and lasts for as long as these actions or omissions are repeated and remain not in conformity with the international obligation' (Article 15(2) of the ILC Articles).

---

[455] *Nicaragua*, para 231. [456] *Nicaragua*, Dissenting Opinion of Judge Schwebel, para 6.
[457] *Case Concerning the Land and Maritime Boundary Between Cameroon and Nigeria (Cameroon v Nigeria: Equatorial Guinea intervening)*, Merits, Judgment, 10 October 2002, ICJ Reports 2002, paras 323–4.
[458] *Oil Platforms*, para 64.
[459] When assessing the attacks by armed groups from the DRC against Uganda, the Court concluded that 'on the evidence before it, *even if this series of deplorable attacks could be regarded as cumulative in character*, they still remained non-attributable to the DRC' (*DRC v Uganda*, para 146; emphasis added).
[460] Jean Salmon, 'Duration of the Breach', in *The Law of International Responsibility*, edited by Crawford, Pellet, and Olleson, p 391.
[461] Ruys, *'Armed Attack'*, p 290. [462] ILC Commentary, p 63.
[463] Tams, 'The Use of Force', p 389.

There are, however, three problems with using the accumulation of events doctrine in the cyber context. First, one of the strategic purposes of the attacker is likely to be that the defender does not realize that the attacks are part of a coordinated strategy.[464] In certain cases, the victim will not even be aware that it is under attack, as low-yield cyber attacks can frequently go undetected: 'successful low-intensity CNA never awaken a sleeping giant'.[465] The problem is neither new nor specific to cyberspace: as Franck highlighted more than 40 years ago with regard to guerrilla warfare, '[w]ith the hit-and-run tactics of wars of national liberation...it is often difficult even to establish convincingly, from a pattern of isolated, gradually cumulative events, when or where the first round began, let alone at whose instigation, or who won it'.[466] Secondly, in order to apply the accumulation of events doctrine, each small-scale attack must be a use of force which is not serious enough to amount to an armed attack.[467] If the cyber attacks, individually considered, are not uses of force but, say, cyber exploitation operations, the doctrine of accumulation of events cannot be applied for the purposes of self-defence. Thirdly, in order to apply the doctrine of the accumulation of events, the attacks must be imputable to the same source: it has already been seen that identification and attribution issues are particularly cumbersome in the cyber context.[468]

V. Chapter VII of the United Nations Charter and the Role of the Security Council

Regardless of its qualification as an armed attack, the state victim of a cyber operation (or any other UN member)[469] could refer the situation to the Security Council under Article 35(1) of the UN Charter and the Council may recommend the appropriate methods to settle the dispute (Article 36(1)). If the Security Council also establishes that the situation amounts to a threat to the peace, breach of the peace, or act of aggression, it could exercise its powers under Chapter VII. The exercise of this competence by the Council also constitutes a limit to the right of individual and collective self-defence by states, as provided in Article 51 of the UN Charter.[470]

---

[464] Antoine Lemay, José M Fernandeza, and Scott Knight, 'Pinprick Attacks, A Lesser Included Case?', in *Conference on Cyber Conflict Proceedings 2010*, edited by Christian Czossek and Karlis Podins (Tallinn: CCDCOE, 2010), p 191.
[465] Watts, 'Low-Intensity Computer Network Attack', p 72.
[466] Franck, 'Who Killed Article 2(4)?', p 820.
[467] Tarcisio Gazzini, *The Changing Rules on the Use of Force in International Law* (Manchester: Manchester University Press, 2006), p 144.
[468] See Chapter 1, Section IV.
[469] A non-member can 'bring to the attention of the Security Council or of the General Assembly any dispute to which it is a party if it accepts in advance, for the purposes of the dispute, the obligations of pacific settlement provided in the present Charter' (Art 35(2)).
[470] According to Art 51, the right of self-defence exists 'until the Security Council has taken measures necessary to maintain international peace and security'. It seems, however, that the

Whether or not cyber operations can be considered breaches of the peace or acts of aggression (Article 3(b) of GA Res 3314 (XXIX), for instance, specifies that 'the use of *any weapons* by a State against the territory of another State' may amount to an act of aggression),[471] they could certainly potentially amount to a 'threat to the peace'. The authors of the cyber operation (be they states or non-state actors), as well as its characterization as a use of force, armed attack, or mere cyber exploitation, would not be a decisive factor in the determination that a threat to the peace exists. Indeed, even though, in the drafters' idea, this notion was limited to the international use of armed force,[472] its scope has been progressively expanded by the Security Council.[473] The General Assembly has repeatedly expressed its concern that cyber technologies 'can potentially be used for purposes that are inconsistent with the objectives of maintaining international stability and security'.[474] Cuba has highlighted that the misuse of information systems and resources for interfering in the domestic affairs of other states and infringing their sovereignty and independence 'may pose a serious threat to international security'.[475] The US *International Strategy for Cyberspace* also notes that '[c]ybersecurity threats can even endanger international peace and security more broadly, as traditional forms of conflict are extended into cyberspace'.[476] Bolivia,[477] China,[478] Estonia,[479] Mexico,[480] Panama,[481] Poland,[482] Russia,[483] Sweden on behalf of the EU member states of the United Nations,[484] and Turkmenistan[485] have all expressed similar concerns. Article 4 of the draft Convention on Information Security proposed by Russia, in particular, lists 11 threats in the information space prejudicial to 'international peace and stability':

1) the use of information technology and means of storing and transferring information to engage in hostile activity and acts of aggression;

Security Council needs to indicate an explicit intention to terminate the right of self-defence of the victim and other states for this limitation to apply (Gill and Ducheine, 'Anticipatory Self-Defense', pp 447–8).

[471] Emphasis added. It has been seen (Chapter 2, Section III.1, pp 71–2) that certain situations envisaged in the Definition could well cover cyber operations. In any case, the list contained in the Definition is not exhaustive (Art 4) and is not binding on the Security Council.

[472] Inger Österdahl, *Threat to Peace. The Interpretation by the Security Council of Article 39 of the UN Charter* (Uppsala: Iustus, 1998), p 85.

[473] It is well-known that the drafters of the Charter deliberately left the notion undefined (United Nations Conference on International Organization, Documents, Vol XII, 1945, p 505).

[474] See eg Preamble, GA Res 66/24, 2 December 2011.

[475] UN Doc A/54/213, 10 August 1999, pp 4–5.

[476] *International Strategy for Cyberspace*, p 4.

[477] UN Doc A/58/373, 17 September 2003, p 2.

[478] UN Doc A/59/116, 23 June 2004, p 4.   [479] Estonia's *Cyber Security Strategy*, p 10.

[480] UN Doc A/59/116/Add.1, 28 December 2004, p 2.

[481] UN Doc A/57/166/Add.1, 29 August 2002, p 5.

[482] UN Doc A/55/140/Add.1, 3 October 2000, p 2.

[483] UN Doc A/C.1/65/PV.15, 20 October 2010, p 20. See also *Conceptual Views Regarding the Activities of the Armed Forces of the Russian Federation in the Information Space*, 9 September 2000, p 8, <http://www.ccdcoe.org/strategies/Russian_Federation_unofficial_translation.pdf> (CCDCOE's unofficial translation).

[484] UN Doc A/56/164, 3 July 2001, p 5.   [485] UN Doc A/66/152/Add.1, 16 September 2011, p 7.

2) purposefully destructive behavior in the information space aimed against critically important structures of the government of another State;
3) the illegal use of the information resources of another government without the permission of that government, in the information space where those resources are located;
4) actions in the information space aimed at undermining the political, economic, and social system of another government, and psychological campaigns carried out against the population of a State with the intent of destabilizing society;
5) the use of the international information space by governmental and non-governmental structures, organizations, groups, and individuals for terrorist, extremist, or other criminal purposes;
6) the dissemination of information across national borders, in a manner counter to the principles and norms of international law, as well as the national legislation of the government involved;
7) the use of an information infrastructure to disseminate information intended to inflame national, ethnic, or religious conflict, racist and xenophobic written materials, images or any other type of presenting ideas or theories that promote, enable, or incite hatred, discrimination, or violence against any individual or group, if the supporting reasons are based on race, skin color, national or ethnic origin, or religion;
8) the manipulation of the flow of information in the information space of other governments, disinformation or the concealment of information with the goal of adversely affecting the psychological or spiritual state of society, or eroding traditional cultural, moral, ethical, and aesthetic values;
9) the use, carried out in the information space, of information and communication technology and means to the detriment of fundamental human rights and freedoms;
10) the denial of access to new information and communication technologies, the creation of a state of technological dependence in the sphere of informatization, to the detriment of another State;
11) information expansion, gaining control over the national information resources of another State.[486]

The problem, however, is whether *any* cyber operation, whatever its nature, scale, and consequences, can be qualified by the Security Council as a threat to the peace in the sense of Article 39 of the Charter. Even though the Council enjoys a broad discretion in determining the existence of such a threat,[487] this *kompetenz-kompetenz* is not unlimited: a threat to the peace could not be 'artificially created as a pretext for the realization of ulterior purposes'.[488] The ICTY

---

[486] Draft Convention on International Information Security (Concept), 2011, Art 4.
[487] See the Dissenting Opinion of Judge Weeramantry in *Questions of Interpretation and Application of the 1971 Montreal Convention arising from the Aerial Incident at Lockerbie (Libyan Arab Jamahiriya v UK; Libyan Arab Jamahiriya v US)*, Order on Request for the indication of Provisional Measures, 14 April 1992, ICJ Reports 1992, p 176: 'the determination under Article 39 of the existence of any threat to the peace, breach of the peace or act of aggression, is one entirely within the discretion of the Council. It would appear that the Council and no other is the judge of the existence of the state of affairs which brings Chapter VII into operation'.
[488] *Legal Consequences for States of the Continued Presence of South Africa in Namibia (South West Africa) notwithstanding Security Council Resolution 276 (1970)*, Advisory Opinion, 21 June 1971, ICJ

made clear that 'the "threat to the peace" is more of a political concept. But the determination that there exists such a threat is not a totally unfettered discretion, as it has to remain, at the very least, within the limits of the Purposes and Principles of the Charter',[489] According to Conforti, the conduct of a state cannot be considered a threat to the peace *when the condemnation is not shared by the opinion of most of the States and their peoples*'.[490] Other commentators refer to the limit of good faith and to the doctrine of abuse of right.[491] It is true that there is no direct judicial control over acts of the Council,[492] but there are indirect ones: the protest by refusal to comply with the resolution by the UN member states, the indirect judicial control when a resolution becomes relevant to decide a case before an international or national tribunal, and, more generally, acceptance of the Security Council's action by the international community.[493]

In order to establish the existence of a threat to the peace, then, the rank of the breached norm or value, the severity of the violation, and its transboundary effects need to be taken into consideration.[494] The assessment would obviously depend on the specific circumstances of each case. For instance, as the US DoD emphasizes, the fact that 'a computer network attack caused widespread damage, economic disruption, and loss of life could well precipitate...action by the Security Council'.[495] Another potential example of threat to the peace in the cyber context is 'any serious CNA conducted by contenders in long-standing global flash-points (e.g., India–Pakistan, Turkey–Greece)'.[496] Iran has also encouraged the Security Council 'to act against those States undertaking cyber

---

Reports 1971, Dissenting Opinion of Judge Fitzmaurice, paras 116–17. The qualification as a threat to the peace of the failure of Libya to extradite the alleged perpetrators of the Lockerbie bombing and to renounce terrorism 'by concrete actions', contained in SC Res 748 (31 March 1992), has for instance been criticized (Susan Lamb, 'Legal Limits to United Nations Security Council Powers', in *The Reality of International Law. Essays in Honour of Ian Brownlie*, edited by Guy S Goodwin-Gill and Stefan Talmon (Oxford: Oxford University Press, 1999), pp 378–9).

[489] *Prosecutor v Tadić*, Case No IT-94-1, Decision on the Defence Motion for Interlocutory Appeal on Jurisdiction, 2 October 1995, para 29.

[490] Benedetto Conforti, *The Law and Practice of the United Nations*, 3rd edn (Leiden and Boston: Nijhoff, 2005), pp 176–7 (emphasis in the original).

[491] See Thomas M Franck, 'Fairness in the International Legal and Institutional System', *Recueil des cours* 240 (1993–III), p 191; Lamb, 'Legal Limits', p 385.

[492] In his Separate Opinion in the *Genocide* case, Judge ad hoc Lauterpacht recalled that the ICJ's power of judicial review 'does not embrace any right of the Court to substitute its [own] discretion for that of the Security Council in determining the existence of a threat to the peace, a breach of the peace or an act of aggression, or the political steps to be taken following such a determination' (*Application of the Convention on the Prevention and Punishment of the Crime of Genocide (Bosnia and Herzegovina v Serbia and Montenegro)*, Further Requests for the Indication of Provisional Measures, Order of 13 September 1993, ICJ Reports 1993, para 99).

[493] Michael Bothe, 'Les limites des pouvoirs du Conseil de sécurité', in *The Development of the Role of the Security Council—Workshop of the Hague Academy of International Law*, edited by René-Jean Dupuy (Dordrecht: Nijhoff, 1992), p 70.

[494] Matthias Herdegen, *Die Befugnisse des UN-Sicherheitsrates: Aufgeklärter Absolutismus im Völkerrecht?* (Heidelberg: Müller, 1998), p 16.

[495] US DoD, *An Assessment*, p 15.

[496] Schmitt, 'Computer Network Attack and the Use of Force', p 928.

attacks and sabotage in the peaceful nuclear facilities'.[497] On the other hand, it has been suggested that 'computer attacks among major Western economic powers (perhaps in the form of economic espionage)...would clearly not threaten the peace if discovered'.[498]

If the Security Council does qualify a cyber operation as a threat to the peace, breach of the peace, or act of aggression, it could make recommendations under Article 39, adopt measures aimed at preventing the worsening of the crisis under Article 40, and, more importantly, adopt coercive measures under Articles 41 and 42. The non-exhaustive list of measures that the Council can recommend or decide under Article 41 includes 'complete or partial interruption of...telegraphic, radio, and other means of communication': the Security Council could thus adopt targeted cyber sanctions or limit the access to the internet of the state responsible for the threat to the peace, breach of the peace, or act of aggression. Member states may be required to prohibit the provision to the targeted state of hardware and software that facilitate connection to the internet and to ensure that webpages are denied access from the domain name of the targeted state.[499] The UN member states may also be required to adopt legislation to implement the sanctions in their domestic legal order, for instance to criminalize certain cyber conduct or to require national ISPs to adopt restrictive measures.[500]

'Should the Security Council consider that measures provided for in Article 41 would be inadequate or have proved to be inadequate',[501] it could authorize UN member states or UN peace forces to conduct cyber attacks amounting to a use of force in order to react against a threat to the peace.[502] It is true that Article 42 only refers to enforcement action 'by air, sea, or land forces': a literal reading of the provision might lead to the conclusion that enforcement in cyberspace is precluded to the Council. The purpose of Article 42, however, was to extend the collective security machinery to all military domains available at the time the Charter was drafted.[503] An evolutionary interpretation of the norm would then include any other military domain that becomes accessible through technological developments, such as outer space and cyberspace.

The Security Council may also 'utilize...regional arrangements or agencies for enforcement action under its authority'.[504] The regional organizations need of course to be competent to undertake peace enforcement operations under their own statutes. The enforcement actions by regional organizations could well include

---

[497] Iranian Foreign Minister's address to the UN Security Council, 28 September 2012, <http://iran-un.org/en/2012/09/28/28-september-2012-2/>.
[498] Schmitt, 'Computer Network Attack and the Use of Force', p 928.
[499] See the sanctions imposed by the United States on Cuba, which also affect access to the internet and use of social networks (UN Doc A/67/167, 23 July 2012, pp 10–11).
[500] Tallinn Manual, p 70.
[501] Article 42 of the UN Charter.
[502] Of course, the Security Council could also authorize the use of traditional force to react against cyber attacks: Jann K Kleffner and Heather A Harrison Dinniss, 'Keeping the Cyber Peace: International Legal Aspects of Cyber Activities in Peace Operations', *International Law Studies* 89 (2013), pp 523–7.
[503] Melzer, *Cyberwarfare*, p 19.
[504] Article 53(1) of the UN Charter. See also Rule 19 of the Tallinn Manual, pp 71–2.

cyber operations: to the extent they amount to a use of force, these operations require the previous authorization of the Security Council under Article 53(1).[505] Furthermore, Article 54 of the Charter requires that the regional organizations must keep the Security Council 'at all times...fully informed' of the cyber operations undertaken for the maintenance of international peace and security, whether or not they amount to a use of force.

VI. Conclusions

In the current absence of specific *jus ad bellum* rules applicable to cyber operations, we are left with the provisions contained in the UN Charter and in customary international law. These rules are flexible enough to be extended to means that did not exist when they were adopted. All in all, as has been seen, it is states themselves that have argued that existing *jus ad bellum* rules apply to at least certain cyber operations.[506]

The main normative conclusions reached in the present Chapter can be summarized in the following points:

(1) A cyber attack by a state against another state is a use of force, and is thus prohibited by Article 2(4) of the UN Charter and its customary counterpart, when it causes or is reasonably likely to cause (a) material damage to property; (b) loss of life or injury to persons; (c) severe disruption of the functioning of critical infrastructures.

(2) Cyber attacks not causing the above consequences may be unlawful interventions in the internal affairs of other states, but are not a use of force.

(3) Cyber exploitation activities may be interferences (when they are not prohibited by international law) or violations of another state's sovereignty, but are never a use of force under Article 2(4) of the UN Charter.

(4) Cyber or kinetic self-defence under Article 51 of the Charter and customary international law may be exercised against a cyber attack by a state or a non-state actor only to the extent that it qualifies as an 'armed attack', ie when the cyber operation amounting to a use of force meets the 'scale and effects' standard identified by the ICJ.

(5) Self-defence against cyber operations not amounting to an armed attack may be exercised exclusively within the limits of the doctrine of the accumulation of events and of anticipatory self-defence. In all other cases, the only lawful response is the adoption of acts of retorsion, non-forcible countermeasures (including cyber attacks below the use of force level) and resort to the UN Security Council.

---

[505] Peace enforcement operations by regional organizations must be distinguished from collective self-defence operations under Art 51 of the UN Charter, which do not require the previous authorization of the Security Council.
[506] See Chapter I, Section III, pp 21–2.

(6) Finally, the standard of proof for claims of self-defence against cyber operations amounting to an armed attack does not differ from that applicable to self-defence against kinetic armed attacks and would normally require 'clear and convincing evidence'.

This Chapter's findings are also recapitulated in the following tables.

Table 2.1 Qualification of different types of cyber operations and cyber-related activities

	Violation of another state's sovereignty	Intervention	Use of force[507]
Cyber attacks causing or reasonably likely to cause material damage to property, loss of life, or bodily injury			✓
Cyber attacks severely disrupting NCIs without physical damage			✓
Other cyber attacks		✓	
Cyber exploitation	✓		
Provision of cyber weapons and training to an armed group acting against another state			✓
Placing a state's own cyber infrastructure at the disposal of another state so that it uses it for perpetrating an act of aggression against a third state			✓

Table 2.2 Remedies against cyber operations

	Acts of retorsion	Non-forcible countermeasures	Self-defence	Resort to the UN Security Council
Cyber exploitation	✓	✓		✓
Cyber attacks causing material damage to property, loss of life, or injury to persons reaching the scale and effects threshold of 'armed attack'	✓	✓	✓	✓
Cyber attacks severely disrupting the functioning of NCIs and reaching the scale and effects threshold of 'armed attack'	✓	✓	✓	✓
Cyber attacks short of 'armed attack'	✓	✓		✓

[507] A use of force is of course *a fortiori* a violation of another state's sovereignty and an unlawful intervention in its internal affairs, and an intervention is *a fortiori* a violation of another state's sovereignty.

3
The Applicability of the *jus in bello* to Cyber Operations

I. Introduction

The *jus in bello* (also known as the law of armed conflict, or laws of war, or—although the expression is technically narrower—international humanitarian law)[1] is the province of international law that regulates *how* hostilities may be conducted in armed conflict and that protects those affected by them. Although rules on the conduct of hostilities can be traced back at least as far as the Old Testament,[2] the backbone of the current *jus in bello* includes the Hague Conventions of 1899 and 1907, which mainly address means and methods of warfare, and the four 1949 Geneva Conventions on the Protection of Victims of War, as supplemented by two additional Protocols in 1977.[3] Many of the provisions contained in these treaties now reflect customary international law: this is certainly true of the Hague Regulations Respecting the Laws and Customs of War on Land annexed to the 1907 Hague Convention IV and of most of the Geneva Conventions, while the status of certain provisions of Additional Protocols I and, even more, II is debatable.[4]

The *jus in bello* is different, and separate, from the *jus ad bellum*, which determines *when* states may use armed force in their international relations. For a state's resort to armed force to be lawful, then, it will have to comply not only with the *jus ad bellum* provisions of the UN Charter and customary international law, but also

---

[1] Although there are some differences of meaning in these expressions, they will be used interchangeably.

[2] Leslie C Green, *The Contemporary Law of Armed Conflict* (Manchester: Manchester University Press, 2008), pp 26–7.

[3] A third Protocol on the Adoption of an Additional Distinctive Emblem was adopted in 2005. The provisions on the conduct of hostilities are mainly contained in the Hague Regulations annexed to the 1907 Hague Convention IV, in Additional Protocol I and, to a lesser extent, in Additional Protocol II. On the other hand, the Geneva Conventions essentially focus on the protection of victims of war. There is also a plethora of treaties on specific warfare issues, such as the legality of certain weapons or the protection of cultural property in time of armed conflict.

[4] See Jean-Marie Henckaerts and Louise Doswald-Beck, *Customary International Humanitarian Law* (Cambridge: Cambridge University Press, 2005). See also the US criticism of the customary status of certain provisions in John Bellinger, III and William J Haynes, II, 'A US Government Response to the International Committee of the Red Cross Study *Customary International Humanitarian Law*', *International Review of the Red Cross* 89 (2007), pp 443 ff.

with the law of armed conflict: breach of or compliance with one does not justify violations of the other. Indeed, the *jus in bello* does not discriminate between an aggressor state and a state using force in self-defence and applies equally to both. The Preamble to Additional Protocol I expressly affirms this principle by making clear that the Protocol applies 'without any adverse distinction based on the nature or origin of the armed conflict, or on the causes espoused by or attributed to the Parties to the conflict'.

If, as has been seen, what triggers the application of the *jus ad bellum* is a 'use of force',[5] the *jus in bello* comes into play only when an 'armed conflict' breaks out, while law enforcement mechanisms as regulated by domestic laws and international human rights law apply to violence short of armed conflict.[6] International human rights law, however, does not cease to apply during armed conflict, and only certain of its provisions may be derogated from in case of national emergency.[7] International humanitarian law and international human rights law, therefore, complement each other, but the former is the *lex specialis* applicable in armed conflict that regulates the conduct of hostilities.[8]

The spatial scope of application of the *jus in bello* includes 'the whole territory of the warring States or, in the case of internal conflicts, the whole territory under the control of a party, whether or not actual combat takes place there'.[9] In *Kunarać*, the ICTY further clarified that '[t]here is no necessary correlation between the area where the actual fighting is taking place and the geographical reach of the laws of war', as long as the act, even if it occurs at a time and place far from where fighting is taking place, is 'closely related to hostilities occurring on other parts of the territories controlled by the parties to the conflict.'[10] This is particularly significant in the cyber context, which defies traditional geographical limitations. As to the temporal scope of application of the *jus in bello*, the *Tadić* decision specified that '[i]nternational humanitarian law applies from the initiation of...armed conflicts

---

[5] See Chapter 2, Section II. It should, however, be recalled that the 1928 Pact of Paris, a *jus ad bellum* instrument, focuses on the renunciation of 'war' as 'as instrument of national policy' (Art I).

[6] Law enforcement is hereby intended as 'all territorial and extraterritorial measures taken by a State to vertically impose public security, law and order or to otherwise exercise its authority or power over individuals in any place or manner whatsoever', excluding those measures that also qualify as 'hostilities' (Nils Melzer, *Targeted Killing in International Law* (Oxford: Oxford University Press, 2008), p 90).

[7] *Legality of the Threat or Use of Nuclear Weapons*, Advisory Opinion, 8 July 1996, ICJ Reports 1996 ('*Nuclear Weapons*'), para 25.

[8] *Nuclear Weapons*, para 25; *Legal consequences of the construction of a wall in the occupied Palestinian territory*, Advisory Opinion, 9 July 2004, ICJ Reports 2005 ('*Legal consequences of the construction of a wall*'), para 106. See also *Report of the International Commission of Inquiry to investigate all alleged violations of international human rights law in the Libyan Arab Jamahiriya*, UN Doc A/HRC/17/44, 1 June 2011, pp 5, 41; *Third Report of the Independent International Commission of Inquiry on the Syrian Arab Republic*, UN Doc A/HRC/21/50, 16 August 2012, p 45.

[9] ICTY, *Prosecutor v Tadić*, Case No IT–94–1, Decision on the Defence Motion for Interlocutory Appeals on Jurisdiction, 2 October 1995, para 70; ICTR, *Prosecutor v Akayesu*, Case No ICTR–96–4–T, Judgment, 2 September 1998, para 635.

[10] ICTY, *Prosecutor v Kunarać, Kovać and Voković*, Case No IT– 96–23 and IT–96–23/1–A, Appeals Chamber Judgment, 12 June 2002, para 57. See Chapter 3, Section II.2, pp 123–5. See also Chapter 4, Section III.1.4.

and extends beyond the cessation of hostilities until a general conclusion of peace is reached; or, in the case of internal conflicts, a peaceful settlement is achieved'.[11] The rules on detention of prisoners of war (POWs) and interned civilians apply until their release or repatriation, while the application of at least certain provisions of the law of occupation continues for as long as the occupation lasts.[12]

With the above in mind, this Chapter will establish when the law of armed conflict applies to cyber operations. As conventional *jus in bello* does not provide a general definition of 'armed conflict' but merely distinguishes between different types of armed conflicts to which different sets of rules apply, these types of conflict will be addressed separately. The next Section will focus on international armed conflicts and will distinguish between several scenarios that might lead to the application of the *jus in bello* to cyber operations. Section III will deal with cyber operations in the context of belligerent occupation, while Section IV will examine cyber operations in and as non-international armed conflicts. Finally, Section V will discuss cyber operations as internal disturbances and tensions.

II. Cyber Operations in and as International Armed Conflicts

Common Article 2(1) of the 1949 Geneva Conventions states that the Conventions apply 'to all cases of declared war or of any other armed conflict which may arise between two or more of the High Contracting Parties, even if the state of war is not recognized by one of them'.[13] It is generally accepted that this provision reflects customary international law and therefore fixes the threshold of application not only of the Geneva Conventions but also of the customary provisions contained in the Hague Conventions.[14]

---

[11] *Tadić*, Decision on the Defence Motion, para 70.
[12] According to Art 42(1) of the Hague Regulations, '[t]erritory is considered occupied when it is actually placed under the authority of the hostile army'. Article 6(3) of Geneva Convention IV fixes a deadline for the application of the Convention to 'one year after the general close of the military operations', with the exception of certain provisions that continue to apply 'for the duration of the occupation, to the extent that [the Occupying] Power exercises the functions of government in such territory'. On the other hand, Art 3(b) of Additional Protocol I provides that the application of the Conventions and the Protocol ceases 'on the termination of the occupation'. According to some commentators, the broader scope of application provided in Additional Protocol I reflects customary international law and, therefore, applies to Geneva Convention IV as well, whether or not the relevant states have ratified the Protocol (Robert Kolb and Richard Hyde, *An Introduction to the International Law of Armed Conflicts* (Oxford and Portland: Hart, 2008), p 104). Others, however, dissent (Yoram Dinstein, *The International Law of Belligerent Occupation* (Cambridge: Cambridge University Press, 2009), p 281).
[13] See also Art 18(1) of the 1954 Hague Convention on the Protection of Cultural Property in the Event of Armed Conflict. Text in Adam Roberts and Richard Guelff, *Documents on the Laws of War* (Oxford: Oxford University Press, 2000), pp 371 ff.
[14] The original scope of application of the 1899 and 1907 Hague Conventions was 'war', not 'armed conflict', but, as a consequence of the development of customary international humanitarian law, they are now applicable to all international armed conflicts, whether or not they amount to 'war'. See Christopher Greenwood, 'The Concept of War in Modern International Law', *International and Comparative Law Quarterly* 36 (1987), p 295; Siobhan Wills, 'The Legal Characterization of the

In the light of the first paragraph of Common Article 2, the Geneva Conventions and customary international humanitarian law would apply to cyber operations between states in three cases: (1) if they are preceded by a declaration of war made through cyber or traditional means of communication; (2) when the cyber operations occur in the context of an already existing international armed conflict and have a nexus with it; and (3) when they amount themselves to an international armed conflict, with or without the concomitant occurrence of kinetic hostilities. These cases will be addressed in turn.

1. Declared war

According to the traditional doctrine of the state of war, '[t]he use of physical force by one state within the territory of another state does not necessarily imply the existence of war in the legal sense'.[15] War in the legal sense is started exclusively by an 'overt act' by which a state manifests its intention to turn a state of peace into a state of war (*animus bellandi*).[16] Not all armed conflicts are therefore 'wars': this occurs only when at least one belligerent has manifested its *animus bellandi*:[17] as the Secretary-General of the League of Nations wrote in 1927, 'measures of coercion, however drastic, which are not intended to create and which are not regarded by the State to which they are applied as creating a state of war, do not legally establish a relation of war between the States concerned'.[18]

Animus bellandi could be implied in certain conduct, in particular the exercise of powers that would only be allowed in a state of war, such as the promulgation of a blockade.[19] A use of force, including by cyber means as described in Chapter 2 of this book, accompanied by an invitation to other states to observe the law of neutrality, as well as certain hostile acts,[20] could also imply an intention to make war.[21] On the other hand, breaking off diplomatic relations does not establish a state of war, but

---

Armed Conflicts in Afghanistan and Iraq: Implications for Protection', *Netherlands International Law Review* 58 (2011), p 177.

[15] George G Wilson, 'Use of Force and War', *American Journal of International Law* 26 (1932), p 327.

[16] Quincy Wright, 'When Does War Exist?', *American Journal of International Law* 26 (1932), p 363. See also John A Cohan, 'Legal War: When Does It Exist and When Does It End?', *Hastings International and Comparative Law Review* 27 (2003–04), p 242.

[17] Marina Mancini, *Stato di guerra e conflitto armato nel diritto internazionale* (Torino: Giappichelli, 2009), p 206.

[18] Reprinted in Ian Brownlie, *International Law and the Use of Force by States* (Oxford: Clarendon, 1963), p 38. As noted by Wright, '[a] state of war may exist without active hostilities, and active hostilities may exist without a state of war' (Wright, 'When Does War Exist?', p 363). Examples of a state of material war not recognized by the belligerents as war was the 1827 Battle of Navarino between the British, French, and Russian forces on one side and the Turkish and Egyptian fleets on the other, as well as the Sino-Japanese conflict that led to the occupation of Manchuria by Japan (p 365).

[19] Avril McDonald, 'Declarations of War and Belligerent Parties: International Law Governing Hostilities Between States and Transnational Terrorist Networks', *Netherlands International Law Review* 54 (2007), p 289; Wright, 'When Does War Exist?', p 364.

[20] Clyde Eagleton, 'The Form and Function of the Declaration of War', *American Journal of International Law* 32 (1938), p 25.

[21] Cohan, 'Legal War', p 254.

is rather a consequence of it.[22] For the parties to the 1907 Hague Convention III Relative to the Opening of Hostilities, the *animus bellandi* must be communicated through a declaration of war. According to its Article 1, '[t]he contracting Powers recognize that hostilities between themselves must not commence without previous and explicit warning, in the form of a declaration of war, giving reasons, or of an ultimatum with a conditional declaration of war'.[23] As Quincy Wright argues, however, '[w]hile this convention undoubtedly imposes a *duty* upon the parties, it is not certain that it restricts their *powers*. Hence if a party to this convention began hostilities with intent to make war without such declaration or ultimatum, while it would clearly be violating its duty to the other parties, nevertheless it is possible that its exercise of power would have legal effect and that a state of war would exist'.[24] By starting an undeclared war, then, the state party would commit an internationally wrongful act in the form of a treaty violation, but the state of war would nevertheless arise if *animus bellandi* is present.

A declaration of war creates a state of war in the legal sense between the belligerents, whether or not hostilities have commenced or actually follow.[25] Several consequences derive from the establishment of a state of war in the legal sense: diplomatic relations are broken off, treaties between belligerents and contracts between their nationals suspended, emergency powers may be exercised at the domestic level and the laws of war and neutrality start to apply.[26] The intention of one belligerent to establish a state of war, manifested through a declaration of war, is sufficient to bring about these consequences: the state to which the declaration is addressed could be dragged into a state of war without having such intention itself.[27] Similarly, if the attacking state acts *sine animo bellandi*, the victim could declare the existence of a state of war and the relevant consequences would ensue, whether or not the attacking state initially so intended.[28]

There is no binding form prescribed for a declaration of war in Hague Convention III, providing that it 'gives reasons', whatever they may be. It cannot be excluded that, in the internet era, the declaration could also be communicated by cyber means, for instance by email. Whether a certain communication amounts to a declaration of war or to an ultimatum containing such declaration, or to a mere threat, depends ultimately not on its form, but on its substance, and in particular on whether *animus bellandi* is present, which should not be 'lightly implied'.[29] The communication must also clearly originate from or be authorized by an organ of

[22] Eagleton, 'The Form', p 24.
[23] It is doubtful whether this provision reflects customary international law (Yoram Dinstein, *War, Aggression and Self-Defence* (Cambridge: Cambridge University Press, 2011), p 32).
[24] Wright, 'When Does War Exist?', p 363 (emphasis in the original).
[25] An example of a legal state of war not followed by actual hostilities is that between certain South American states and Germany during the First and Second World Wars (Dinstein, *War, Aggression and Self-Defence*, p 9).
[26] Wright, 'When Does War Exist?', p 363; Mancini, *Stato di guerra*, pp 24 ff.
[27] Wright, 'When Does War Exist?', p 363.
[28] Wright, 'When Does War Exist?', p 365; Cohan, 'Legal War', p 255. On the role of third states, see Mancini, *Stato di guerra*, pp 18–19.
[29] Lord McNair and Arthur D Watts, *The Legal Effects of War*, 4th edn (Cambridge: Cambridge University Press, 1966), p 8. According to the EECC, for instance, 'the essence of a declaration of

the state that is constitutionally competent to declare war and be addressed to the target state.[30]

Both cyber attacks and cyber exploitation operations conducted by the belligerents against each other after a declaration of war are regulated by the relevant *jus in bello* provisions, whether or not kinetic hostilities occur: the rules on the conduct of attacks will only apply to cyber attacks amounting to acts of violence, while other rules (for instance, those on war espionage) may apply to cyber exploitation. It is, however, a fact that declarations of war have not been issued in recent conflicts.[31] Indeed, the use of these semantics has become largely rhetoric and symbolic: as will be seen in the next Section, the very notion of 'war' as a legal concept has been replaced by that of 'armed conflict'. This has led Sir Christopher Greenwood to write that 'the state of war has become an empty shell which international law has already discarded in all but name'.[32] In the information age, declarations of war, as an 'element of sportsmanlike warning',[33] are even more unlikely: the incessant flow of information through mass media, the internet and social networks, as well as the use of surveillance technology such as satellites, will in most cases make sure that the victim state will have some awareness that a kinetic attack is about to take place even without a declaration of war. Requiring a declaration of 'cyber war' would also be unrealistic as it is scarcely reconcilable with the surprise and plausible deniability factors that constitute two of the main advantages of cyber operations.

2. Cyber operations 'in the context of' an existing international armed conflict

Cyber operations may be employed during an existing traditional international armed conflict as 'force multipliers'. The definition of 'cyber war' contained in a 2010 Report by the Italian Parliamentary Committee for the Security of the Republic, for instance, describes it as 'a proper conflict scenario between nations, fought by systematically destroying critical protection defences of the adversary's security, or through the disruption or shutting down of strategic communication networks, *and the integration of such activities with the properly belligerent ones*'.[34]

---

war is an explicit affirmation of the existence of a state of war between belligerents' (EECC, *Partial Award*, Jus ad bellum, *Ethiopia's Claims no 1-8*, 19 December 2005, RIAA, Vol XXVI, p 467). The Commission, therefore, denied that the resolution adopted by the Ethiopian Council of Ministers and Parliament condemning the Eritrean invasion and demanding the unconditional and immediate withdrawal of Eritrean forces from Ethiopian territory amounted to a declaration of war (p 467).

[30] Dinstein, *War, Aggression and Self-Defence*, p 33.

[31] Mancini, *Stato di guerra*, pp 195–7. See, however, the case of the alleged US declaration of war on Al-Qaeda in 2001 (McDonald, 'Declarations of War', pp 279 ff).

[32] Greenwood, 'The Concept of War', p 305. See also UK Ministry of Defence, *The Manual of the Law of Armed Conflict* (Oxford: Oxford University Press, 2004), p 29.

[33] Eagleton, 'The Form', p 29.

[34] COPASIR, *Relazione sulle possibili implicazioni e minacce per la sicurezza nazionale derivanti dall'utilizzo dello spazio cibernetico*, Doc XXXIV, no 4, 7 July 2010, p 17, <http://www.parlamento.it/documenti/repository/commissioni/bicamerali/COMITATO%20SICUREZZA/Doc_XXXIV_n_4.pdf> (emphasis added). The translation is mine. The original text of the definition of 'cyber war' contained in the Report reads as follows: 'scenario relativo ad un vero e proprio conflitto tra Nazioni,

An example of this scenario is the 2008 armed conflict between Georgia and the Russian Federation, where Georgia's governmental and media websites were taken off-line or defaced during the initial phases of the conflict allegedly by Russian hackers, thus affecting Georgia's ability to communicate and possibly also the operability of its armed forces.[35] Russia's responsibility, however, has not been conclusively established.

Rule 20 of the Tallinn Manual makes the generally accepted point that '[c]yber operations executed in the context of an armed conflict are subject to the law of armed conflict', whether or not the operations amount to resort to armed force themselves.[36] The Manual, however, fails to explain what 'in the context of' means and the Commentary limits itself to note that it could refer to either the fact that the operations are conducted by a belligerent against an adversary or that they are carried out to contribute to a belligerent's military effort.[37] Neither explanation, however, is, on its own, entirely satisfactory. To address the issue, it is useful to refer to the notion of 'belligerent nexus' developed by the ICRC in relation to the notion of direct participation in hostilities.[38] According to the ICRC, the belligerent nexus requires that the act must be 'specifically designed to directly cause the required threshold of harm in support of a party to the conflict and to the detriment of another'.[39] It is submitted that, if the cyber operations are conducted by a belligerent against another and cause or are reasonably likely to cause the required threshold of harm to the adversary,[40] the nexus is established and the cyber operations, as acts of hostilities, would fall under the scope of the law of international armed conflict. The operations may be attributable to a belligerent under the secondary rules on attribution contained in the ILC Articles on State Responsibility, already examined in Chapter 1, or under Article 91 of Additional Protocol I, which more broadly provides that '[a] Party to the conflict...shall be

combattuto attraverso il sistematico abbattimento delle barriere di protezione critica della sicurezza dell'avversario, ovvero attraverso il disturbo o lo «spegnimento» delle reti di comunicazione strategica, e l'integrazione di queste attività con quelle propriamente belliche'.

[35] Report of the Independent Fact-Finding Mission on the Conflict in Georgia, September 2009, Vol II, pp 217–18, <http://www.ceiig.ch/Report.html>; Eneken Tikk, Kadri Kaska, Kristel Rünnimeri, Mari Kert, Anna-Maria Talihärm, and Liis Vihul, *Cyber Attacks Against Georgia: Legal Lessons Identified* (CCDCOE, November 2008), pp 7–12, <http://www.carlisle.army.mil/DIME/documents/Georgia%201%200.pdf>. The CCDCOE Report concludes, however, that 'it is highly problematic to apply Law of Armed Conflict to the Georgian cyber attacks—the objective facts of the case are too vague to meet the necessary criteria of both state involvement and gravity of effect' (p 23). The CCDCOE Report is correct to say that there is no conclusive evidence that the cyber operations were attributable to Russia. On the other hand, the gravity of the effect does not seem relevant in the Georgian scenario, at least for those cyber operations that occurred after the beginning of the armed conflict.

[36] *Tallinn Manual on the International Law Applicable to Cyber Warfare* (Cambridge: Cambridge University Press, 2013), p 75.

[37] Tallinn Manual, p 76.

[38] The requirements for direct participation in hostilities will be further explored in Chapter 4, Section III.1.3.d.

[39] Recommendation V(3), in ICRC, *Interpretive Guidance on the Notion of Direct Participation in Hostilities under International Humanitarian Law*, Geneva, 2009 (prepared by Nils Melzer), p 58, <http://www.icrc.org/eng/resources/documents/publication/p0990.htm> (emphasis in the original).

[40] On the threshold of harm, see Chapter 4, Section III.1.3.d, pp 204–6.

responsible for *all* acts committed by persons forming part of its armed forces.'[41] The belligerent nexus, however, does not necessarily require attribution to a belligerent, being sufficient that the operations support a party to the conflict to the detriment of another. A hacktivist group that spontaneously carries out cyber attacks causing the required threshold of harm to a belligerent in support of another belligerent, without being an organ or agent of the latter, would therefore conduct acts of hostilities that would also fall under the remit of the *jus in bello*.

It should be noted that, with specific regard to individual criminal responsibility for violations of international humanitarian law, the ICTY has argued that the nexus with an armed conflict exists whenever the acts are 'closely related' to the hostilities.[42] The ICTY nexus is not necessarily identical to that identified by the ICRC in the context of direct participation in hostilities.[43] In *Kunarac et al*, the Appeals Chamber found that

> [w]hat ultimately distinguishes a war crime from a purely domestic offence is that a war crime is shaped by or dependent upon the environment—the armed conflict—in which it is committed.... The armed conflict need not have been causal to the commission of the crime, but the existence of an armed conflict must, at a minimum, have played a substantial part in the perpetrator's ability to commit it, his decision to commit it, the manner in which it was committed or the purpose for which it was committed. Hence, if it can be established... that the perpetrator acted *in furtherance of or under the guise of the armed conflict*, it would be sufficient to conclude that his acts were closely related to the armed conflict.[44]

The ICTY nexus requirement distinguishes war crimes from domestic offences committed during an armed conflict, although the distinction might not always be easy to make, especially in the cyber context.[45] The requirement that the perpetrator must have acted 'in furtherance of or under the guise of the armed conflict' should be narrowly construed: as the International Criminal Tribunal for Rwanda (ICTR) suggests, 'the expression "under the guise of the armed conflict" does not mean simply "at the same time of an armed conflict" and/or "in any circumstances created in part by the armed conflict"'.[46] Therefore, '[p]arasitical criminality that opportunistically uses the cover of the armed conflict does not, in principle, satisfy the requirement of nexus'.[47] What is necessary is that 'a person must be part of—or closely related to—the military power apparatus that has been established to fight

---

[41] Article 91 of Additional Protocol I (emphasis added). Article 3 of Hague Convention IV has a virtually identical formulation.

[42] See ICTY, *Prosecutor v Miodrag Jokić*, Case No IT-01-42/1-S, Trial Chamber Sentencing Judgment, 18 March 2004, para 12; ICTY, *Prosecutor v Blagojević and Jokić*, Case No IT-02-60-T, Trial Chamber Judgment, 17 January 2005, para 536.

[43] ICRC, *Interpretive Guidance*, pp 58–9.

[44] *Kunarać, Kovać and Voković*, para 58 (emphasis added).

[45] Guénaël Mettraux, 'Nexus with Armed Conflict', in *The Oxford Companion to International Criminal Justice*, edited by Antonio Cassese (2009), p 435.

[46] ICTR, *Rutaganda v Prosecutor*, Case No ICTR–96–3–A, Appeals Chamber Judgment, 26 May 2003, para 570. As Mettraux explains, 'there should be no presumption or fiction that, because a crime is committed in time of war, it therefore automatically constitutes a war crime. No such presumption exists under international law' (Guénaël Mettraux, *International Crimes and the* ad hoc *Tribunals* (Oxford: Oxford University Press, 2005), p 42).

[47] Mettraux, *International Crimes*, p 44.

an international or internal enemy. Moreover, he must have access, and be able, to employ the methods and means of warfare'.[48] The nexus, however, does not necessarily require that the perpetrators are combatants or have a 'special relationship' with one of the belligerents,[49] or that the crimes were 'part of a policy or of a practice officially endorsed or tolerated by one of the parties to the conflict, or that the act be in actual furtherance of a policy associated with the conduct of war or in the actual interest of a party to the conflict'.[50] It also does not require that the conduct could have only been committed because of the existence of an armed conflict: in *Kunarać*, the ICTY Appeals Chamber explained that the 'proposition that the laws of war only prohibit those acts which are specific to an actual wartime situation is not right.... The laws of war do not necessarily displace the laws regulating a peacetime situation; the former may add elements requisite to the protection which needs to be afforded to victims in a wartime situation'.[51]

The ICTY Appeals Chamber has suggested certain indicators to be considered in order to establish the existence of the nexus requirement for individual criminal responsibility purposes, including 'the fact that the perpetrator is a combatant; the fact that the victim is a non-combatant; the fact that the victim is a member of the opposing party; the fact that the act may be said to serve the ultimate goal of a military campaign; and the fact that the crime is committed as part of or in the context of the perpetrator's official duties'.[52] None of these indicators, however, is on its own conclusive.[53] Should no nexus with an armed conflict exist, the acts (including cyber operations) would not qualify as violations of international humanitarian law giving rise to individual criminal responsibility, but rather as ordinary crimes committed during an armed conflict.

3. Cyber operations without concurrent kinetic hostilities

This scenario involves states conducting cyber operations against each other without concurrent traditional hostilities. According to the ICRC, whether isolated cyber operations will be regarded as amounting to an armed conflict 'will probably be determined in a definite manner only through future state practice'.[54]

---

[48] Harmen Van der Wilt, 'War Crimes and the Requirement of a Nexus with an Armed Conflict', *Journal of International Criminal Justice* 10 (2012), p 1127.

[49] ICTR, *Prosecutor v Akayesu*, Case No ICTR–96–4, Appeals Chamber Judgment, 1 June 2001, para 444.

[50] ICTY, *Prosecutor v Tadić*, Case No IT–94–1–T, Opinion and Judgment, Trial Chamber, 7 May 1997, para 573. See also *Kunarać, Kovać and Voković*, para 58.

[51] *Kunarać, Kovać and Voković*, para 60. As Mettraux explains, '[t]he criminality of war...overlaps a great deal with peacetime criminality and many of those acts that would qualify as war crimes (such as murder or rape) would often qualify as domestic offences too if committed in peacetime, so that the fact that certain acts or conduct may fall in one category does not exclude that they would also fall in the other' (Mettraux, *International Crimes*, p 40).

[52] *Kunarać, Kovać and Voković*, para 59. See also ICTY, *Prosecutor v Limaj*, Case No IT–03–66–T, Trial Chamber Judgment, 30 November 2005, para 91.

[53] Mettraux, 'Nexus with Armed Conflict', p 436.

[54] ICRC, *International Humanitarian Law and the Challenges of Contemporary Armed Conflicts*, October 2011, p 37, <http://www.icrc.org/eng/assets/files/red-cross-crescent-movement/31st-int ernational-conference/31-int-conference-ihl-challenges-report-11-5-1-2-en.pdf>.

The following pages, however, will demonstrate that it is already possible to draw at least some conclusions.

The 1949 Geneva Conventions deliberately replace the use of the term 'war' employed in the Hague Conventions with 'armed conflict' with the intent of broadening the material scope of application of the Conventions and make the application of international humanitarian law dependent on a factual assessment.[55] Although the Conventions do not define the term 'international armed conflict', Pictet's Commentary of Common Article 2 clarifies that '[a]ny difference arising between two States and leading to the intervention of members of the armed forces is an armed conflict within the meaning of Article 2, even if one of the Parties denies the existence of a state of war'.[56] Similarly, the Commentary to Article 1 of Additional Protocol I on international armed conflicts, affirms that 'humanitarian law also covers any dispute between two States involving the use of their armed forces'.[57] No reference is made to *animus bellandi* or any other intention of the belligerents as an element for the existence of an international armed conflict, which, according to the Commentary, is instead based only on two objective factors: a dispute between at least two states and the intervention of their armed forces.[58] The former element excludes that accidental border incursions or bombings or armed intervention on the territory of a state with its consent qualify as international armed conflicts. If taken literally, the latter element is problematic for several reasons. First, there is no universally agreed definition of 'armed forces'.[59] Secondly states may rely on paramilitary forces or proxies in order to escape the application of international humanitarian law. In several jurisdictions the armed forces can also be used for law enforcement and disaster reaction purposes, and

---

[55] Jean S Pictet (ed), *Commentary on the Geneva Conventions of 12 August 1949* (Geneva: International Committee of the Red Cross, 1952–60), Vol 4, p 20. The expression 'armed conflict' also appears in the Preambles of the 1899 Hague Convention II with Respect to the Laws and Customs of War on Land and of the 1907 Hague Convention IV, but not in relation to the scope of application of the Conventions.

[56] Pictet (ed), *Commentary*, Vol 4, p 20. It is generally understood that the Conventions apply not only if one of the parties does not recognize the state of war, but also if neither do (Dapo Akande, 'Classification of Armed Conflicts: Relevant Legal Concepts', in *International Law and the Classification of Conflicts*, edited by Elizabeth Wilmshurst (Oxford: Oxford University Press, 2012), p 40).

[57] Yves Sandoz, Christophe Swinarski, and Bruno Zimmermann (eds), *Commentary on the Additional Protocols of 8 June 1977 to the Geneva Conventions of 12 August 1949* (Dordrecht: Nijhoff, 1987), para 62. Article 1(3) of Additional Protocol I clarifies that the Protocol 'shall apply in the situations referred to in Article 2 common' of the Geneva Conventions.

[58] Schindler also argues that an armed conflict in the sense of Common Art 2 exists 'when parts of the armed forces of two States clash with each other' (Dietrich Schindler, 'The Different Types of Armed Conflicts According to the Geneva Conventions and Protocols', *Recueil des cours* 163 (1979–II), p 131). The UK *Manual of the Law of Armed Conflict* adopts the same approach and states that '[t]he law of armed conflict applies in all situations when the armed forces of a state are in conflict with those of another state or are in occupation of territory' (*The Manual of the Law of Armed Conflict*, p 27). The Manual defines 'armed forces' as 'all organized armed forces, groups and units which are under a command responsible to that Party for the conduct of its subordinates, even if that Party is represented by a government or an authority not recognized by an adverse Party' (p 39).

[59] Lindsay Moir, *The Law of Internal Armed Conflict* (Cambridge: Cambridge University Press, 2002), p 39.

not exclusively for military operations.[60] In reading the Commentary, though, we should not lose sight of the fact that, when it was drafted, armed conflicts were conducted mainly, if not exclusively, by the armed forces of states. What is relevant for the determination of the existence of an international armed conflict, then, is not *who* carries out the activity on behalf of the states involved, but *what* activity is typically associated with the armed forces, ie resort to armed force.[61] It is not surprising that the reference to the intervention of the state's armed forces has been dropped in subsequent attempts to define an international armed conflict. Indeed, in a 2008 opinion paper, the ICRC itself did not refer to it, but merely reproduced the definition famously conceived by the ICTY Appeals Chamber in *Tadić*, where the Tribunal succinctly defined an (international) armed conflict as 'a resort to armed force between States'.[62] The Commentary to Rule 22 of the Tallinn Manual confirms that, if civilian intelligence agents conduct cyber operations amounting to armed force, an armed conflict might exist and that, on the other hand, if the armed forces only engage in espionage activities, no such conflict comes into existence in the absence of concurrent hostilities.[63]

In light of the *Tadić* definition, then, the two components of an international armed conflict are: (1) a dispute between at least two states; (2) and their 'resort to armed force' against each other. Starting from the latter, these two elements will now be applied in the cyber context in order to verify when cyber operations not accompanied by the use of traditional weapons can give rise to an international armed conflict.

3.1 'resort to armed force'

In *Tadić*, the ICTY did not explain what 'resort to armed force' means. It particular, it did not clarify whether it has the same meaning as 'use of force' in the *jus ad*

---

[60] Charles Garraway, 'War and Peace: Where is the Divide?', *International Law Studies* 88 (2012), p 104.

[61] Michael N Schmitt, 'Wired Warfare: Computer Network Attack and the *Jus in Bello*', *International Law Studies* 76 (2002), p 191.

[62] *Tadić*, Decision on the Defence Motion, para 70. Similarly, the 1992 German Military Manual states that '[a]n international armed conflict exists if one party uses force of arms against another party' (The Federal Ministry of Defence of the Federal Republic of Germany, *Humanitarian Law in Armed Conflicts*, ZDv 15/2, 1992, Section 202, <http://www.humanitaeres-voelkerrecht.de/ManualZDv15.2.pdf>). The *Tadić* definition is also incorporated in Art 2(b) of the ILC's Draft Articles on the Effects of Armed Conflict on Treaties (*Yearbook of the International Law Commission*, 2011, Vol II, Part Two, p 173). The definition contained in the 1985 Resolution of the Institute of International Law on the Effects of Armed Conflicts on Treaties is phrased differently: an armed conflict is 'a state of war or an international conflict which involve armed operations which by their nature or extent are likely to affect the operation of treaties between States parties to the armed conflict or between States parties to the armed conflict and third States, regardless of a formal declaration of war or other declaration by any or all of the parties to the armed conflict' (Resolution adopted on 28 August 1985 at the Helsinki Session, Art 1, <http://www.idi-iil.org/idiE/resolutionsE/1985_hel_03_en.PDF>). Although the language is different, there is no reference to the intervention of armed forces, and 'armed operations' can be taken as equivalent to 'armed force'.

[63] Tallinn Manual, p 83.

bellum sense and whether it implies a minimum level of intensity for an international armed conflict to exist.

3.1.1 'Use of force' and 'resort to armed force'

The notion of 'armed force' has already been discussed in the context of Article 2(4) of the UN Charter.[64] The question is whether any 'use of [armed] force' in the *jus ad bellum* sense also amounts to a 'resort to armed force' that determines the existence of an international armed conflict under the *jus in bello*, or whether the former is a broader concept than the latter. In Beckett's view, the questions of whether a cyber operation is a breach of Article 2(4) of the UN Charter or an armed conflict are 'essentially synonymous', as 'any use of force is regulated by IHL [International Humanitarian Law]'.[65] This view finds support in Judge Shahabuddeen's Separate Opinion in the *Tadić* Appeals Judgment, where he contends that 'an armed conflict involves a use of force' and therefore the question of whether there is an international armed conflict between two states depends on whether a state has used (armed) force against another.[66] Ziolkowski also maintains that an international armed conflict exists and the *jus in bello* applies whenever a state uses force in the sense of the *jus ad bellum* against another state, 'with the possible exception of quick, discrete and "surgical" use of force in the meaning of Article 2(4) UN Charter without further retort by the victim (as e.g. the bombardment of the Iraqi nuclear reactor in Osirak/Iraq by Israeli Air Force in 1981)'.[67]

On the other hand, in the *Nicaragua* Judgment the ICJ held that '[c]learly, use of force may *in some circumstances* raise questions of [international humanitarian] law',[68] implying that not always does a use of armed force amount to an armed conflict and thus trigger the application of the *jus in bello*. This can be explained in light of the ICJ's expansive notion of 'use of [armed] force', which includes indirect uses such as arming and training of armed groups:[69] these uses of force in the *jus ad bellum* sense do not amount to 'resort to armed force' under the *jus in bello*.[70] The *Tadić* Appeals Judgment confirms that the provision of military equipment, training or funding by a state to rebels fighting against another state is not sufficient

---

[64] Chapter 2, Section II.1.
[65] Jason Beckett, 'New War, Old Law: Can the Geneva Paradigm Comprehend Computers?', *Leiden Journal of International Law* 13 (2000), p 41.
[66] ICTY, *Prosecutor v Tadić*, Case No IT–94–1–A, Appeals Chamber Judgment, 15 July 1999, Separate Opinion of Judge Shahabuddeen, para 7.
[67] Katharina Ziolkowski, 'Computer Network Operations and the Law of Armed Conflict', *Military Law and Law of War Review* 49 (2010), p 68.
[68] *Military and Paramilitary Activities in and against Nicaragua (Nicaragua v US)*, Merits, Judgment, 27 June 1986, ICJ Reports 1986 ('*Nicaragua*'), para 216 (emphasis added).
[69] *Nicaragua*, para 228.
[70] The *Dalmia* Arbitration Award also suggests that 'use of force' in the *jus ad bellum* is broader than 'armed conflict' in the *jus in bello* ('if the Members of the Organisation must be presumed not to intend to use *force* (except within the narrow limits allowed by the Charter), they must *a fortiori* be presumed not to intend to resort to *war*'): *Dalmia Cement Ltd v National Bank of Pakistan*, International Chamber of Commerce, Arbitration Tribunal (Professor Pierre Lalive, Sole Arbitrator), 18 December 1967, *International Law Reports* 67 (1984), pp 619–20 (emphasis in the original).

to internationalize the conflict.[71] Dinstein also claims that the mere supply of arms to rebels does not bring about a state of war in the material sense.[72] In his view, however, if the state is not only arming the rebels, but also training them, it would be 'waging warfare' against the state fought by the rebels.[73] The German Military Manual, currently under revision, provides that support for a third party's 'acts of war' will be considered an act of war of the supporting state only 'if it is directly, i.e. closely related in space and time to measures harmful to the adversary. Cooperation in arms production or other activities to support the armed forces will not suffice'.[74] The General Assembly's Declaration on the Definition of Aggression provides other examples of use of armed force under the *jus ad bellum* that do not necessarily entail 'resort to armed force' under the *jus in bello*: the 'use of armed forces of one State which are within the territory of another State with the agreement of the receiving State, in contravention of the conditions provided for in the agreement or any extension of their presence in such territory beyond the termination of the agreement' and '[t]he action of a State in allowing its territory, which it has placed at the disposal of another State, to be used by that other State for perpetrating an act of aggression against a third State'.[75] Other violations of Article 2(4), such as measures involving the threat but not the use of armed force (for instance a 'quarantine'), also do not initiate, in themselves, an international armed conflict.

According to Melzer, the existence of an international armed conflict does not depend on whether a state simply uses force against another in the sense of Article 2(4), but rather on the occurrence of 'belligerent hostilities', which is a narrower concept.[76] This also appears to be the approach of the Tallinn Manual, Rule 22 of which states that '[a]n international armed conflict exists whenever there are hostilities, which may include or be limited to cyber operations, occurring between two or more States',[77] where 'hostilities' is intended as 'the collective application of

---

[71] *Tadić*, Appeals Chamber Judgment, para 137. In *Rajić*, the ICTY Trial Chamber also found that the internal conflict between the Bosnian Croats and their government in central Bosnia became international 'as a result of the significant and continuous military intervention of the Croatian Army in support of the Bosnian Croats' (ICTY, *Prosecutor v Rajić*, Case No IT–95–12–R61, Review of the Indictment pursuant to Rule 61 of the Rules of Procedure and Evidence, 13 September 1996, para 21).

[72] Dinstein, *War, Aggression and Self-Defence*, p 10.

[73] Dinstein, *War, Aggression and Self-Defence*, p 10.

[74] German Ministry of Defence, *Humanitarian Law in Armed Conflict*, Section 214.

[75] Definition of Aggression, GA Res 3314 (XXIX), 14 December 1974, Art 3(f).

[76] Nils Melzer, *Cyberwarfare and International Law*, UNIDIR, 2011, p 24, <http://www.isn.ethz.ch/Digital-Library/Publications/Detail/?lng=en&id=134218>. Conventional international humanitarian law does not define 'hostilities'. The Commentary to Art 51(3) of Additional Protocol I defines 'hostile acts' as 'acts which by their nature and purpose are intended to cause actual harm to the personnel and equipment of the armed forces' (Sandoz, Swinarski, and Zimmermann (eds), *Commentary*, para 1942). Similarly, the Commentary to Art 13 of Additional Protocol II defines 'hostilities' as 'acts of war that by their nature or purpose struck at the personnel and *matériel* of enemy armed forces' (para 4788). These definitions are too restrictive, as they do not take into account harm to civilians and civilian objects (see *Public Committee Against Torture in Israel et al v The Government of Israel et al*, Israel's Supreme Court, HCJ 769/02, 11 December 2005 ('*Targeted Killings*'), para 33 (per Judge Barak)).

[77] Tallinn Manual, p 79.

means and methods of warfare'.[78] The ICRC *Interpretive Guidance on the Notion of Direct Participation in Hostilities* also defines 'hostilities' as 'the (collective) resort by the parties to the conflict to means and methods of injuring the enemy'.[79] As will be seen more in-depth in Chapter 4,[80] cyber capabilities and operations can indeed be means and methods of warfare. It is sufficient to recall here that, according to the ICRC, '[i]f a cyber operations [*sic*] is used against an enemy in an armed conflict in order to cause damage, for example by manipulation of an air traffic control system that results in the crash of a civilian aircraft, it can hardly be disputed that such an attack is in fact a method of warfare and is subject to prohibitions under IHL'.[81] The Commentary to the HPCR Manual on Air and Missile Warfare confirms that '[m]eans of warfare include non-kinetic systems, such as those used in EW and CNAs. The means would include the computer and computer code used to execute the attack, together with all associated equipment'.[82] It further specifies that loss of life, injury, damage or destruction 'need not result from physical impact...since the force used does not need to be kinetic. In particular, CNA hardware, software and codes are weapons that can cause such effects through transmission of data streams.'[83] The fact that existing *jus in bello* instruments were drafted with kinetic weapons in mind, then, does not entail that the capacity to cause an explosion is an essential characteristic of weaponry: in its Advisory Opinion on the *threat or use of nuclear weapons*, the ICJ held that international humanitarian law 'applies to all forms of warfare and to all kinds of weapons, those of the past, those of the present and those of the future', in spite of the 'qualitative as well as quantitative difference' that there might be with traditional weapons.[84]

In any case, in order to qualify as a 'means of warfare', the software must be able to 'injure the enemy':[85] if the program is designed solely for the purpose of infiltrating a computer and stealing information, it would not be a 'weapon' in the sense highlighted above and its use between states would not trigger an international armed conflict.[86] What does 'injuring' entail then? In the context of the

---

[78] Tallinn Manual, p 82.   [79] ICRC, *Interpretive Guidance*, p 43.
[80] Chapter 4, Section II.
[81] ICRC, *International Humanitarian Law and the Challenges*, p 37.
[82] HPCR, *Manual on International Law Applicable to Air and Missile Warfare* (Cambridge: Cambridge University Press, 2013), p 31.
[83] HPCR Manual, p 49.
[84] *Nuclear Weapons*, para 86. The Court was of course referring to the applicability of the Hague and Geneva laws to nuclear weapons.
[85] ICRC, *Interpretive Guidance*, p 43.
[86] Thomas Rid and Peter McBurney, 'Cyber-Weapons', *RUSI Journal* 157, no 1 (February 2012), p 11. The definition of 'information war' put forward by the Russian Federation, ie '[c]onfrontation between States in the information field, with a view to damaging information systems, processes and resources and vital structures, and undermining another State's political and social systems, as well as the mass psychological manipulation of a State's population and the destabilization of society', is therefore too broad, as it is not limited to the conduct of hostilities (UN Doc A/54/213, 10 August 1999, p 10). The definition is also contained almost verbatim in Annex I of the Agreement between the Governments of the Member States of the Shanghai Cooperation Organization on Cooperation in the Field of International Information Security of 16 June 2009.

jus ad bellum, it was submitted that not only cyber operations causing or reasonably likely to cause physical damage to property, loss of life or injury to persons, but also those that *severely* disrupt the adversary's military capabilities or the functioning of civilian critical infrastructure are a use of force prohibited by Article 2(4) of the UN Charter.[87] The question is whether the same conclusion holds true for the purpose of establishing the existence of an international armed conflict.[88] Both the above-mentioned ICRC and HPCR documents seem to imply that only destructive cyber operations may give rise to an armed conflict. Dinstein also claims that violence is an essential ingredient of 'hostilities', which should be understood in terms of consequences, ie destruction or damage to property or mental or physical harm to individuals.[89] He subsequently admits, however, that hostilities also include 'certain non-violent acts, provided that they are directly connected to military operations against the enemy (e.g., logistics or the gathering of intelligence about the enemy)'.[90] According to Michael Schmitt, international humanitarian law applies 'whenever computer network attacks can be ascribed to a State, are more than merely sporadic and isolated incidents and are either intended to cause injury, death, damage, or destruction (or analogous effects), or such consequences are foreseeable'.[91] Louise Doswald-Beck also limits the applicability of international humanitarian law to situations when the 'CNA is undertaken by official sources and is intended to, or does, result in physical damage to persons, or damage to objects that goes beyond the bit of computer program or data attacked'.[92] On the other hand, Melzer argues that 'state-sponsored cyber operations would give rise to an international armed conflict if they are designed to harm another state not only by directly causing death, injury or destruction, but also by directly adversely affecting its military operations or military capacity'.[93] Similarly, Ziolkowski opines that not only cyber operations that cause deaths, injuries, or destruction of property qualify as armed conflict, but also those resulting in the medium to long-term disruption of NCIs.[94] These latter views have been supported by certain states. The AIV/CAVV Report on cyber warfare claims that 'if an organised cyber attack (or series of attacks) leads to the destruction of or substantial or long-lasting damage

---

[87] Chapter 2, Section II.1.2.

[88] The Group of Experts that drafted the Tallinn Manual could not reach consensus on this point, although the issue was apparently only discussed in the context of the requirement of intensity for non-international armed conflicts (Tallinn Manual, p 88).

[89] Yoram Dinstein, *The Conduct of Hostilities under the Law of International Armed Conflict*, 2nd edn (Cambridge: Cambridge University Press, 2010), p 1.

[90] Dinstein, *The Conduct of Hostilities*, p 2.

[91] Schmitt, 'Wired Warfare', p 192.

[92] Louise Doswald-Beck, 'Some Thoughts on Computer Network Attack and the International Law of Armed Conflict', *International Law Studies* 76 (2002), p 165.

[93] Melzer, *Cyberwarfare*, p 24.

[94] Ziolkowski, 'Computer Network Operations', p 75. Brown argues that, although cyber attacks 'do not accomplish the work of explosives, they do inflict harm on the class of people that LOAC [Law of Armed Conflict] is meant to protect, regardless of whether the harm is intentional or simply a collateral effect of an attack on some other target. International law should therefore govern the use of computers for these purposes as well' (Davis Brown, 'A Proposal for an International Convention to Regulate the Use of Information Systems in Armed Conflict', *Harvard International Law Journal* 47 (2006), p 188; footnote omitted).

to computer systems managing critical military or civil infrastructure, it could conceivably be considered an armed conflict and international humanitarian law would apply. The same is true of a cyber attack that seriously damages the state's ability to perform essential tasks, causing serious and lasting harm to the economic or financial stability of that state and its people.'[95] The UK Under-Secretary for Security and Counter-terrorism also declared that a cyber attack that took out a power station would be considered an act of war.[96] These views all imply a compelling point: the dependency of modern societies on computers, computer systems, and networks has made it possible to cause significant damage to states and persons through non-destructive means. From this perspective, it is significant that Panama noted that misuse of information and telecommunication systems is 'a new form of violence'.[97]

3.1.2 Does the 'resort to armed force' need to reach a minimum level of intensity to initiate an international armed conflict?

Whether or not the resort to armed force between states needs to reach a minimum level of intensity in order to qualify as an international armed conflict is the subject of much controversy. The *Tadić* decision leaves the point unclear: if its definition of armed conflict attaches the 'protracted' and 'organization' requirements only to non-international armed conflicts while an international armed conflict exists 'whenever' there is a resort to armed force between states, it then refers to 'the intensity requirements applicable to *both* international and internal armed conflicts' and concludes that, in the case before the Court, there had been '*protracted, large-scale* violence between the armed forces of different States'.[98] In the subsequent *Mucić* case, however, the ICTY found that, unlike non-international armed conflicts, in international armed conflicts 'the existence of armed force between States is sufficient of itself to trigger the application of international humanitarian law', whether or not it reaches a certain intensity.[99]

The ILA Final Report on the Meaning of Armed Conflict maintains that low-intensity engagements are usually treated by states as incidents, and not as armed conflicts.[100] This conclusion finds support in the US President's certification of 25 April 1997, according to which there are 'situations in which the U.S. is not engaged in a use of force of a scope, duration, and intensity that would

---

[95] AIV/CAVV, *Cyber Warfare*, No 77, AIV/No 22, CAVV, December 2001, p 24, <http://www.aiv-advies.nl/ContentSuite/upload/aiv/doc/webversie__AIV77CAVV_22_ENG.pdf>. The example offered in the Report is that of 'a coordinated and organised attack on the entire computer network of the financial system (or a major part of it) leading to prolonged and large-scale disruption and instability that cannot easily be averted or alleviated by normal computer security systems'.
[96] Jamie Doward, 'Britain fends off flood of foreign cyber-attacks', *The Observer*, 7 March 2010, <http://www.theguardian.com/technology/2010/mar/07/britain-fends-off-cyber-attacks>.
[97] UN Doc A/57/166/Add.1, 29 August 2002, p 5.
[98] *Tadić*, Decision on the Defence Motion, para 70 (emphasis added).
[99] ICTY, *Prosecutor v Mucić*, Case No IT–96–21–T, Trial Chamber Judgment, 16 November 1998, para 184.
[100] *Final Report on the Meaning of Armed Conflict in International Law*, in ILA, Report of the Seventy-Fourth Conference (The Hague, 2010), p 708.

trigger the laws of war with respect to the U.S. forces'.[101] Such activities include, inter alia, peace operations, humanitarian assistance, recovery operations, arms control, and counter-terrorism operations.[102] The AIV/CAVV Report also argues that 'hostilities must reach a sufficient level of intensity' that goes beyond 'border skirmishes or isolated incidents in the air or at sea' to qualify as international armed conflicts.[103] The Italian Military Penal Code of War defines armed conflict as 'conflict where at least one of the parties uses weapons against the other party in a militarily organized and prolonged manner for the conduct of belligerent operations'.[104] Similarly, the French Manual of the Law of Armed Conflict provides that '[i]l faut un certain seuil de violence pour qualifier une situation de conflit armé. En deçà de ce seuil, on parle seulement de troubles et de tensions internes. Les émeutes, les actes isolés et sporadiques de violence et autres actes analogues ne sont pas des conflits armés', without distinguishing between international and non-international armed conflicts.[105] In his analysis of Oppenheim's definition of 'war', Dinstein recalls that there are incidents involving the use of force that states usually do not consider as 'war', such as exchanges of fire between border patrols of neighbouring states, the interception of planes and the sinking of foreign vessels by naval units.[106] A 'comprehensive' use of force, on the other hand, necessarily qualifies as 'war' in the material sense. According to Dinstein, the comprehensive character of the use of force should be assessed: '(i) spatially, across sizeable tracts of land or far-flung corners of the ocean; (ii) temporally, over a protracted period of time; (iii) quantitatively, entailing massive military operations or a high level of firepower; (iv) qualitatively, inflicting extensive human casualties and destruction of property'.[107]

On the other hand, the ICRC's position is that there is no minimum threshold of intensity for international armed conflicts: international humanitarian law applies to any shot fired, which prevents controversies about whether the intensity threshold has been reached.[108] The Commentary to Common Article 2 clearly states that '[i]t makes no difference how long the conflict lasts, or how much slaughter takes place, or how numerous are the participating forces'.[109] The Commentary

---

[101] Cited in Derek I Grimes, John Rawcliffe, and Jeannine Smith (eds), *Operational Law Handbook* (2006), p 20, <http://www.fas.org/irp/doddir/army/law2006.pdf>.
[102] *Joint Doctrine for Military Operations Other Than War*, Joint Publication 3–07 (1995), p III–1, <http://www.bits.de/NRANEU/others/jp-doctrine/jp3_07.pdf>.
[103] AIV/CAVV, *Cyber Warfare*, p 23.
[104] Article 165, as replaced by Law no 6 of 31 January 2002 (in *Gazzetta Ufficiale*, 2 February 2002, no 28) and amended by Law no 1 of 27 February 2002 (in *Gazzetta Ufficiale*, 27 February 2002, no 49). The translation is mine.
[105] Ministère de la Défense, *Manuel de droit des conflits armés*, 2001, Definition of 'Guerre', <http://www.defense.gouv.fr/content/download/77498/693317/file/Manuel_de_droit_des_conflits_armes.pdf>.
[106] Dinstein, *War, Aggression and Self-Defence*, p 11.
[107] Dinstein, *War, Aggression and Self-Defence*, p 12.
[108] ICRC, *International Humanitarian Law and the Challenges*, p 7. The Commentary of Rule 22 of the Tallinn Manual argues first that isolated incidents and sporadic clashes do not amount to international armed conflicts, and then that 'it would be prudent to treat the threshold of international armed conflict as relatively low', without however identifying such threshold (Tallinn Manual, p 83).
[109] Pictet (ed), *Commentary*, Vol 3, p 23.

suggests that international law applies '[e]ven if there has been no fighting, [as] the fact that persons covered by the Convention [III] are detained is sufficient for its application'.[110] The Commentary to Article 1 of Additional Protocol I confirms that '[n]either the duration of the conflict, nor its intensity, play a role: the law must be applied to the fullest extent required by the situation of the persons and the objects protected by it'.[111] This finds support in several military manuals. The Royal Australian Air Force's *Operation Law for RAAF* (Royal Australian Air Force) Commanders, for instance, specifies that, in international armed conflicts, '[t]he duration and intensity of the conflict are not relevant to whether an armed conflict exists'.[112] Louise Doswald-Beck also speaks for numerous scholars where she argues highlights the risk that the threshold approach 'would lead to the need for evaluations that would create inevitable uncertainties and ultimately to the same problems faced when establishing whether "war" existed without a formal declaration'.[113] Schindler opines that '[t]he existence of an armed conflict within the meaning of Article 2 common to the Conventions can always be assumed when parts of the armed forces of two States clash with each other. Even a minor frontier incident is sufficient. Any kind of use of arms between two States brings the Conventions into effect'.[114] Finally, Kleffner maintains that 'requiring a certain level of intensity for the resort to armed force between states to amount to an international armed conflict bears the risk of creating an international legal vacuum or of depriving certain categories of persons of the protections that international humanitarian law provides'.[115]

Both the contextual and teleological criteria of interpretation suggest that this latter view is preferable.[116] Indeed, it should be recalled that, under a literal reading of Common Article 2 of the Geneva Conventions, not only is a minimum level of intensity of hostilities not required, but even their total absence does not preclude the application of the Conventions, in case of a declared war or an occupation that meets with no armed resistance. As to the teleological criterion of interpretation, the purpose of Common Article 2 is to extend the protection of victims of war as widely as possible. The risk of abuses highlighted by the ICRC, therefore,

---

[110] Pictet (ed), *Commentary*, Vol 3, p 23.
[111] Sandoz, Swinarski, and Zimmermann (eds), *Commentary*, para 62.
[112] Royal Australian Air Force, *Operations Law for RAAF Commanders*, AAP 1003, 2nd edn, May 2004, p 42, <http://airpower.airforce.gov.au/Publications/Details/156/AAP1003-Operations-Law-for-RAAF-Commanders-2nd-Edition.aspx>.
[113] Doswald-Beck, 'Some Thoughts', p 164. See also Thilo Marauhn and Zacharie F Ntoubandi, 'Armed Conflict, Non-International', *Max Planck Encyclopedia of Public International Law* (2012), Vol I, p 627; Eric David, *Principes de droit des conflits armés*, 3rd edn (Bruxelles: Bruylant, 2002), p 109 ('tout *affrontement armé* entre forces des Etats parties aux C.G. de 1949 (et éventuellement au 1er P.A. de 1977) relève de ces instruments, quelle que soit l'ampleur de cet affrontement: une escarmouche, un incident de frontière entre les forces armées des Parties suffisent à provoquer l'application des Conventions (et du 1er Protocol, s'il lie les Etats) à cette situation)'; Kolb and Hyde, *An Introduction*, p 76.
[114] Schindler, 'The Different Types', p 131.
[115] Jann K Kleffner, 'Scope of Application of International Humanitarian Law', in *The Handbook of International Humanitarian Law*, edited by Dieter Fleck, 3rd edn (Oxford: Oxford University Press, 2013), p 45.
[116] Article 31 of the 1969 Vienna Convention on the Law of Treaties.

is a compelling argument: if the belligerents could deny the existence of an international armed conflict by claiming that hostilities have not reached a minimum level of intensity, the application of international humanitarian law would in fact depend—again—on the intention of the parties. Of course, if the resort to armed force is limited as in border incidents or 'surgical' attacks on certain facilities, only a very limited number of international humanitarian law provisions will come into play, in particular those on targeting. But the point is that, whenever belligerent hostilities occur between states, at least some *jus in bello* rules become applicable.[117]

Cyber operations causing or likely to cause material damage to property, loss of life or injury to persons, should, however, be distinguished from merely disruptive ones. With regard to the former, by analogy with kinetic operations, no *de minimis* threshold of property damage or personal injury is needed for an international armed conflict to exist. If—and it is a big 'if'—Stuxnet was attributable to states and did cause material damage to the gas centrifuges at Natanz, it would have triggered the application of the law of targeting between the responsible state(s) and Iran, in the same way as if the operation had been carried out by kinetic means.[118] This reasoning should, of course, not lead to unreasonable results: it is common sense that a cyber operation that causes the destruction of one computer would not as such determine the application of international humanitarian law, in the same way as soldiers throwing stones at each other or killing a cow across the border would not initiate an international armed conflict.[119] In a nutshell, 'the greater the damage, the more likely the situation will be treated as an armed conflict',[120] whatever means are employed to cause that damage.

As to cyber operations resulting in loss of functionality but not physical damage, however, it is only those operations that exceed mere inconvenience and *significantly* disrupt the correct functioning of military or civilian critical infrastructures that can potentially qualify as 'resort to armed force' and thus initiate an international armed conflict, as it is only in these cases that the effects of disruption can be equated to those of destruction caused by traditional armed force.[121] This minimum threshold is not for the resort to armed force to be an international armed conflict, but for the cyber operation to *be* a resort to armed force. Hence, even if it were demonstrated that Russia was behind the operations, the 2007 DDoS attacks on Estonia would not qualify as an international armed conflict between the two states: although they targeted critical infrastructures (banking and

---

[117] Kleffner, 'Scope of Application', p 45.
[118] Michael N Schmitt, 'Classification of Cyber Conflict', *Journal of Conflict and Security Law* 17 (2012), pp 251–2. *Contra*, see Cordula Droege, 'Get Off My Cloud: Cyber Warfare, International Humanitarian Law, and the Protection of Civilians', *International Review of the Red Cross* 94 (2012), p 548.
[119] Schmitt, 'Classification of Cyber Conflict', p 252.
[120] Knut Dörmann, *Applicability of the Additional Protocols to Computer Network Attacks*, p 3, <http://www.icrc.org/eng/assets/files/other/applicabilityofihltocna.pdf>.
[121] Droege distinguishes between cyber operations that manipulate a banking system and cause economic loss and those targeting vital infrastructures like electricity or water supply systems, that, if prolonged, could have prejudicial consequences on the population, and argues that only the latter can amount to armed force (Droege, 'Get Off My Cloud', pp 548–9).

communications), no property damage or personal injury occurred and no serious disruption ensued. A different conclusion may be reached with regard to a cyber attack that takes down the national grid for a prolonged time, with severe negative repercussions on the provision of medical services, transport, financial markets, and security.

In light of the above analysis, it can be concluded that cyber operations amount to 'resort to armed force' when they entail the direct use of cyber means or methods of warfare in support of a belligerent to the detriment of another. Such use must result or be reasonably likely to result in harm to the adversary in the form of physical damage to property, loss of life or injury to persons, or serious disruption of critical infrastructures. If such resort to armed force takes place 'between states', there is an international armed conflict: this element will be examined next.

3.2 'between states'

In order to amount to an international armed conflict, the resort to armed force must be between at least two states that are acting against each other. Without clear attribution, challenging as it might be in the cyber context, it is impossible to establish whether there is an armed conflict and its international nature, and therefore to determine if and what *jus in bello* provisions apply. The following analysis will therefore assume that the cyber operations have been conclusively attributed to certain states.

The states involved must be resorting to armed force *against each other*. There is no international armed conflict when the extraterritorial resort to armed force is the result of error or when the state has consented to the armed operations on its territory:[122] as the UK *Manual of the Law of Armed Conflict* explains, 'an accidental border incursion by members of the armed forces would not, in itself, amount to an armed conflict, nor would the accidental bombing of another country'.[123] This is particularly important in the cyber context, because of the likelihood that malware will spread inadvertently to non-belligerents' infrastructures.

When the cyber operations amount to resort to armed force, the governmental or private character of the targeted infrastructure is not relevant to the determination of the existence of an international armed conflict (as has been seen, most NCIs are not owned by the government, but by the private sector). It is also not necessary that a state react against the initial resort to armed force by another state for an international armed conflict to exist.[124] This conclusion finds support in the fact that, according to Common Article 2, the Geneva Conventions also apply if the state of war is not recognized by *one* of the parties to the conflict and in case

---

[122] Sylvain Vité, 'Typology of Armed Conflicts in International Humanitarian Law: Legal Concepts and Actual Situations', *International Review of the Red Cross* 91 (2009), pp 72–3.

[123] UK Ministry of Defence, *The Manual of the Law of Armed Conflict*, p 29.

[124] But see, *contra*, Mary Ellen O'Connell, 'Combatants and the Combat Zone', *University of Richmond Law Review* 43 (2009), p 111; Michael Bothe, Karl J Partsch, and Waldemar A Solf, *New Rules for Victims of Armed Conflicts: Commentary on the Two 1977 Protocols Additional to the Geneva Conventions of 1949* (The Hague, Boston, and London: Nijhoff, 1982), p 46.

of occupation which is not met with armed resistance. The fact that a reaction to the initial resort to armed force is not needed is further confirmed by some military manuals. The Canadian Forces' *Law of Armed Conflict Manual,* for instance, defines an armed conflict as 'a conflict between states in which at least one party has resorted to the use of armed force to achieve its aims'.[125] Similarly, Section 202 of the 1992 German Military Manual determines that an international armed conflict exists 'if one party uses force of arms against another party'.[126] Arguing that a reaction is necessary for an international armed conflict to occur would lead to the absurd result that the attacker would be free from targeting constraints under the *jus in bello* until the attacked state reacts, and would therefore be encouraged to carry out an overwhelming first strike to prevent the application of international humanitarian law.

Not all extraterritorial resort to armed force is an international armed conflict. Four scenarios must be distinguished: (1) the resort to armed force by state A against an armed group located in the territory of state B but operating against state A; (2) the resort to armed force by state A in support of an armed group that fights against state B, in whose territory it is located; (3) the resort to armed force by state A in support of state B against an armed group located in state B; (4) the resort to armed force by an armed group under the 'overall control' of state A against state B, in whose territory the group is located. With regard to the second scenario, if state A conducts kinetic or cyber operations amounting to armed force in support of the armed group, the conflict is mixed: the law of international armed conflict will apply to the hostilities between states A and B, while the law of non-international armed conflict will continue to regulate the hostilities between state B and the insurgents.[127] As the Commission of Inquiry on Libya has stated, the 'international armed conflict is legally separate to the continuing non-international armed conflict' and is thus a 'co-existing international armed conflict'.[128] On the other hand, if state A conducts kinetic or cyber operations amounting to armed force in order to support the government of state B to quell the insurgency (as in the third scenario), the conflict would remain non-international, as there is no resort to armed force *between* states.[129]

---

[125] Office of the Judge Advocate General, *Law of Armed Conflict Manual: At the Operational and Tactical Levels*, Joint Doctrine Manual, B–GJ–005–104/FP–02, 2001, p GL–2, <http://www.fichl.org/uploads/media/Canadian_LOAC_Manual_2001_English.pdf>. Very similar language is used in the Australian Defence Doctrine, Executive Series ADDP 06.4, 11 May 2006, p 13, <http://www.defence.gov.au/adfwc/Documents/DoctrineLibrary/ADDP/ADDP06.4-LawofArmedConflict.pdf>.

[126] German Ministry of Defence, *Humanitarian Law in Armed Conflict*, Section 202. It is interesting that a non-international armed conflict requires a 'confrontation' (Section 210).

[127] See Giulio Bartolini, 'Air Targeting in Operation *Unified Protector* in Libya. Jus ad bellum and IHL Issues: An External Perspective', in *Legal Interoperability and Ensuring Observance of the Law Applicable in Multinational Deployments*, edited by Stanislas Horvat and Marco Benatar (Brussels: International Society for Military Law and the Law of War, 2013), pp 260–1.

[128] *Report of the International Commission of Inquiry on Libya* (2011), p 31. In *Nicaragua*, the ICJ famously found that an international armed conflict between the United States and Nicaragua existed simultaneously with a non-international armed conflict between Nicaragua and the *contras* (*Nicaragua*, para 219).

[129] Andreas Paulus and Mindia Vashakmadze, 'Asymmetrical War and the Notion of Armed Conflict—A Tentative Conceptualization', *International Review of the Red Cross* 91 (2009), p 101.

According to the ICTY Appeals Chamber in the *Tadić* judgment, a conflict is internationalized if state A exercises 'overall control' over an armed group fighting against state B, providing that the group is organized and hierarchically structured: to meet this level of control, it is sufficient that the state '*has a role in organising, coordinating or planning the military actions* of the military group, in addition to financing, training and equipping or providing operational support to that group [...] regardless of any specific instructions by the controlling State concerning the commission of each of those acts'.[130] Other forms of support such as 'the mere provision of financial assistance or military equipment or training' are not sufficient, in themselves, to internationalize the conflict.[131] In his Separate Opinion, Judge Shahabuddeen criticized the majority judges and maintained that what internationalizes a conflict is not the fact that an external state has overall control of the insurgents or even less that the conduct of the insurgents is attributable to it, but rather that the external state is using force through the insurgents against the state that the insurgents are fighting.[132] He then recalls that the *Nicaragua* judgment qualified the arming and training of armed groups as a threat or use of force, but not the mere supply of funds, which is only a violation of the principle of non-intervention:[133] as Yugoslavia did much more than merely supplying the Bosnian Serbs with money, it used force against Bosnia and Herzegovina and thus there was an international armed conflict between the two states, irrespective of whether the actions of the Bosnian Serbs could be attributed to Yugoslavia under the law of state responsibility.[134] With respect, it is submitted that this view, if intriguing, cannot be shared. Judge Shahabuddeen is correct when he argues that it is not the secondary rules on attribution provided in the law of state responsibility, be they general or special, that determine the nature of the conflict, but rather the primary rules themselves: as has been observed, 'attribution *suffices*, but it need not be *necessary*, for internationalization'.[135] However, the relevant primary rules here are not contained, as Judge Shahabuddeen maintains, in the *jus ad bellum*, but rather in the *jus in bello*. From this point of view, the overall control standard developed by the ICTY might well be the applicable one, but not as an attribution rule, as the majority of the Court in *Tadić* seem to suggest, but as part of the primary rule containing the definition of international(ized) armed conflict. This conclusion is supported by the ICJ's findings in the 2007 *Genocide* case, where the Court held that the overall

---

[130] *Tadić*, Appeals Chamber Judgment, para 137 (emphasis in the original). For the case of a 'private individual who is engaged by a State to perform some specific illegal acts in the territory of another State (for instance, ... carrying out acts of sabotage)' and of unorganized, non-military and non-hierarchical groups of individuals, the ICTY retains the effective control test, ie the need to prove the issue of specific instructions concerning the commission of the illegal act or the state's public retroactive approval of the individual's actions (para 118).
[131] *Tadić*, Appeals Chamber Judgment, para 137.
[132] *Tadić*, Appeals Chamber Judgment, Separate Opinion of Judge Shahabuddeen, para 7.
[133] *Tadić*, Appeals Chamber Judgment, Separate Opinion of Judge Shahabuddeen, paras 8–12.
[134] *Tadić*, Appeals Chamber Judgment, Separate Opinion of Judge Shahabuddeen, para 14.
[135] Marko Milanović and Vidan Hadzi-Vidanović, 'A Taxonomy of Armed Conflict', in *Research Handbook of International Conflict and Security Law*, edited by Nigel D White and Christian Henderson (Cheltenham: Elgar, 2013), p 295 (emphasis in the original).

control test might well be 'applicable and suitable' to the determination of the existence of an international armed conflict, as

> logic does not require the same test to be adopted in resolving the two issues, which are very different in nature: the degree and nature of a State's involvement in an armed conflict on another State's territory which is required for the conflict to be characterized as international, can very well, and without logical inconsistency, differ from the degree and nature of involvement required to give rise to that State's responsibility for a specific act committed in the course of the conflict.[136]

The Court, thus, did not exclude that different standards might apply to different regimes and that the overall control test might be the correct one for the qualification of conflicts, although it rejected it as an attribution standard.

With regard to cyber operations amounting to the resort to armed force by a state *against* an armed group on the territory of another state (as in the above first scenario), the Supreme Court of Israel famously held that an international armed conflict is one that 'crosses the borders of the state—whether or not the place in which the armed conflict occurs is subject to belligerent occupation'.[137] With due respect to the Court, this conclusion is not correct. Whenever the spillover violence, if sporadic, has a nexus with an existing non-international armed conflict occurring in another state, it can be subsumed under that conflict and thus falls under the same regulatory framework, even if it partly occurs in the territory of another state.[138] Even when there is no internal armed conflict in the territory of the intervening state and the armed group operates exclusively from abroad, as in the case of Hezbollah in southern Lebanon, the conflict between the intervening state and the non-state actors remains non-international, since what defines the international or non-international nature of an armed conflict is not the location of the hostilities, but rather the parties involved:[139] indeed, non-state actors would not be able to comply with many of the provisions contained in the law of international armed conflict.[140] This conclusion is confirmed by Articles 1 and 7 of the Statute of the ICTR, that empower the Tribunal to apply the law of non-international armed conflict even though the Rwandan conflict had spread to neighbouring Burundi and DRC.[141] In *Lubanga*, the ICC Trial Chamber also

---

[136] *Application of the Convention on the Prevention and Punishment of the Crime of Genocide (Bosnia and Herzegovina v Serbia and Montenegro)*, Merits, Judgment, 26 February 2007, ICJ Reports 2007, paras 404, 405.

[137] *Targeted Killings*, para 18 (per Judge Barak).

[138] ICRC, *International Humanitarian Law and the Challenges*, pp 9–10; Milanović and Hadzi-Vidanović, 'A Taxonomy', p 291.

[139] Liesbeth Zegveld, *Accountability of Armed Opposition Groups in International Law* (Cambridge: Cambridge University Press, 2002), p 136; Milanović and Hadzi-Vidanović, 'A Taxonomy', p 275 ('the fact that the hostilities between a non-state actor and a state cross state borders does not justify treating that non-state actor as being on par with the state in terms of the IAC [international armed conflict]/NIAC [non-international armed conflict] distinction').

[140] Noam Lubell, 'The War (?) against Al-Qaeda', in *International Law*, edited by Wilmshurst, p 434.

[141] ICRC, 'How is the Term "Armed Conflict" Defined in International Humanitarian Law?', Opinion paper, March 2008, p 5, <http://www.icrc.org/eng/assets/files/other/opinion-paper-armed-conflict.pdf>.

held that '[i]t is widely accepted that when a State enters into a conflict with a non-governmental armed group located in the territory of a neighbouring State and the armed group is acting under the control of its own state' it is an international armed conflict between the two states.[142] On the other hand, 'if the group is not acting on behalf of a government, in the absence of two States opposing each other, there is no international armed conflict'.[143] The US Supreme Court in the *Hamdan* case also found that Common Article 3 applies to the 'conflict' between the United States and Al-Qaeda in spite of its transnational character.[144]

It should finally be recalled that in two cases the law of international armed conflict applies to an armed conflict not 'between states', but between a state and an armed group: when the insurgents have been recognized as belligerents by the government against which they fight and in the situations envisaged in Article 1(4) of Additional Protocol I. Recognition of belligerency by the government, which had the main consequence of making the laws of war applicable to the hostilities between the government and the insurgents,[145] has not been practised for over a century and has in fact been replaced by Common Article 3 of the Geneva Conventions and by the practice of unilaterally declaring the application of at least certain international humanitarian law provisions by the parties to the non-international armed conflict.[146] Article 1(4) of Additional Protocol I submits 'wars of national liberation' to the law of international armed conflict, even though hostilities are between a state and non-state actors. Cyber operations amounting to resort to armed force conducted by a national liberation movement in their struggle against colonial domination, racist régimes or alien occupation would then potentially fall under the scope of application of Additional Protocol I for the states parties to it.[147] For the application of the

---

[142] ICC, *Prosecutor v Thomas Lubanga Dyilo*, Case No ICC–01/04–01/06–2842, Trial Chamber I Judgment, 14 March 2012, para 541. According to the ICC, the test that applies to establish whether the external state has control of the armed group is that of 'overall control' (para 541).

[143] *Lubanga*, Trial Chamber I Judgment, para 541.

[144] *Hamdan v Rumsfeld et al* 548 US 557, 126 S Ct 2749 (2006), pp 67–8, <http://www.giurcost.org/links/guantanamo.pdf>. The Supreme Court, however, did not clearly classify the conflict between the United States and Al-Qaeda or said that there was a conflict at all: it held that the qualification was unnecessary as, in any case, Common Art 3 would apply as 'minimal protection' (p 68). It also did not specify whether Common Art 3 applied between the United States and Al-Qaeda as treaty law or customary international law.

[145] It is not clear whether, if recognition of belligerency is granted, the conflict will be regulated by the Geneva Conventions in their entirety or only by Common Art 3. On the basis of the *travaux préparatoires*, Moir suggests that only Common Art 3 will apply (Moir, *The Law*, pp 40–2). See *contra* Dinstein, *The International Law*, p 34. Recognition of belligerency by third states entailed the prohibition for those states to militarily support either side, but did not affect the conduct of hostilities between the belligerents (Schindler, 'The Different Types', p 145).

[146] Schinder, 'The Different Types', p 146. It appears that the American Civil War was the last uncontroversial case of recognition of belligerency, although the debate arose in other subsequent conflicts (Yair M Lootsteen, 'The Concept of Belligerency in International Law', *Military Law Review* 166 (2000), p 110). This non-use, however, does not mean that recognition of belligerency has fallen into desuetude: see Ian Scobbie, 'Gaza', in *International Law*, edited by Wilmshurst, pp 303–4.

[147] It does not seem that Art 1(4) has achieved customary status (Anthony Cullen, *The Concept of Non-International Armed Conflict in International Humanitarian Law* (Cambridge: Cambridge University Press, 2010), p 85). On the definition of colonial domination, alien occupation and racist

Protocol, however, a *de minimis* threshold needs to be reached. Article 96 of Additional Protocol I states that the Conventions and the Protocol apply only when there is an 'authority representing a people' which, by issuing a declaration, is able to assume the same rights and obligations of a contracting party: this necessarily implies that the national liberation movement possesses a quasi-state level of organization which would exclude groups that exist and operate merely online.[148] As to the intensity of the hostilities, the United Kingdom has attached a declaration when signing Additional Protocol I, stating that '[i]n relation to Article 1 [of Additional Protocol I]... the term "armed conflict" of itself and in its context implies a certain level of intensity of military operations which must be present before the Conventions or the Protocol are to apply to any given situation, and that this level of intensity cannot be less than that required for the application of Protocol II, by virtue of Article 1 of that Protocol, to internal conflicts'.[149] At the moment of ratification in 1998, the United Kingdom rephrased the declaration and specified that 'the term "armed conflict" of itself and in its context denotes a situation of a kind which is not constituted by the commission of ordinary crimes including acts of terrorism whether concerted or in isolation'.[150]

III. Cyber Operations During Partial or Total Belligerent Occupation

According to their Common Article 2(2), the Geneva Conventions also apply 'to all cases of partial or total occupation of the territory of a High Contracting Party, even if the said occupation meets with no armed resistance'.[151] The Conventions do not define 'occupation', but Article 42 of the 1907 Hague Regulations provides that 'territory is considered occupied when it is actually placed under the authority of the hostile army'.[152] 'Unlike 'war' and like 'armed conflict', then, occupation is an exclusively factual situation that cannot exist only in legal terms following a declaration not substantiated by effectiveness.[153]

---

régime, see the ICRC Commentary of Art 1(4) of Additional Protocol I, in Sandoz, Swinarski, and Zimmermann (eds), *Commentary*, para 112.

[148] On the organization requirement, see Chapter 3, Section IV.1.2.
[149] Text in UNTS, Vol 1125, p 432. On Additional Protocol II, see Chapter 3, Section IV.2.
[150] Text in Roberts and Guelff, *Documents*, p 510.
[151] Denmark was for instance occupied by Germany in 1940 with no or only minimal resistance (Adam Roberts, 'What is a Military Occupation?', *British Year Book of International Law* 55 (1984), p 252).
[152] 'Belligerent' occupation (*occupatio bellica*) refers to the territory of a state which is under the effective control of another state without its consent as a consequence of hostilities (Eyal Benvenisti, 'Occupation, Belligerent', in *Max Planck Encyclopedia of Public International Law* (2012), Vol VII, p 920). Occupation can, however, also be established peacefully (Roberts, 'What is a Military Occupation?', pp 273–6). A case of *occupatio pacifica* was, for instance, that of Iceland by the British and American forces during the Second World War (Dinstein, *The International Law*, p 35). It is also possible that the presence of foreign troops in the territory of a state, initially authorized by the territorial state, turns into a situation of occupation if the troops do not withdraw after consent is revoked (Kolb and Hyde, *An Introduction*, p 230).
[153] SC Res 1483 (22 May 2003), however, modified the application of Art 42 of the Hague Regulations in that it did not qualify Poland as an Occupying Power even though a small area in

Apart from exceptional cases,[154] there must be land forces on the ground for a territory to be occupied, as otherwise 'it is hard to conceive of the manner in which an occupier with no ground presence could realistically be expected to execute its obligations under *jus in bello*'.[155] If a territory cannot be occupied simply by establishing control of its airspace,[156] then, *a fortiori* exercising authority over the cyberspace of a state (assuming that this is technically possible) would not determine that the state is under occupation. Occupation differs from invasion as it implies some elements of stability.[157] According to the US Military Tribunal in the 1948 *Hostages* case, '[t]he term invasion implies a military operation while an occupation indicates the exercise of governmental authority to the exclusion of the established government'.[158] The ICTY Trial Chamber has suggested some non-cumulative indicators in order to establish whether actual authority over a territory has been established:

- the occupying power must be in a position to substitute its own authority for that of the occupied authorities, which must have been rendered incapable of functioning publicly;
- the enemy's forces have surrendered, been defeated or withdrawn. In this respect, battle areas may not be considered as occupied territory. However, sporadic local resistance, even successful, does not affect the reality of occupation;
- the occupying power has a sufficient force present, or the capacity to send troops within a reasonable time to make the authority of the occupying power felt;
- a temporary administration has been established over the territory;
- the occupying power has issued and enforced directions to the civilian population.[159]

southern Iraq was 'actually placed under the authority' of Polish troops. Similarly, SC Res 1546 (8 June 2004), considered the occupation of Iraq to be terminated as at the end of June 2004 even though little had changed on the ground.

[154] Tristan Ferraro, 'Determining the Beginning and End of an Occupation Under International Humanitarian Law', *International Review of the Red Cross* 94 (2012), pp 157–8.

[155] Yuval Shany, 'Faraway, So Close: The Legal Status of Gaza After Israel's Disengagement', *Yearbook of International Humanitarian Law* 8 (2005), p 380. According to the ICTY, it is not necessary that the occupier exercises territorial control in the form of governmental powers, or that it has established an administration to exercise those powers, as long as it is in a position to do so (ICTY, *Prosecutor v Naletilić and Martinović*, Case No IT–98–34–T, Trial Chamber Judgment, 31 March 2003, para 217). Ferraro identifies three constitutive criteria for the establishment of occupation: 'the unconsented-to presence of foreign forces, the foreign forces' ability to exercise authority over the territory concerned in lieu of the local sovereign, and the related inability of the latter to exert its authority over the territory' (Ferraro, 'Determining', p 142). See also Michael Bothe, 'Effective Control During Invasion: A Practical View on the Application Threshold of the Law of Occupation', *International Review of the Red Cross* 94 (2012), pp 39–40.

[156] Dinstein, *The International Law*, p 48.

[157] Natalino Ronzitti, *Diritto internazionale dei conflitti armati* 4th edn (Torino: Giappichelli, 2011), p 255. Dinstein also argues that 'as long as fighting goes on, and the local population is not in thrall of the invading force, one cannot seriously talk of belligerent occupation' (Dinstein, *The International Law*, p 42). See similarly Bothe, 'Effective Control', p 39.

[158] *United States of America v Willem List et al*, Case No 7, 19 February 1948, in *Trials of War Criminals Before the Nuremberg Military Tribunals under Control Council Law No. 10*, Vol 11, Government Printing Office, Washington, 1950, p 1243.

[159] *Naletilić and Martinović*, para 217 (footnotes omitted).

Once the Occupying Power has actually established its authority, the law of occupation applies to that territory even if hostilities continue elsewhere. Furthermore, as stated by Israel's Supreme Court, 'if the military force gained effective and practical control over a certain area, it is immaterial that its presence in the territory is limited in time or that the intention is to maintain only temporary military control'.[160] Similarly, the EECC held that 'where combat is not occurring in an area controlled even for just a few days by the armed forces of a hostile Power,...the legal rules applicable to occupied territory should apply'.[161]

It is well known that, according to Jean Pictet's Commentary on Article 6(1) of Geneva Convention IV,[162] 'occupation' in the Geneva law has a broader meaning than under Article 42 of the Hague Regulations. Adopting a teleological interpretation of the rules of Geneva Convention IV, Pictet replaces control over territory with control over protected persons as the main criterion for the application of the law of occupation,[163] and argues that, while the Hague Regulations find application only when the situation is stable and territory is under the effective control of the Occupying Power, Geneva Convention IV applies to 'all persons who find themselves in the hands of a Party to the conflict or an Occupying Power of which they are not nationals'.[164] In Pictet's view, then, there is no distinction between occupation and invasion in the Geneva law, as any successful invasion triggers the application of the law of occupation. This entails that

[t]he relations between the civilian population of a territory and troops advancing into that territory, whether fighting or not, are governed by the present Convention. There is no intermediate period between what might be termed the invasion phase and the inauguration of a stable regime of occupation. Even a patrol which penetrates into enemy territory without any intention of staying there must respect the Conventions in its dealings with the civilians it meets.[165]

Pictet's interpretation has been applied in the ICTY's *Naletilić and Martinović* Trial Judgment, which, referring to the Commentary, held that 'the application of

---

[160] *Tsemel et al v Minister of Defence et al*, HCJ 102/82 (1983), quoted in Eyal Benvenisti, *The International Law of Occupation* (Oxford: Oxford University Press, 2012), p 200.

[161] EECC, *Partial Award, Central Front–Eritrea's Claims 2, 4, 6, 7, 8, and 22*, 28 April 2004, RIAA, Vol XXVI, Part IV, p 136.

[162] According to this provision, the Convention 'shall apply from the outset of any conflict or occupation mentioned in Article 2'.

[163] Marten Zwanenburg, 'Challenging the Pictet Theory', *International Review of the Red Cross* 94 (2012), p 32.

[164] Pictet (ed), *Commentary*, Vol 4, p 60. An intermediate position sees Art 42 as the main threshold for the application of the law of occupation, but certain provisions of Geneva Convention IV are also considered to apply in the invasion phase where effective control over the territory has not consolidated yet. It may, however, be difficult to distinguish what provisions of the Geneva Convention IV apply to occupation only and what also to the invasion phase (Zwanenburg, 'Challenging', p 35; Kenneth Watkin, 'Use of Force During Occupation: Law Enforcement and Conduct of Hostilities', *International Review of the Red Cross 94* (2012), p 272).

[165] Pictet (ed), *Commentary*, Vol 4, p 60.

the law of occupation as it affects "individuals" as civilians protected under Geneva Convention IV does not require that the occupying power have actual authority. For the purposes of those individuals' rights, a state of occupation exists upon their falling into "the hands of the occupying power." Otherwise civilians would be left, during an intermediate period, with less protection than that attached to them once occupation is established.'[166] Nothing in the *travaux préparatoires* of the Geneva Conventions, however, suggests that the drafters intended to reject Article 42 as the occupation threshold: on the contrary, Article 154 of Geneva Convention IV expressly recognizes that the Convention does not replace, but rather is supplementary to the Hague Regulations.[167] The distinction between invasion and occupation also results clearly from Article 4(A)(6) of Geneva Convention III on the *levée en masse*, which expressly refers to a situation of invasion in non-occupied territory. Indeed, applying the law of occupation to the invasion phase would require the Occupying Power to comply with obligations that it would be impossible for it to fulfil: most of Part III, Section III of Geneva Convention IV implies that the belligerent has established some degree of authority over enemy territory.[168] '[I]n the absence of a definition of "occupation" in the Geneva Conventions', the ICTY Trial Chamber itself eventually referred 'to the Hague Regulations and the definition provided therein, bearing in mind the customary nature of the Regulations', although it mentioned the Hague test for occupation only when dealing with property in occupied territory, while it used Pictet's test with regard to individuals.[169] In both the *Palestinian Wall* Advisory Opinion and the *DRC v Uganda* Judgment, the ICJ confirmed that Article 42 reflects customary international law and, in the latter case, found that mere stationing of Ugandan troops in certain locations was not sufficient to conclude that the areas in question were occupied, being it necessary that Uganda had substituted its own authority to that of the DRC government.[170] The 1992 German Military Manual further confirms that '[o]ccupied territory does not include battle areas, i.e. areas which are still embattled and not subject to permanent occupational authority (area of invasion, withdrawal area). The general rules of international humanitarian law shall be applicable here.'[171]

Once the territory is 'actually placed under the authority of the hostile army', the cyber operations of the Occupying Power in relation to the occupied territory fall under the scope of the provisions of the law of occupation that regulate the rights and duties of the occupier. Notices to the population in occupied territory could, for instance, be issued by the Occupying Power by cyber means.[172] Article 51(1) of Geneva Convention IV would also prohibit the Occupying Power from conducting online propaganda aimed 'at securing voluntary enlistment' of

---

[166] *Naletilić and Martinović*, para 221. [167] Ferraro, 'Determining', p 136.
[168] Zwanenburg, 'Challenging', p 34.
[169] *Naletilić and Martinović*, paras 215, 222.
[170] *Legal Consequences of the Construction of a Wall*, para 78; *Armed Activities on the Territory of the Congo (DRC v Uganda)*, Judgment, 19 December 2005, ICJ Reports 2005, paras 172–3. See also Shany, 'Faraway, So Close', pp 374 ff; Benvenisti, *The International Law of Occupation*, pp 4–5.
[171] German Ministry of Defence, *Humanitarian Law in Armed Conflicts*, Section 528.
[172] Tallinn Manual, p 239.

protected persons. Similarly, cyber attacks by the Occupying Power that destroy 'real or personal property belonging individually or collectively to private persons, or to the State, or to other public authorities, or to social or co-operative organizations' are prohibited by Article 53 of Geneva Convention IV, 'except where such destruction is rendered absolutely necessary by military operations'. If interpreted in a manner that takes into account the radically increasing informatization of modern societies, the provision would extend to cyber operations that shut down infrastructures without destroying them. Cyber operations disrupting medical and hospital establishments and services, public health and hygiene in occupied territory (for instance by corrupting medical data) are prohibited by Article 56(1) of Geneva Convention IV, while cyber operations aimed at taking possession 'of cash, funds, and realizable securities' which are the property of protected persons are a violation of Article 53(1) of the Hague Regulations. On the other hand, '[a]ll appliances...for the transmission of news...may be seized, even if they belong to private individuals, but must be restored and compensation fixed when peace is made' (Article 53(2) of the Hague Regulations). The Commentary of Rule 90 of the Tallinn Manual suggests that this provision may be extended to include taking remote control of privately-owned cyber infrastructure, providing that the Occupying Power can use the property for its own purposes and the owner is denied its use.[173]

It may be, however, that the Occupying Power also conducts cyber operations that go beyond the exercise of policing or governance powers and amount to resort to armed force for counter-insurgency purposes. It might also be that armed groups in occupied territory resort to cyber operations in order to challenge the authority of the Occupier. The issue of the legal framework regulating the use of armed force during *occupatio bellica* (hostilities or law enforcement) is a controversial one.[174] The following scenarios can be envisaged in the cyber context: (a) cyber operations conducted between the Occupying Power and the Occupied State; (b) cyber operations between the Occupying Power and militias

---

[173] Commentary to Rule 90, Tallinn Manual, p 247.

[174] The majority of occupation law scholars and the ICRC opine that a distinction should be made between situations of 'calm' occupation and those of 'troubled' occupation, where hostilities have continued or resumed (see eg Louise Doswald-Beck, 'The Right to Life in Armed Conflict: Does International Humanitarian Law Provide All the Answers?', *International Review of the Red Cross* 88 (2006), p 892; ICRC, *Occupation and Other Forms of Administration of Foreign Territory*, 2012, pp 114–15, <http://www.icrc.org/spa/assets/files/publications/icrc-002-4094.pdf>; Andreas Paulus, 'The Use of Force in Occupied Territory: The Applicable Legal Framework', in ICRC, *Occupation*, p 142; Jelena Pejic, 'Conflict Classification and the Law Applicable to Detention and the Use of Force', in *International Law*, edited by Wilmshurst, p 109). According to this view, in cases of 'calm' occupation the law enforcement model applies and the use of force is governed by occupation law and, subordinately, international human rights law. On the other hand, in cases of 'troubled' occupation, ie in the presence of 'the sort of hostilities that characterize active armed conflict' (ICRC, *Occupation*, pp 120–1), the use of force falls under the remit of the law on the conduct of hostilities, even though international human rights law continues to apply and complements international humanitarian law as *lex generalis* (p 119). The expression 'troubled occupation' has, however, an oxymoronic connotation: it seems preferable to refer to the threshold of non-international armed conflict in order to establish whether the use of force during occupation falls under the hostilities or law enforcement paradigms.

or other resistance movements 'belonging to a Party to the conflict';[175] (c) cyber operations between the Occupying Power and armed groups in the occupied state that are not resistance movements 'belonging to a Party to the conflict'; (d) cyber operations between different armed groups not belonging to a party to the conflict conducted in occupied territory; and (e) cyber operations conducted by individuals who are sympathetic with one of the belligerents but do not belong to the parties to the conflict and whose actions are not coordinated.

As to (a) and (b), the situation is a continuation of the international armed conflict that determined the situation of belligerent occupation. The cyber operations will therefore fall under the relevant provisions of the law of international armed conflict: according to the ICRC, '[a]ny degree of armed violence involving the occupying power on one side, and the armed forces and other organized armed groups belonging to the occupied State on the other, would...justify resort to the IHL rules governing the conduct of hostilities', whether or not the territory remains occupied.[176] As to scenario (b), conventional international humanitarian law does not clarify what 'belonging to a Party to the conflict' means. The ICRC Report on *Occupation and Other Forms of Administration of Foreign Territory* cautiously concludes that 'organized armed groups linked to the occupied State, either through the overall control exerted by the latter or as regular members of its armed forces, would be involved in an international armed conflict when fighting the occupying forces'.[177] The 'belonging to a Party to the conflict' requirement will be discussed in Chapter 4.[178]

As to (c) and (d), and assuming that the armed violence is protracted and the armed group organized, it is controversial whether a conflict between a state (the Occupying Power) and a non-state actor or between non-state actors in occupied territory qualify as an international or a non-international armed conflict. In the *Targeted Killings* judgment, the Supreme Court of Israel adopted the former view and qualified the conflict between Israel and terrorist groups in the West Bank and Gaza as an international armed conflict relying on the situation of belligerent occupation of the Palestinian Territories.[179] According to the ICRC, however, 'a confrontation between an occupying power and organized armed groups not belonging to the occupied State should be viewed as detached from the original international armed conflict and construed as a new armed conflict

---

[175] Article 4(A)(2) of Geneva Convention III.
[176] ICRC, *Occupation*, p 121. Dinstein also maintains that '[w]hen hostilities go on in an occupied territory, the law of belligerent occupation does not disappear. It has to be applied in contiguity with LOIAC [Law of International Armed Conflict] norms relating to combat' (Dinstein, *The International Law*, p 100). The distinguished author specifies that the Occupying Power will have to comply with occupation law *vis-à-vis* civilians in occupied territory, and with the law on hostilities *vis-à-vis* combatants and civilians taking direct part in hostilities (p 100).
[177] ICRC, *Occupation*, p 127.
[178] Chapter 4, Section III.1.3.b, pp 195–6.
[179] *Targeted Killings*, para 18 (per Judge Barak). See also ICC, *Prosecutor v Thomas Lubanga Dyilo*, Case No ICC–01/04–01/06–803, Pre-Trial Chamber I Decision on the confirmation of charges, 29 January 2007, para 220; ICC, *Katanga and Chui*, ICC–01/04–01/07–717, Pre-Trial Chamber I Decision on the confirmation of charges, 30 September 2008, para 240.

of a non-international character, provided the IHL requirements of organization and intensity are met', whether or not occupation continues.[180] This view has been supported by the ICC in the *Lubanga* Judgment, where the Trial Chamber found that 'although there is evidence of direct intervention on the part of Uganda [in the DRC], this intervention would only have internationalised the conflict between the two states concerned (*viz.* the DRC and Uganda)'.[181] If the threshold of armed conflict has not been reached, either because the cyber operations do not amount to 'protracted' armed violence or because the involved groups or individuals are not organized (as in scenario (e)), the law on the conduct of hostilities will not apply to them and the use of armed force will fall under the law enforcement framework.

In the specific context of the Palestinian Territories, both Israel and Palestinian hackers have conducted a variety of cyber operations after the beginning of the second *intifada* in 2000.[182] It has been argued that the situation in Gaza is still one of occupation and that the conflict between Israel and Hamas is of an international character, as a consequence of the occupation and of the recognition of Hamas as belligerents following the 2009 naval blockade of the Gaza Strip by Israeli forces.[183] If this conclusion is correct, any cyber operation between the belligerents and any other cyber operation that has a nexus with the occupation falls under the relevant *jus in bello* rules. This conclusion would not change if the conflict between Israel and Hamas were deemed to be of a non-international character, although, in this case, it may be more difficult to determine the nexus between the cyber operation and the conflict. It can be doubted, for instance that Anonymous's cyber operations in the context of the Middle East conflict fell under the remit of the *jus in bello*. Even though the belligerent nexus does not necessarily require attribution to a belligerent (neither is Anonymous a party to the conflict, nor does it belong to one), it has not been convincingly established that the operations caused, or were intended to cause, the required threshold of harm to a belligerent to the detriment of another.[184] They are, therefore, better qualified as ordinary (cyber) crimes committed during an armed conflict.

---

[180] ICRC, *Occupation*, p 127. See also Doswald-Beck, 'The Right to Life', p 894; Yutaka Arai-Takahashi, *The Law of Occupation. Continuity and Change of International Humanitarian Law, and its Interaction with International Human Rights Law* (Leiden: Nijhoff, 2009), p 299; Pejic, 'Conflict Classification', p 109.

[181] *Lubanga*, Trial Chamber I Judgment, para 563. In literature, for the application of the law of international armed conflict, including the law of targeting, to uprisings in occupied territory, see Akande, 'Classification', pp 47–8.

[182] See Chapter I, Section I, p 8. See also William A Owens, Kenneth W Dam, and Herbert S Lin, *Technology, Policy, Law, and Ethics Regarding U.S. Acquisition and Use of Cyberattack Capabilities* (Washington: The National Academies Press, 2009), p 278.

[183] Scobbie, 'Gaza', pp 300–2.

[184] On the belligerent nexus and the threshold of harm, see Section II.2 of this Chapter and Section III.1.3.d of Chapter 4. It should be noted that the cyber operations that have been conducted so far in the context of the conflict in the Middle East, including those during Operation Pillar of Defense, do not qualify as cyber attacks amounting to resort to armed force, as they were mainly conducted for defacement and propaganda purposes.

IV. Cyber Operations in and as Non-International Armed Conflicts

Non-international armed conflicts are armed conflicts that do not take place between states, although they might spill over into or from other countries: 'internal' armed conflicts are therefore a sub-category of non-international armed conflicts, limited to those that occur within the territory of one state only. When conducted in the context of an already initiated traditional non-international armed conflict, cyber operations will fall under the scope of the *jus in bello* applicable to this type of conflict to the extent that they have a nexus with the conflict:[185] the same considerations developed in relation to international armed conflicts apply here as well, although the nexus with a non-international one will likely be more difficult to prove and, therefore, it will be more challenging to distinguish between cyber operations amounting to acts of hostilities and ordinary crimes.[186] The following Sections will verify whether the *jus in bello* instruments applicable to armed conflicts of a non-international character apply to cyber operations when no kinetic hostilities concomitantly occur. As each instrument has a different threshold for its application, they will be discussed separately. What should be pointed out from the beginning is that '[t]he exact same amount of violence may produce an IAC [international armed conflict] if perpetrated between states, but might not qualify as a NIAC [non-international armed conflict] if committed by non-state actors'.[187]

---

[185] It appears, for instance, that both during the first (1994) and second (1999–2000) Chechen wars, Russia and the separatist group conducted cyber operations against each other (Tikk, Kaska, Rünnimeri, Kert, Talihärm, and Vihul, *Cyber Attacks Against Georgia*, p 5). In 1997, the Tamil Tigers also conducted cyber operations to shut down the servers and email systems of Sri Lankan embassies (Dino Kritsiotis, 'Enforced Equations', *European Journal of International Law* 24 (2013), p 146). The Syrian government has used cyber capabilities against the insurgents since early 2012. The operations are, however, essentially cyber exploitation aimed to obtain access to personal information. The opposition forces have conducted defacement operations (Justin Salhani, 'In Syria, the Cyberwar Intensifies', *Defense News*, 18 January 2013, <http://www.defensenews.com/article/20130118/C4ISR01/301180018/In-Syria-Cyberwar-Intensifies>).

[186] See Chapter 3, Section II.2. Another problem is that, unlike states and international organizations, there are no rules for the attribution of acts and omissions to armed groups (Liesbeth Zegveld, 'Accountability of Organized Armed Groups', in *Non-state Actors and International Humanitarian Law*, edited by Marco Odello and Gian Luca Beruto (Milano: Franco Angeli, 2010), pp 111–12): it is, therefore, not clear what individuals impute their conduct to the armed group as a separate legal entity. At least some provisions of the law of state responsibility may be applied by analogy to armed groups that have a state-like structure. For other armed groups, it has been suggested that effective control by the group over individual members is what determines the group's responsibility (pp 112–13). A role may also be played by the notion of 'continuous combat function' developed by the ICRC in relation to direct participation in hostilities (see Chapter 4, Section III.1.3.c, pp 200–2).

[187] Milanović and Hadzi-Vidanović, 'A Taxonomy', p 274.

1. Article 3 Common to the 1949 Geneva Conventions

Common Article 3 of the 1949 Geneva Conventions is the first provision adopted to apply '[i]n the case of armed conflict not of an international character occurring in the territory of one of the High Contracting Parties'.[188] As Common Article 3 is generally understood as reflecting customary international law, the limitation to states parties to the Geneva Conventions is now obsolete. The reference to 'one' contracting party, which would make Common Article 3 applicable to internal conflicts only, should be interpreted as 'any' contracting party, so that the provision also applies to non-international armed conflicts that spread to the territory of other states: such evolutionary interpretation of Common Article 3 takes into account contemporary practice and the increased frequency of the 'spillover' scenario.[189] The fact that cyber operations occur in and through cyberspace, and not strictly speaking in the 'territory' of a state, does not prevent the application of Common Article 3: as has been seen, cyber operations can be seen as 'the reduction of information to electronic format and the actual movement of that information between physical elements of cyber infrastructure'.[190]

Common Article 3 does not refer to any intensity threshold that the conflict needs to reach for the provision to apply: no mention is made, for instance, of the traditional requirements for recognition of belligerency. The Commentary, however, suggests certain non-binding criteria in order to establish the existence of a Common Article 3 conflict:

(1) That the Party in revolt against the *de jure* Government possesses an organized military force, an authority responsible for its acts, acting within a determinate territory and having the means of respecting and ensuring respect for the Convention.
(2) That the legal Government is obliged to have recourse to the regular military forces against insurgents organized as military and in possession of a part of the national territory.
(3) (a) That the *de jure* Government has recognized the insurgents as belligerents; or
    (b) That it has claimed for itself the rights of a belligerent; or
    (c) That it has accorded the insurgents recognition as belligerents for the purposes only of the present Convention; or
    (d) That the dispute has been admitted to the agenda of the Security Council or the General Assembly of the United Nations as being a threat to international peace, a breach of the peace, or an act of aggression.
(4) (a) That the insurgents have an organization purporting to have the characteristics of a State.
    (b) That the insurgent civil authority exercises *de facto* authority over the population within a determinate portion of the national territory.

---

[188] In *Nicaragua*, the ICJ found that Common Art 3 reflects 'elementary considerations of humanity' and constitutes 'a minimum yardstick' applicable to all armed conflicts (*Nicaragua*, para 218).
[189] Jelena Pejic, 'The Protective Scope of Common Article 3: More Than Meets the Eye', *International Review of the Red Cross 93* (2011), pp 199–205.
[190] Melzer, *Cyberwarfare*, p 5. See Chapter 1, Section III.1, pp 23–4.

(c) That the armed forces act under the direction of an organized authority and are prepared to observe the ordinary laws of war.

(d) That the insurgent civil authority agrees to be bound by the provisions of the Convention.[191]

The *travaux préparatoires* also reveal that the expression 'armed conflict of a non-international character' was originally understood as 'civil war' under the doctrine of belligerency.[192] As it was initially conceived, then, Common Article 3 applied to 'armed conflicts, with "armed forces" on either side engaged in "hostilities"—conflicts, in short, which are in many respects similar to an international war, but take place within the confines of a single country'.[193] Having said that, the scope of Common Article 3 has subsequently expanded and the significance of the above-mentioned indicators has been progressively reduced.[194] After all, the Commentary itself specifies that even a conflict that does not possess any of the above characteristics could still fall under the scope of Common Article 3, the application of which should be 'as wide as possible'.[195] As has been observed,

[t]he legal interests contemplated by common Article 3 GCs [Geneva Conventions] are different from those envisaged by the rules on conduct of hostilities. Common Article 3 GCs provides the minimum standards of treatment and procedure for the persons captured in armed conflict, without succumbing to the countervailing notion of military necessity. On the other hand, the rules on the conduct of hostilities are purported to minimise loss, injuries and damage. They are susceptible to an intrinsic balance that must be struck against the varying standard of military necessity.[196]

While not that of a 'civil war', it is, however, generally agreed that a situation of internal armed violence needs to reach a minimum threshold in order to amount to a non-international armed conflict in the sense of Common Article 3. The ICTY Appeals Chamber has famously defined non-international armed conflicts as '*protracted* armed violence between governmental authorities and *organized* armed groups or between such groups within a State'.[197] This definition has proved to be very successful and is now considered 'the most authoritative formulation of the threshold associated with common Article 3':[198] in addition to being consistently referred to by the ICTY and other international criminal tribunals in subsequent decisions as well as by international commissions of inquiry, expert reports and military manuals,[199] it has been incorporated in Article 8(2)(e) of the Statute of the ICC and in the ILC Draft Articles on the Effects of Armed Conflict on Treaties (although with some adaptations). In a 2008 opinion paper, the ICRC itself adopted a definition of non-international armed conflict clearly inspired by

---

[191] Pictet (ed), *Commentary*, Vol 3, p 36.
[192] Cullen, *The Concept*, p 49.
[193] Pictet (ed), *Commentary*, Vol 3, p 37.
[194] Cullen, *The Concept*, p 50.
[195] Pictet (ed), *Commentary*, Vol 3, p 36.
[196] Arai-Takahashi, *The Law of Occupation*, p 299.
[197] *Tadić*, Decision on the Defence Motion, para 70 (emphasis added).
[198] Cullen, *The Concept*, p 122.
[199] See the list in Cullen, *The Concept*, pp 120–2.

the *Tadić* language.[200] Rule 23 of the Tallinn Manual also uses the ICTY definition and adjusts it to the cyber context.[201]

The elements of a non-international armed conflict that allow to distinguish 'an armed conflict from banditry, unorganized and short-lived insurrections, or terrorist activities, which are not subject to international humanitarian law',[202] are therefore: (1) armed violence; (2) which is protracted; (3) and occurring between governmental authorities and armed groups, or between armed groups; (4) providing that the armed groups are 'organized'. A group is 'armed' if it detains and is able to use weapons, ie in our context, 'if it has the capacity of undertaking cyber attacks'.[203] 'Armed violence' can be interpreted as a synonymous of 'armed force': the comments made above in the context of international armed conflicts apply here as well.[204] The third element makes clear that what characterizes a non-international armed conflict is the nature of the belligerents, and not where the hostilities take place. The second and fourth requirements are cumulative: if the armed violence is protracted but there is no organized armed group, the situation does not qualify as a non-international armed conflict, but rather as internal disturbances.[205] Similarly, if there is an organized armed group but their actions are sporadic or isolated, there is no armed conflict of a non-international character. On the other hand, the motives, political or not, of the armed group are irrelevant.[206] Therefore, the fact that those behind the 2007 DDoS attacks on Estonia may have had political motives did not turn per se the situation into an armed conflict, unless it was proved that the hackers had a sufficient level of organization and the operation amounted to 'protracted' armed violence.

In the following pages, the requirements of the protracted character of armed violence and of the organization of the armed group will be applied to the cyber context.

---

[200] ICRC, 'How is the Term', p 5 (*protracted armed confrontations* occurring between governmental armed forces and the forces of one or more armed groups, or between such groups arising on the territory of a State [party to the Geneva Conventions]. The armed confrontation must reach *a minimum level of intensity* and the parties involved in the conflict must show *a minimum of organisation*'; emphasis in the original). There are, however, some minor differences between the ICRC and the *Tadić* definitions: in the former, 'armed violence' is replaced with 'armed confrontations' and 'governmental authorities' is replaced with 'governmental armed forces'. Also, all parties to the conflict, ie including the government, must show a minimum level of organization, not only the armed groups.
[201] Tallinn Manual, p 84.   [202] *Tadić*, Opinion and Judgment, para 562.
[203] Tallinn Manual, p 88.
[204] See Chapter 3, Section II.3.1. Indeed, Art 2(b) of the ILC Draft Articles on the Effects of Armed Conflict on Treaties replaces 'armed violence' with 'armed force' in its definition of 'armed conflict'.
[205] The intensity and organization requirements have also been expressly incorporated in Rule 210 of the German Military Manual, according to which '[a] non-international armed conflict is a confrontation between the existing governmental authority and groups of persons subordinate to this authority, which is carried out by force of arms within national territory and reaches the magnitude of an armed riot or a civil war' (German Ministry of Defence, *Humanitarian Law in Armed Conflicts*, Section 210). The *Third Report of the Independent International Commission of Inquiry on the Syrian Arab Republic* also assessed the intensity of the hostilities and the level of organization of the Free Syrian Army and other opposition groups involved in order to establish the applicability of Common Art 3 (*Third Report of the Independent International Commission of Inquiry on the Syrian Arab Republic*, p 45).
[206] ICRC, *International Humanitarian Law and the Challenges*, p 11. See *Limaj*, para 170.

1.1 'protracted' armed violence

The ICTY has clarified that 'protracted armed violence' has to be interpreted as referring to the intensity of the conflict, and not to its duration.[207] Indeed, intensity does not necessarily mean continuous: it implies, but is not limited to, a time element.[208] Obviously, the longer the duration the easier it will be to prove that the conflict is 'intense', but a short duration does not automatically entail that the situation does not amount to an armed conflict. The decision of the Inter-American Commission on Human Rights on the *Abella* case is exemplary from this perspective, as it qualified as a non-international armed conflict the events at La Tablada military base despite their very brief duration. According to the Commission, '[w]hat differentiates the events at the La Tablada base from [internal disturbances and tensions] are the concerted nature of the hostile acts undertaken by the attackers, the direct involvement of governmental armed forces, and the nature and level of the violence attending the events in question. More particularly, the attackers involved carefully planned, coordinated and executed an armed attack, i.e. a military operation, against a quintessential military objective—a military base'.[209] On the other hand, as the Venice Commission explained, 'sporadic bombings and other violent acts which terrorist networks perpetrate in different places around the globe and the ensuing counter-terrorism measures, even if they are occasionally undertaken by military units, cannot be said to amount to an "armed conflict" in the sense that they trigger the applicability of International Humanitarian Law'.[210] The fact that the government employs the armed forces to restore law and order does not per se mean that armed violence is intense, although it might be an indicator. In fact, as has already been noted, in several states the armed forces can be used for law enforcement purposes, and not exclusively for military operations.[211]

The ICTY has suggested some indicative factors in order to assess whether the armed violence is sufficiently intense: 'the number, duration and intensity of individual confrontations; the type of weapons and other military equipment used; the number and calibre of munitions fired; the number of persons and type of forces partaking in the fighting; the number of casualties; the extent of material destruction; and the number of civilians fleeing combat zones', as well as the involvement of the UN Security Council.[212] Some of these indicators can be transposed by analogy of consequences to cyber operations causing or reasonably likely to cause material damage to property, loss of life or injury of persons in order

---

[207] ICTY, *Prosecutor v Ramush Haradinaj*, Case No IT–04–84–T, Trial Chamber I Judgment, 3 April 2008, para 49.
[208] Commentary to Rule 23, Tallinn Manual, p 87.
[209] *Juan Carlos Abella v Argentina*, Case 11.137, Report No 55/97, Inter-American Commission of Human Rights, OEA/Ser.L/V/II.95, Doc 7 rev, 18 November 1997, para 155.
[210] European Commission for Democracy through Law (Venice Commission), Opinion on the International Legal Obligations of Council of Europe Member States in Respect of Secret Detention Facilities and Inter-State Transport of Prisoners, 17 March 2006, Opinion no 363/2005, CDL–AD (2006)009, p 19, <http://www.venice.coe.int/webforms/documents/CDL-AD(2006)009.aspx>.
[211] Garraway, 'War and Peace', p 104.
[212] *Haradinaj*, para 49. See also *Limaj*, paras 135–70; ICTY, *Prosecutor v Boškoski and Tarčulovski*, Case No IT–04–82–T, Trial Chamber II Judgment, 10 July 2008, paras 177–8.

to assess whether the intensity requirement has been met or not. Their application would likely lead to conclude that, even if it was proved that it did damage a certain number of centrifuges, operations like Stuxnet, if conducted by or against an armed group, would not reach the minimum intensity required for the existence of a non-international armed conflict.[213] According to the Tallinn Manual, however, '[f]requent, albeit not continuous, cyber attacks occurring within a relatively defined period may be characterized as protracted':[214] whether multiple attacks can cumulatively reach a minimum level of intensity depends on the circumstances of each case.[215] With regard to disruptive cyber operations not resulting in physical consequences, it is likely that only multiple coordinated cyber operations seriously disrupting the functioning of several or all critical infrastructures of a heavily digitally reliant state for a prolonged time may potentially be considered by states to reach the intensity requirement needed for the application of the law of non-international armed conflict in the absence of associated kinetic hostilities.

It is worth noting that, unlike in international armed conflicts, only a 'confrontation' establishes a non-international armed conflict, and not unilateral armed acts that are not met with an armed reaction.[216] Indeed, 'isolated and sporadic acts of violence' are expressly disqualified as non-international armed conflicts by Article 1(2) of Additional Protocol II, which also applies, *qua* customary international law, to Common Article 3.[217]

1.2 The organization requirement

The armed group must also possess a certain level of organization for Common Article 3 to apply.[218] The organization requirement excludes 'private wars' from the scope of the law of armed conflict: a group, however small, must exist and be organized, while a single individual or several non-coordinated individuals cannot be a party to the armed conflict, even if they inflict severe damage.[219] This is particularly relevant in the cyber context: although even individual hackers could potentially cause significant damage on state infrastructure, the absence of an organized group would preclude the applicability of the *jus in bello*.

---

[213] Robin Geiss, 'Cyber Warfare: Implications for Non-international Armed Conflicts', *International Law Studies* 89 (2013), p 633.
[214] Commentary to Rule 23, Tallinn Manual, p 88.
[215] As Akande points out, 'prolonged violence may suffice even though the individual confrontations do not result in extensive casualties or destruction and are mere "pin-pricks"' (Akande, 'Classification', p 53). On the doctrine of 'accumulation of events' in the *jus ad bellum* context, see Chapter 2, Section IV, pp 108–10.
[216] The 1992 German Military Manual is explicit in this regard: while Rule 202 provides that an international armed conflict exists 'if one party uses force of arms against another party', a non-international armed conflict requires a 'confrontation' (German Ministry of Defence, *Humanitarian Law in Armed Conflicts*, Sections 202 and 210).
[217] Geiss, 'Cyber Warfare', p 633.
[218] Moir, *The Law*, p 36.   [219] Schindler, 'The Different Types', p 147.

The organization requirement, however, should not be understood too rigidly.[220] In particular, it does not necessarily entail a level of organization analogous to that of state armed forces. In *Limaj*, the ICTY argued that

[t]he KLA [Kosovo Liberation Army] was effectively an underground organisation, operating in conditions of secrecy out of concern to preserve its leadership, and under constant threat of military action by the Serbian forces. The members of the General Staff did not meet regularly because of the security situation and identified themselves not by their names but by numbers for the same reason. In these circumstances it is of no surprise that the organisational structure and the hierarchy of the KLA were confusing, or not known, to outside observers, and that, to some, this suggested a state of confusion.[221]

The Court found that the above did not affect the existence of a non-international armed conflict as '*some* degree of organization by the parties' is enough.[222] In the above-mentioned *Abella* case, the Inter-American Court of Human Rights also held that 'confrontations between *relatively* organized armed forces' are potentially sufficient to trigger the application of Common Article 3, as long as the attacks are 'carefully planned, coordinated and executed'.[223] The ILA *Initial Report on the Meaning of Armed Conflict* sees the requirement of organization as inversely related to that of intensity: 'the higher the level of organisation the less degree of intensity may be required and vice versa'.[224]

The ICTY has suggested certain indicators in order to establish whether a group is sufficiently organized, including 'the existence of a command structure and disciplinary rules and mechanisms within the group; the existence of a headquarters; the fact that the group controls a certain territory; the ability of the group to gain access to weapons, other military equipment, recruits and military training; its ability to plan, coordinate and carry out military operations, including troop movements and logistics; its ability to define a unified military strategy and use military tactics; and its ability to speak with one voice and negotiate and conclude agreements such as cease-fire or peace accords'.[225] Whether or not the armed group complies with international humanitarian law is, on the other hand, not significant: 'so long as the armed group possesses the organisational *ability* to comply with the obligations of international humanitarian law, even a pattern of such type of violations would not necessarily suggest that the party did not possess the level of organisation required to be a party to an armed conflict'.[226] The above indicators

---

[220] See Marco Sassòli, 'Implementation of International Humanitarian Law', *Yearbook of International Humanitarian Law 10* (2007), pp 56–7.

[221] *Limaj*, para 132 (footnotes omitted).   [222] *Limaj*, para 89 (emphasis added).

[223] *Abella*, paras 152, 155 (emphasis added).

[224] *Initial Report on the Meaning of Armed Conflict in International Law*, in ILA, Report of the Seventy-Third Conference (Rio de Janeiro, 2008), p 840.

[225] *Haradinaj*, para 60. See also *Limaj*, paras 94–134 and the five groups of factors listed in *Boškoski and Tarčulovski*, paras 199–203. The *Report of the International Commission of Inquiry on Libya* has also indicated some factors to assess the organization of the group: 'whether there is a hierarchical command structure, the extent to which it is able to carry out organized operations (e.g. organises into zones of responsibility, means of communication); discipline systems, the nature of logistical arrangements and how the group presents itself (e.g. whether it is capable of involvement in negotiations)' (*Report of the International Commission of Inquiry on Libya* (2011), p 30).

[226] *Boškoski and Tarčulovski*, para 205 (emphasis in the original).

are not binding, not exhaustive and not cumulative, and none of them is more important than the others.

Several scenarios can be identified in relation to cyber operations conducted by groups according to their level of organization:

(a) The Hezbollah scenario: a traditional armed group, with physical infrastructure and contact among its members, which has a hierarchical organization and the capacity to enforce its decisions towards the members of the group (with or without control of territory).

(b) The Al-Qaeda scenario: a network where members physically meet, share an ideology and/or a mission but lack a proper structure capable of planning and coordinating the group's activities and of ensuring compliance with its directives.[227]

(c) The 'Anonymous' scenario: a loose collective whose members communicate exclusively online, for instance through chatrooms and social networks, with no physical infrastructure and whose activities are at times coordinated but not organized.[228]

(d) The Georgia 2008 scenario: individual hacktivists that pursue the same goal but that meet neither in person nor online. What links them is for instance the fact that they follow instructions contained in, or use malware downloaded from, blogs, forums, and websites, and/or respond to an appeal to conduct cyber operations against a certain target.[229]

(e) The 'botnet' scenario: multiple users that participate in the attack, often in great numbers, but who are unaware of their participation, as their computers have been taken over by others in a DDoS attack.

(f) The RBN scenario: a cybercrime firm hired to conduct cyber operations in return for financial gain.

Scenario (a) is the most likely to meet the organization requirement for the application of Common Article 3: if such armed group conducts standalone cyber operations of sufficient intensity or cyber operations in connection with kinetic hostilities against the adversary, the operations would fall under the law of non-international armed conflict. On the other hand, it would be difficult to conclude that scenarios (b) and, even more, (c) and (d) satisfy any of the indicators suggested by the ICTY in order to assess whether the group has some degree of organization: in fact, they are not even groups, but 'collectives', whose activities are at best coordinated (when they are at all).[230] Anonymous, for instance, has

---

[227] McDonald, 'Declarations', p 309; Lubell, 'The War (?) against Al-Qaeda', p 436.
[228] Thomas Rid, *Cyber War Will Not Take Place* (London: Hurst & Co, 2013), pp 128–31.
[229] Gary Brown, 'Law at Cyberspeed: Answering Military Cyber Operators' Legal Questions', in *International Humanitarian Law and New Weapons Technologies*, edited by Wolff Heintschell von Heinegg and Gian Luca Beruto (Milano: Franco Angeli, 2012), p 167; Tikk, Kaska, Rünnimeri, Kert, Talihärm, and Vihul, *Cyber Attacks Against Georgia*, pp 9–10.
[230] Laurie R Blank, 'International Law and Cyber Threats from Non-State Actors', *International Law Studies* 89 (2013), p 425. The Commentary to the Tallinn Manual also concludes that 'if a

'no command structure at all, not even an articulated set of goals, no formal membership structure, and...no formal deliberative process. These are groups whose organisational structure is almost entirely automated, more like a queuing system and a repository and a mailing list than a political movement or even a street-gang'.[231] In similar situations, conclusively determining group membership is likely to be an impossible task and the group's means to enforce discipline would be extremely limited.[232] As to scenario (e), the computers hijacked to create a botnet are not a 'group'. However, the element of organization could be assessed with regard to the botnet controller: an individual, an organized group, or not, and the situation would therefore fall within one of the other scenarios.[233] The same conclusion applies to scenario (f): the organization requirement should be evaluated with regard to those who hired the firm to conduct the cyber operations: in that, the situation is not different from the use of private military security companies in traditional armed conflicts. If the cyber crime firm is not acting on behalf of someone else but on its own for financial gain, it will have to be sufficiently organized for its actions to potentially fall under the scope of the *jus in bello*, the motivations, political or not, of the group being irrelevant for the qualification of conflicts.[234]

To be clear, even if the group has a sufficient degree of organization, the other elements of a non-international armed conflict need to be present for Common Article 3 to apply in the absence of concurrent kinetic hostilities: if the cyber operations they conduct do not amount to 'armed violence' in the sense highlighted above, ie if they are cyber exploitation operations or cyber attacks that do not cause material damage to property, loss of life or injury to persons or severe disruption of the functioning of critical infrastructures, and if such armed violence is not 'protracted', there is no 'armed conflict' and the matter falls under the scope of domestic criminal laws.[235] In fact, even though it cannot be excluded that the situation may change in the future, cyber technologies have so far played a limited

---

website offers malware and a list of potential cyber targets, those who independently use the site to conduct attacks would not constitute an organized armed group' (Commentary to Rule 23, Tallinn Manual, p 90).

[231] Cory Doctorow, 'Disorganised but effective: how technology lowers transaction costs', *The Guardian*, 21 June 2012, <http://www.guardian.co.uk/technology/2012/jun/21/how-technology-lowers-transaction-costs>. Schmitt, however, does not rule this out and argues that whether the group meets the organization criterion in this case 'should depend on the nature of their collective action. Is there an identifiable "leader" responsible for setting up and maintaining the site? Do they only attack targets found on the site?' (Michael N Schmitt, 'Classification in Future Conflict', in *International Law*, edited by Wilmshurst p 463).

[232] Geiss, 'Cyber Warfare', p 636.

[233] There could be a further layer between the botnet handler and the bots, which is used to control multiple independent botnets similarly to a regiment/battalion/company structure in traditional military forces (William A Owens, Kenneth W Dam, and Herbert S Lin, *Technology, Policy, Law, and Ethics Regarding U.S. Acquisition and Use of Cyberattack Capabilities* (Washington: The National Academies Press, 2009), p 94).

[234] ICRC, *International Humanitarian Law and the Challenges*, p 11.

[235] Melzer, therefore, concludes that cyber operations by an armed group are likely to qualify as hostilities 'when they occur repeatedly over a certain duration and emanate from territory where the attacked state cannot exercise its law enforcement authority, and where the local authority is unwilling or unable to intervene' (Melzer, *Cyberwarfare*, p 24).

role in non-international armed conflicts, as armed groups often do not have the capacity to conduct significant cyber attacks and do not have a cyber infrastructure to attack in retaliation.[236] Until now, cyberspace has been used by non-state actors essentially for cyber exploitation and for propaganda.[237]

2. Additional Protocol II

Common Article 3 is supplemented, but not replaced, by the 1977 Additional Protocol II. Indeed, the expression 'armed conflict not of an international character', which is the scope of application of Common Article 3, only appears in the Preamble to the Protocol in reference to Common Article 3, but not in the main text. This is because, in the words of Article 1 of Protocol II, '[the Protocol]...develops and supplements Article 3 common to the Geneva Conventions of 12 August 1949 without modifying its existing conditions of application'.

Under its Article 1, in order for Additional Protocol II to apply, the armed conflict must meet the following requirements: (1) it must not be covered by Article 1 of Additional Protocol I, ie be an international armed conflict or a 'war of national liberation'; (2) it must take place in the territory of a High Contracting Party; (3) it must be between the government's armed forces on the one hand and 'dissident armed forces or other organized armed groups' on the other; (4) the opposition forces must be 'under responsible command'; and (5) 'exercise such control over a part of its territory as to enable them to carry out sustained and concerted military operations and to implement this Protocol'; (6) the situation must not amount to 'internal disturbances and tensions, such as riots, isolated and sporadic acts of violence and other acts of a similar nature'. Even though the threshold is lower than that required for the traditional recognition of belligerency,[238] when compared to that of Common Article 3 the scope of application of Additional Protocol II is rather restrictive and only applies in case of high magnitude conflicts.[239] Common Article 3, then, continues to be the only provision that regulates non-international armed conflicts not reaching the higher Additional Protocol II threshold, as well as conflicts between different armed groups with no involvement of governmental forces. Common Article 3 also continues to apply as a 'minimum standard of humanity' to all other armed conflicts, including those falling under the scope of application of Additional Protocol II.

The first and sixth requirements fix the highest and lowest thresholds of the scope of application of Additional Protocol II. The second requirement is a reminder that treaties only apply to the states that have ratified them.[240] The third requirement excludes the application of the Protocol to conflicts between different groups of

---

[236] Geiss, 'Cyber Warfare', pp 642–3.   [237] Geiss, 'Cyber Warfare', p 643.
[238] Cullen, *The Concept*, pp 106–7.
[239] A recent example of an Additional Protocol II conflict is that developed in Libya in late February 2011, after the insurgents established *de facto* authority over part of the national territory (*Report of the International Commission of Inquiry on Libya* (2011), p 31).
[240] Paulus and Vashakmadze, 'Asymmetrical War', p 118.

insurgents without the involvement of governmental forces. Furthermore, unlike in Common Article 3 conflicts, the armed group must be under a 'responsible command' for Additional Protocol II to apply:[241] the ICRC Commentary specifies that the existence of a responsible command 'does not necessarily mean that there is a hierarchical system of military organization similar to that of regular armed forces. It means an organization capable, on the one hand, of planning and carrying out sustained and concerted military operations, and on the other, of imposing discipline in the name of a de facto authority'.[242] As to the fifth requirement, the extent of controlled territory may be modest, but control must be exercised by the armed group with 'some degree of stability...for them to be capable of effectively applying the rules of the Protocol'.[243] The ICRC Commentary to Additional Protocol II clarifies that

'[s]ustained' (in French the reference is to '*opérations continues*') means that the operations are kept going or kept up continuously. The emphasis is therefore on continuity and persistence. 'Concerted' (in French: '*concertées*') means agreed upon, planned and contrived, done in agreement according to a plan. Thus we are talking about military operations conceived and planned by organized armed groups.[244]

According to the ICTY, the degree of organization required to carry out 'sustained and concerted military operations' is higher than that required to engage in 'protracted violence' for the purposes of Common Article 3.[245] Indeed, 'Common Article 3 reflects basic humanitarian protections, and a party to an armed conflict only needs a minimal degree of organisation to ensure their application'.[246] The higher degree of organization that the armed group must possess under Additional Protocol II derives from the more detailed and demanding rules contained therein, which require more stable territorial control to be applicable.[247]

If the above requirements are applied in the cyber context, one has to conclude that Additional Protocol II is extremely unlikely to ever extend to standalone cyber operations between a government and an armed group that exists only online. In particular, it is hard to see how such group could be able to impose discipline and obtain territorial control by cyber means only.[248] The only situations in which the Protocol could potentially apply to cyber operations are, then, those of traditional armed groups meeting the Additional Protocol II requirements that (also) engage in such operations against the government. Having said that, it should be recalled that, according to the ICRC *Study on Customary International Humanitarian Law*, 'the difference in thresholds [between Common Article 3 and Additional Protocol II] is rendered largely irrelevant because most, if not all, of

---

[241] Cullen, *The Concept*, p 158.
[242] Sandoz, Swinarski, and Zimmermann (eds), *Commentary*, para 4463.
[243] Sandoz, Swinarski, and Zimmermann (eds), *Commentary*, para 4467.
[244] Sandoz, Swinarski, and Zimmermann (eds), *Commentary*, para 4469 (footnote omitted).
[245] *Boškoski and Tarčulovski*, para 197.
[246] *Boškoski and Tarčulovski*, para 197 (footnote omitted).
[247] *Boškoski and Tarčulovski*, para 197. See also Sandoz, Swinarski, and Zimmermann (eds), *Commentary*, para 4467.
[248] Droege, 'Get Off My Cloud', p 550. It has been maintained, however, that with modern technological developments in warfare, the ability to conduct 'sustained and concerted' military

the rules under AP II [Additional Protocol II] have been transposed to CA3 [Common Article 3] NIACs under customary law'.[249] Should one accept this view, the customary rules of Additional Protocol II would apply to cyber operations under the same broader conditions highlighted above with regard to Common Article 3.[250]

It is worth recalling that, when defining its scope of application, Article 8(2)(e) of the 1998 Statute of the ICC refers to 'armed conflicts that take place in the territory of a State when there is protracted armed conflict between governmental authorities and organized armed groups or between such groups'.[251] It is doubtful whether this provision introduces a third category of non-international armed conflicts narrower than Common Article 3 but broader than Additional Protocol II conflicts (as the conflict needs be protracted but without requiring control of territory by the armed group), or simply reformulates the scope of application of Common Article 3 as seems preferable.[252] Be that as it may, the purpose of Article 8(2)(e) is to define the jurisdiction of the ICC and not the scope of application of international humanitarian law:[253] it is, therefore, beyond the scope of this Chapter.

V. Cyber Operations as 'Internal Disturbances and Tensions'

Armed violence below the threshold of Article 3 Common to the Geneva Conventions does not amount to an armed conflict and falls within the scope of international human rights law, domestic laws, and law enforcement mechanisms. According to the ICRC, 'there are *internal disturbances*, without being an armed conflict, when the State uses armed force to maintain order; there are *internal tensions*, without being internal disturbances, when force is used as a preventive

---

operations does not necessarily depend on control of land (William H Boothby, *The Law of Targeting* (Oxford: Oxford University Press, 2012), p 555; Heather Harrison Dinniss, *Cyber Warfare and the Laws of War* (Cambridge: Cambridge University Press, 2012), p 136).

[249] Milanović and Hadzi-Vidanović, 'A Taxonomy', p 286. See also Kolb and Hyde, *An Introduction*, p 79; Elizabeth Wilmshurst, 'Conclusions', in *International Law*, edited by Wilmshurst, pp 482–3. The Report of the International Fact-Finding Commission on the Conflict in Georgia goes even further and maintains that 'it is generally recognised that the same IHL customary law rules generally apply to all types of armed conflicts' (Report of the International Fact-Finding Commission on the Conflict in Georgia, Vol II, p 304). This may be correct with regard to the rules on the conduct of hostilities, but it is debatable with regard to other rules, such as those on detention.

[250] A commentator has warned, however, against the risk of overloading that might result from expanding normative content and at the same time lowering the threshold of application: '[t]o increase the obligations to such an extent that the armed group is unable to meet them serves no useful purpose' and could actually lead to 'violations spiralling out of control' (Sandesh Sivakumaran, 'Re-envisaging the International Law of Internal Armed Conflict', *European Journal of International Law* 22 (2011), pp 254–5).

[251] Article 8(2)(f) of the ICC Statute.

[252] Cullen favours the latter option, as Art 8(2)(e) refers to 'armed conflicts of a non-international character', which is the expression used in Common Art 3 but not in Additional Protocol II (Cullen, *The Concept*, p 182). See also Pejic, 'The Protective Scope', p 193.

[253] Vité, 'Typology', p 83.

measure to maintain respect for law and order'.[254] Internal disturbances have been then defined as 'confrontation within the country, which is characterized by a certain seriousness or duration and which involves acts of violence' but which does not 'necessarily degenerate[s] into open struggle'.[255] However, extensive police or even armed forces are employed to restore internal order and there might be a high number of victims, which necessitates the application of 'a minimum of humanitarian rules'.[256] As to 'internal tensions', they consist of 'situations of serious tension (political, religious, racial, social, economic, etc.), but also the sequels of armed conflict or of internal disturbances' and can be characterized by large-scale arrests, high number of political prisoners, ill-treatment or inhumane conditions of detention, suspension of fundamental judicial guarantees, alleged disappearances.[257] In its 1973 Commentary on the Draft Additional Protocols to the Geneva Conventions, the ICRC provided other examples of internal disturbances and tensions not reaching the threshold of a non-international armed conflict:

– riots, that is to say, all disturbances which from the start are not directed by a leader and have no concerted intent;
– isolated and sporadic acts of violence, as distinct from military operations carried out by armed forces or organized armed groups;
– other acts of a similar nature which incur, in particular, mass arrests of persons because of their behavior or political opinion.[258]

In light of the above, the use of force by a government against cyber attacks amounting to 'acts of violence' by non-organized hackers may qualify as internal disturbances. The same conclusion holds true of the use of force by a government against cyber attacks amounting to 'acts of violence' by an organized armed group if such attacks do not reach the intensity required to be an armed conflict. If the use of force by the government is a preventive measure to maintain law and order, the situation would be better qualified as internal tensions. The 2007 DDoS attacks against Estonia cannot be ascribed to either situation, as the attacks were not 'acts of violence' in the sense described above and there was no use of kinetic or cyber force by the Estonian government to react against them: passive cyber defences and criminal investigations were used instead. Cyber exploitation operations would also not determine a situation of 'internal disturbances and tensions', as they do not involve violence, but focus on intelligence collection, surveillance, or reconnaissance.

Even though they escape the application of international humanitarian law, the 1995 Turku Declaration recommends the application of most, if not all, Common Article 3 rights and duties to internal disturbances and tensions.[259] This is consistent

---

[254] Sandoz, Swinarski, and Zimmermann (eds), *Commentary*, para 4477 (emphasis added).
[255] Sandoz, Swinarski, and Zimmermann (eds), *Commentary*, para 4475.
[256] Sandoz, Swinarski, and Zimmermann (eds), *Commentary*, para 4475.
[257] Sandoz, Swinarski, and Zimmermann (eds), *Commentary*, para 4476.
[258] Quoted in Marco Sassòli, Antoine A Bouvier, and Anne Quintin, *How Does Law Protect in War?*, 3rd edn (Geneva: ICRC, 2011), Vol III, p 2. See also *Abella*, para 149.
[259] Declaration of Minimum Humanitarian Standards (Turku Declaration), Report of the Sub-Commission on Prevention of Discrimination and Protection of Minorities on its Forty-sixth

with the ICRC's view that the application of the provision should be 'as wide as possible'.[260] In fact, Common Article 3 provides minimum humanitarian standards that it would be difficult to deny even to common criminals.[261]

VI. Conclusions

In light of the analysis conducted in this Chapter, some conclusions can be drawn with regard to the applicability of the *jus in bello* to cyber operations. In general, cyber operations fall under the remit of the law of armed conflict when they occur in the context of an international or non-international armed conflict *and* qualify as acts of hostilities, or when they amount themselves to an international or non-international armed conflict. More specifically, the relevant *jus in bello* rules would apply to cyber operations in the following cases:

(1) if the cyber operations between belligerents are preceded by a declaration of war, whether or not kinetic hostilities follow;

(2) if the cyber operations are conducted by the belligerents against each other in an already existing international or non-international armed conflict or are otherwise conducted in support of a party to the conflict to the detriment of another, and cause the required threshold of harm;

(3) if the exchange of cyber operations between states amounts in itself to 'resort to armed force', ie they entail the use of cyber means or methods of warfare resulting in material damage to property, loss of life or bodily injury, or serious disruption of critical infrastructures;

(4) if an organized armed group conducts cyber operations amounting to protracted armed violence against a state or against another organized armed group;[262]

(5) if the cyber operations are conducted by the occupying state in the exercise of its policing and governance powers in occupied territory, or are part of the mounted resistance by the local population to the exercise of such powers;

(6) if the cyber operations accompany the resumption or continuation of kinetic hostilities in occupied territory and have a nexus with such hostilities, or amount themselves to the initiation, resumption, or continuation of an international or non-international armed conflict in occupied territory.[263]

Session, Commission on Human Rights, 51st Session, Provisional Agenda Item 19, UN Doc E/CN.4/1995/116, 31 January 1995, p 4. The Declaration was mentioned in the *Tadić*, Decision on the Defence Motion, para 119.

[260] Pictet (ed), *Commentary*, Vol 3, p 36.   [261] Pictet (ed), *Commentary*, Vol 3, pp 36–7.
[262] Whether the conflict falls under Common Art 3 of the Geneva Conventions or Additional Protocol II depends on their respective thresholds.
[263] In case of non-international armed conflicts in occupied territory, the threshold for the existence of this type of conflict must be met, whether by kinetic or cyber means.

The *jus in bello* classification of standalone cyber operations is summarized in the following table.

Table 3.1 Jus in bello *classification of standalone cyber operations*

	Cyber attacks causing material damage to property, loss of life or injury to persons	Cyber attacks causing severe disruption of critical infrastructures	Other cyber attacks	Cyber exploitation
Between states	IAC	IAC	No armed conflict	No armed conflict
Between a state and an organized armed group	NIAC (if the threshold of intensity is reached)	NIAC (if the threshold of intensity is reached)	No armed conflict	No armed conflict
Between a state and an organized armed group in the territory of another state	NIAC (if the threshold of intensity is reached)	NIAC (if the threshold of intensity is reached)	No armed conflict	No armed conflict
Between organized armed groups on the territory of one state	NIAC (if protracted enough)	NIAC (if protracted enough)	No armed conflict	No armed conflict
Between the Occupying Power and the Occupied State and/or a resistance movement belonging to a party to the conflict	IAC	IAC	No resumption or continuation of hostilities	No resumption or continuation of hostilities
Between the Occupying Power and an organized armed group in occupied territory	NIAC (if the threshold of intensity is reached)	NIAC (if the threshold of intensity is reached)	No armed conflict	No armed conflict
Between organized armed groups in occupied territory	NIAC (if the threshold of intensity is reached)	NIAC (if the threshold of intensity is reached)	No armed conflict	No armed conflict

Table 3.1 (Continued)

	Cyber attacks causing material damage to property, loss of life or injury to persons	Cyber attacks causing severe disruption of critical infrastructures	Other cyber attacks	Cyber exploitation
Between states and individuals or non-organized groups, or between non-organized groups	No armed conflict or resumption or continuation of hostilities	No armed conflict or resumption or continuation of hostilities	No armed conflict or resumption or continuation of hostilities	No armed conflict or resumption or continuation of hostilities

4
Cyber Operations and the Conduct of Hostilities

I. Introduction

This Chapter addresses the legal issues arising from the use of cyber warfare in armed conflict. As the *jus in bello* applies differently to international and non-international armed conflicts, the correct qualification of the conflict, as discussed in Chapter 3, is of fundamental importance in order to determine the applicable rules. International armed conflicts are regulated by the 1899 and 1907 Hague Conventions, the 1949 Geneva Conventions on the Protection of the Victims of War and their 1977 Additional Protocol I. On the other hand, non-international armed conflicts fall under the scope of application of Article 3 Common to the Geneva Conventions and of Protocol II Additional to the Geneva Conventions. There are also treaties on specific means of warfare and on the protection of specific objects, some of which expressly apply to both international and non-international armed conflicts.[1]

Under customary international law, however, while significant differences still exist with regard to the application of the law of detention, the legal regimes for international and non-international armed conflicts in relation to the conduct of hostilities have become more and more similar. Already in its first case, the ICTY affirmed that

> customary rules have developed to govern internal strife. These rules...cover such areas as protection of civilians from hostilities, in particular from indiscriminate attacks, protection of civilian objects, in particular cultural property, protection of all those who do not (or no longer) take active part in hostilities, as well as prohibition of means of warfare proscribed in international armed conflicts and ban of certain methods of conducting hostilities.[2]

With regard to the legality of weapons, in particular, the Court noted that '[w]hat is inhumane, and consequently proscribed, in international wars, cannot

---

[1] See, for instance, the 1954 Hague Cultural Property Convention and its Second Protocol of 1999, the 1997 Ottawa Convention on Anti-Personnel Mines (Art 1), the 1993 Chemical Weapons Convention (Art 1(1)), the 2008 Cluster Weapons Convention (Art 1(1)), the 1980 Convention on Certain Conventional Weapons (as amended in 2001), and Additional Protocol II to the Conventional Weapons Convention (Art 1(2)).

[2] ICTY, *Prosecutor v Tadić*, Case No IT–94–1, Decision on the Defence Motion for Interlocutory Appeals on Jurisdiction, 2 October 1995, para 127.

Introduction

but be inhumane and inadmissible in civil strife'.[3] More recently, the 2005 ICRC Study on *Customary International Humanitarian Law* identified 148 rules (out of 161) that apply to both international and non-international armed conflicts, as, according to the Study, state practice does not distinguish between the two types of conflict in the application of such rules.[4] As the ICTY itself warned, however, the extension of rules conceived for international armed conflicts to non-international ones 'has not taken place in the form of a full and mechanical transplant of those rules to internal conflict; rather, the general essence of those rules, and not the detailed regulation they may contain, has become applicable to internal conflicts'.[5] Throughout the present Chapter, therefore, the situations where a difference in legal regulation still remains will be highlighted and discussed separately.

While computers can be used in support of the application of the law of armed conflict (eg to maintain target data, estimate the best targeting route or weapon, or calculate collateral damage),[6] the high dependency of modern societies on computers, computer systems, and networks has also dramatically expanded the means and methods at the disposal of the belligerents to inflict damage on each other. As a consequence, the use of cyber operations in contemporary warfare is fast becoming strategically as important as airpower in traditional conflicts.[7] Of course, the rules on the conduct of hostilities apply to cyber operations only to the extent that they qualify as 'acts of hostilities'. The notion of 'hostilities' has already been discussed in Chapter 3.[8] It is sufficient to recall here that the ICRC *Interpretive Guidance on the Notion of Direct Participation in Hostilities* defines 'hostilities' as 'the (collective) resort by the parties to the conflict to means and methods of injuring the enemy'.[9] Hence, they include the resort to means or methods of cyber warfare in support of a belligerent that result, or are reasonably likely to result, in harm to another belligerent above the required threshold.[10] In particular, hostilities embrace not only violent acts, ie 'attacks',[11] but also any act that negatively affects the military

---

[3] *Tadić*, Decision on the Defence Motion, para 119.

[4] Jean-Marie Henckaerts and Louise Doswald-Beck, *Customary International Humanitarian Law* (Cambridge: Cambridge University Press, 2005), Vol I, p xxxv.

[5] *Tadić*, Decision on the Defence Motion, para 126.

[6] Michael N Schmitt, Heather A Harrison Dinniss, and Thomas C Wingfield, *Computers and War: The Legal Battlespace*, 2004, p 1, <http://www.hpcrresearch.org/sites/default/files/publications/schmittetal.pdf>. See Human Rights Watch, *Off Target. The Conduct of the War and Civilian Casualties in Iraq*, 2003, p 19, <http://www.hrw.org/reports/2003/12/11/target> ('U.S. air forces carry out a collateral damage estimate using a computer model designed to determine the weapon, fuze, attack angle, and time of day that will ensure maximum effect on a target with minimum civilian casualties').

[7] Robin Geiss and Henning Lahmann, 'Cyber Warfare: Applying the Principle of Distinction in an Interconnected Space', *Israel Law Review* 45 (2012), p 384.

[8] Chapter 3, Section II.3.1.1, p 129 ff.

[9] ICRC, *Interpretive Guidance on the Notion of Direct Participation in Hostilities under International Humanitarian Law*, Geneva, 2009 (prepared by Nils Melzer), p 43, http://www.icrc.org/eng/resources/documents/publication/p0990.htm. The Tallinn Manual also defines 'hostilities' as 'the collective application of means and methods of warfare' (*Tallinn Manual on the International Law Applicable to Cyber Warfare* (Cambridge: Cambridge University Press, 2013), p 82).

[10] On cyber capabilities and operations as means and methods of warfare, see Section II of this Chapter. On the threshold of harm, see Section III.1.3.d of this Chapter, pp 204–6.

[11] See Section III.1.1 of this Chapter, pp 129–32.

capabilities of the adversary, including 'operations designed to enhance one's own capabilities',[12] such as the operation of cyber defences to protect military objectives and the collection of tactical intelligence.

It is worth recalling that the ICJ found that the law of armed conflict 'applies to all forms of warfare and to all kinds of weapons, those of the past, those of the present and those of the future', in spite of the 'qualitative as well as quantitative difference' that there might be with traditional weapons,[13] and that the ICRC, the 'guardian' of international humanitarian law, recalled 'the obligation of all parties to conflicts to respect the rules of international humanitarian law if they resort to means and methods of cyberwarfare, including the principles of distinction, proportionality and precaution'.[14] The United States also emphasized that the principles of distinction, indiscriminate attacks and proportionality play an 'important role' in establishing the legality of cyber attacks in the context of an armed conflict and that 'targeting analysis would have to be conducted for information technology attacks just as it traditionally has been conducted for attacks using kinetic (conventional and strategic) weapons'.[15]

A problem specific to Additional Protocol I is that its Section I of Part IV on the general protection against the effects of hostilities only applies *qua* treaty law to 'land, air or sea warfare which may affect the civilian population, individual civilians or civilian objects on land' and 'to all attacks from the sea or from the air against objectives on land'.[16] The provision only expressly mentions land, air, or sea warfare: a literal reading may lead to conclude that cyber warfare is excluded. The purpose of Article 49(3), however, was to clarify whether air attacks on objectives on land fell within air or land warfare, and not to list exhaustively the domains of warfare to which the Protocol applies and exclude future scenarios: an evolutionary interpretation of the norm would then lead to include any other military domain that becomes available through technological developments, such as outer space and cyberspace, at least when civilians or civilian property on land are affected.[17]

---

[12] Michael N Schmitt, 'Cyber Operations and the *Jus in Bello*: Key Issues', *International Law Studies* 87 (2011), p 101.

[13] *Legality of the Threat or Use of Nuclear Weapons*, Advisory Opinion, 8 July 1996, ICJ Reports 1996 ('*Nuclear Weapons*'), para 86. The Court was of course referring to the applicability of the Hague and Geneva laws to nuclear weapons.

[14] UN General Assembly First Committee, 9th meeting, UN Doc A/C.1/66/PV.9, 11 october 2011, p 21.

[15] Comments submitted by the United States to the UN Secretary-General, UN Doc A/66/152, 15 July 2011 11 October 2011, p 19.

[16] Article 49(3). See Wolff Heintschel von Heinegg, 'The Current State of the Law of Naval Warfare: A Fresh Look at the *San Remo Manual*', *International Law Studies* 82 (2006), pp 278–9.

[17] According to the HPCR Manual on Air and Missile Warfare, however, 'as a general principle, the same legal regime applies equally in all domains of warfare (land, sea or air)' (HPCR, Manual on International Law Applicable to Air and Missile Warfare (Cambridge: Cambridge University Press, 2013), p 142). On the law of naval warfare, see *San Remo Manual on International Law Applicable to Armed Conflicts at Sea* (text in Adam Roberts and Richard Guelff, *Documents on the Laws of War*, 3rd edn (Oxford: Oxford University Press, 2000), pp 573 ff).

Having established that Article 49(3) of Additional Protocol I does not prevent the application of the Protocol in the cyber context, another possible obstacle could result from declarations such as that made by the United Kingdom on ratification of the Protocol, according to which 'the rules introduced by the Protocol apply exclusively to conventional weapons without prejudice to any other rules of international law applicable to other types of weapons'.[18] However, it is submitted that this and similar declarations, that focus on weapons rather than on domains of warfare, are irrelevant as far as means of cyber warfare are concerned, for several reasons. First, 'conventional weapons' does not mean 'non-traditional', but must rather be interpreted in opposition to weapons of mass destruction, in particular nuclear weapons, as it becomes clear if one reads the UK declaration in its entirety.[19] Secondly, the declaration only applies to the new rules introduced by Additional Protocol I: however, most of the provisions of the Protocol, and certainly its fundamental principles, now reflect customary international law. Finally, as has been seen, the United Kingdom and several other states have expressly declared that they accept that the law of armed conflict, including Additional Protocol I, applies to cyber warfare.[20]

The application in the cyber context of the existing law, conceived with kinetic weaponry in mind, presents however 'new and unique challenges that will require consultation and cooperation among nations'.[21] Several characteristics of cyber operations are likely to affect the application of the law on the conduct of hostilities: they are conducted remotely, they often produce effects almost instantaneously, they use essentially dual-use infrastructures, their effects on the infrastructures controlled by the information systems are often more relevant than the direct effects on the information itself, and their technology is easily accessible to anyone, not only to the military. On the basis of these characteristics, not all of which are necessarily unique to cyber operations, this Chapter will investigate how the existing law on the conduct of hostilities applies to cyber operations in the following manner. The analysis will first focus on whether cyber capabilities and operations are lawful means and methods of warfare for the conduct of hostilities. Section III will discuss how the law of targeting applies to and limits cyber operations amounting to 'attack' in armed conflict, while Section IV will examine cyber operations as acts of hostilities short of 'attack'. Finally, Section V will discuss cyber operations as a remedy against violations of the *jus in bello* by the belligerents. To be clear, the present Chapter will only deal with the use of force as such, and not with internment and detention. Furthermore, although it is accepted that international human rights law also applies as *lex generalis* in armed conflict, especially in non-international

---

[18] Roberts and Guelff, *Documents*, p 510. See also France's declaration upon ratification of Additional Protocol I, 11 April 2001, para 2, <http://www.icrc.org/applic/ihl/ihl.nsf/Notification.xsp?action=openDocument&documentId=D8041036B40EBC44C1256A34004897B2>.

[19] In the *Nuclear Weapons* Advisory Opinion, the ICJ made clear that the rules of Additional Protocol I that codify pre-existing customary international law apply 'to all means and methods of combat including nuclear weapons (*Nuclear Weapons*, para 84).

[20] See Chapter I, Section III.1, pp 21–2.

[21] UN Doc A/66/152, 15 July 2011, p 19.

ones and in situations of occupation,[22] the Chapter will exclusively focus on the application of the *lex specialis*, ie the *jus in bello*.

II. The Legality of Means and Methods of Cyber Warfare

The law of armed conflict takes a two-pronged approach to the regulation of hostilities: on the one hand, it prohibits certain means and methods of warfare, on the other it regulates the use of the lawful ones. The latter aspect will be discussed in Section III: it is only after establishing that a certain method or means of warfare is not inherently unlawful that it becomes necessary to investigate the legality of the modalities of its use. Methods of cyber warfare should be distinguished from means of cyber warfare. 'Methods of cyber warfare' are defined in Rule 41 of the Tallinn Manual as 'cyber tactics, techniques, and procedures by which hostilities are conducted'.[23] According to the Manual, methods 'denote more than those operations that rise to the level of an "attack"' and would, for instance, also include operations that disrupt enemy communications.[24] Examples are 'flood' attacks,[25] attacks aimed at taking over a system in order to assume control of a network and/or modulate connectivity, privileges or service, deception, psychological warfare or tactical intelligence gathering through cyber exploitation, cyber 'blockades'.

'Means of cyber warfare' are 'cyber weapons and their associated systems' and include 'any cyber device, materiel, instrument, mechanism, equipment, or software used, designed, or intended to be used to conduct a cyber attack'.[26] The Commentary to Rule 41 of the Tallinn Manual further explains that 'cyber weapons' are 'cyber means of warfare that are by design, use, or intended use capable of causing either (i) injury to, or death of, persons; or (ii) damage to, or destruction of, objects, that is, causing the consequences required for qualification of a cyber operation as an attack'.[27] Injury is considered as including 'serious illness and severe mental suffering'.[28] As has been seen, however, this definition is too narrow, as it does not include cyber tools causing loss of functionality without physical consequences (unless they also cause severe mental suffering) and is at odds, for instance, with US Air Force Instruction 51–402, which defines 'weapons' as 'devices designed to kill, injure, disable or *temporarily incapacitate* people,

---

[22] *Legal consequences of the construction of a wall in the occupied Palestinian territory*, Advisory Opinion, 9 July 2004, ICJ Reports 2005 ('*Legal Consequences of the Construction of a Wall*'), para 106.
[23] Tallinn Manual, p 141.   [24] Tallinn Manual, p 142.
[25] As already seen, 'flood attacks' aim to inundate the target with excessive calls, messages, enquiries, or requests in order to overload it and force its shut down.
[26] Tallinn Manual, pp 141, 142.
[27] Tallinn Manual, pp 141–2. Boothby also defines cyber weapons narrowly as 'any computer equipment or computer device that is designed, intended or used, in order to have violent consequences, that is, to cause death or injury to persons or damage or destruction of objects' (William H Boothby, 'Methods and Means of Cyber Warfare', *International Law Studies* 89 (2013), p 389).
[28] Tallinn Manual, p 108.

or destroy, damage or *temporarily incapacitate* property or materiel'.[29] A more accurate definition of 'cyber weapon' is 'computer code that is used, or designed to be used, with the aim of threatening or causing physical, functional, or mental harm to structures, systems, or living beings'.[30] Malware such as worms, viruses, botnet codes, time and logic bombs can be used as weapons in order to cause harm. On the other hand, means of warfare do not include cyberspace, ie the digital domain through which malware is delivered.[31] Trap doors and sniffers are more accurately qualified as cyber weapon systems, as they do not directly cause the harm but allow a cyber weapon to do so.[32]

Both means and methods of cyber warfare can produce multiple effects.[33] This has important consequences on the application of the principles of distinction and proportionality. As has been seen,[34] the primary effects of a cyber operation are the immediate results on the attacked computer system or network, ie the deletion, corruption or alteration of data, or system disruption through a DDoS attack or other cyber attacks. The secondary effects are those on the machine or infrastructure operated by the attacked system (if any), ie its physical damage or incapacitation. Tertiary effects are those on the persons affected by the destruction or incapacitation of the attacked system or infrastructure eg military or civilians that benefit from the electricity produced by a power plant incapacitated by a cyber operation. These effects can be permanent (if the operation results in data loss or physical damage), temporary (if data recovery is possible and functionality can be restored) or transient (when normal functioning resumes immediately after the end of the attack through rebooting or resetting the system).[35] The only cyber operation that is known to have caused secondary effects in the form of physical damage to a NCI is Stuxnet. On the other hand, the DDoS attack on Estonia and the attacks on Georgia only caused limited primary and tertiary effects. As to the 2012 cyber attack on Saudi Aramco, it caused significant primary effects, as it

---

[29] Air Force Instruction 51-402, 27 July 2011, p 6 <http://www2.gwu.edu/nsarchiv/NSAEBB/NSAEBB424/docs/Cyber-053.pdf>.

[30] Thomas Rid and Peter McBurney, 'Cyber-Weapons', *RUSI Journal* 157, no 1 (February 2012), p 7.

[31] Tallinn Manual, p 142.

[32] The US DoD defines a weapon system as '[a] combination of one or more weapons with all related equipment, materials, services, personnel, and means of delivery and deployment (if applicable) required for self-sufficiency' (US DoD, *Dictionary of Military and Associated Terms*, Joint Publication 1–02, 8 November 2010 (As Amended Through 15 August 2013), p 303).

[33] William A Owens, Kenneth W Dam, and Herbert S Lin, *Technology, Policy, Law, and Ethics Regarding U.S. Acquisition and Use of Cyberattack Capabilities* (Washington: The National Academies Press, 2009), p 80. See also Pia Palojärvi, *A Battle in Bits and Bytes: Computer Network Attacks and the Law of Armed Conflict* (Helsinki: Erik Castrén Institute of International Law, 2009), p 32; Boothby, 'Methods and Means', p 390.

[34] See Chapter II, Section II.1.1, p 52.

[35] Robert Fanelli and Gregory Conti, 'A Methodology for Cyber Operations Targeting and Control of Collateral Damage in the Context of Lawful Armed Conflict', in *2012 4th International Conference on Cyber Conflict* (2012), edited by Christian Czosseck, Rain Ottis, and Katharina Ziolkowski (Tallinn: CCDCOE, 2012), pp 323–4.

wiped off three-quarters of the data stored in the company's corporate computers.[36] Saudi Aramco was also forced to shut down its internal network to prevent the virus from spreading.

According to the ICRC, 'means and methods of warfare which resort to cyber technology are subject to IHL just as any new weapon or delivery system has been so far when used in an armed conflict by or on behalf of a party to such conflict'.[37] The Commentary to the HPCR *Manual on Air and Missile Warfare* confirms that '[m]eans of warfare include non-kinetic systems, such as those used in EW and CNAs. The means would include the computer and computer code used to execute the attack, together with all associated equipment'.[38] It further specifies that death, injury, damage or destruction 'need not result from physical impact... since the force used does not need to be kinetic. In particular, CNA hardware, software and codes are weapons that can cause such effects through transmission of data streams'.[39] In April 2013, the US Air Force upgraded six cyber capabilities to 'weapon' status.[40] Other states that have expressed the view that malware can be used as a weapon include Belarus, Cuba, Kazakhstan, Panama, Russia, Spain, and the United Kingdom.[41]

To the extent that they are means or methods of warfare, cyber capabilities and operations fall under the scope of application of Article 36 of Additional Protocol I.[42] The norm provides that

[i]n the study, development, acquisition or adoption of a new weapon, means or method of warfare, a High Contracting Party is under an obligation to determine whether its employment would, in some or all circumstances, be prohibited by this Protocol or by any other rule of international law applicable to the High Contracting Party.[43]

---

[36] Nicole Perlroth, 'In Cyberattack on Saudi Firm, U.S. Sees Iran Firing Back', *The New York Times*, 23 October 2012, <http://www.nytimes.com/2012/10/24/business/global/cyberattack-on-saudi-oil-firm-disquiets-us.html?pagewanted=all&_r=0>.

[37] ICRC, *International Humanitarian Law and the Challenges of Contemporary Armed Conflicts*, October 2011, pp 36–7, <http://www.icrc.org/eng/assets/files/red-cross-crescent-movement/31st-international-conference/31-int-conference-ihl-challenges-report-11-5-1-2-en.pdf>.

[38] HPCR Manual, p 31.

[39] HPCR Manual, p 49.

[40] Andrea Shalal-Esa, 'Six U.S. Air Force cyber capabilities designated "weapons"', Reuters, 8 April 2013, <http://www.reuters.com/article/2013/04/09/net-us-cyber-airforce-weapons-idUSBRE93801B20130409>.

[41] See Chapter 2, Section II.1, pp 50–1.

[42] See also Harold Koh, 'International Law in Cyberspace', Speech at the USCYBERCOM Inter-Agency Legal Conference, 18 September 2012, in CarrieLyn D Guymon (ed), *Digest of United States Practice in International Law*, 2012, p 596. See also Rule 48 of the Tallinn Manual, p 153. The Experts did not reach consensus on whether para (a) of the Rule extends to methods of cyber warfare (p 154).

[43] See also UK Ministry of Defence, *The Manual of the Law of Armed Conflict* (Oxford: Oxford University Press, 2004), p 119; The Federal Ministry of Defence of the Federal Republic of Germany, *Humanitarian Law in Armed Conflict*, ZDv 15/2, 1992, Section 405, <http://www.humanitaeres-voelkerrecht.de/ManualZDv15.2.pdf>. '[A]ny other rule of international law applicable to the High Contracting Party' refers to 'any agreement on disarmament concluded by the Party concerned, or any other agreement related to the prohibition, limitation or restriction on the use of a weapon or a particular type of weapon, concluded by this Party.... Naturally, it also includes the rules which form part of international customary law' (Yves Sandoz, Christophe Swinarski, and Bruno Zimmermann (eds), *Commentary on the Additional Protocols of 8 June 1977 to the Geneva Conventions*

Although it has not been included in the rules identified in the ICRC Study, it has been suggested that this obligation now reflects customary international law.[44] In fact, even states not parties to Additional Protocol I, like the United States, undertake legal reviews of new means and methods of warfare, including cyber ones.[45] There is no express obligation under conventional international humanitarian law to review new means and methods of warfare for use in non-international armed conflicts, but there are hardly any weapons that can be used only in one type of conflict: weapons being developed for use in a non-international armed conflict could also be potentially used in an international one, leading then to the application of Article 36.

The obligation to review new means and methods of warfare arises well before their use in armed conflict, ie as early as they are being 'studied'. In most cases, the review will be conducted in peacetime[46] and 'will include those items of equipment which, whilst they do not constitute a weapon as such, nonetheless have a direct impact on the offensive capability of the force to which they belong'.[47] There is no obligation on how to conduct such review or to disclose its results and the determinations of one state do not bind another.[48]

In order to establish the legality of means and methods of warfare, one has first to verify whether specific rules, either customary or conventional, allow or prohibit them.[49] It has already been seen in Chapter 1 that this is not the case of means and methods of cyber warfare.[50] In the absence of specific provisions, one has to

---

of 12 August 1949 (Dordrecht: Nijhoff, 1987), para 1472). Procedures to review the legality of weapons have been established in several states, including Australia, Belgium, Canada, France, Germany, the Netherlands, Norway, Sweden, the United Kingdom, and the United States (Boothby, 'Methods and Means', p 400). In 2006, the ICRC published a Guide on procedures to be undertaken in order to review new weapons as required by Art 36 of Additional Protocol I (ICRC, *A Guide to the Legal Review of New Weapons, Means, and Methods of Warfare: Measures to Implement Article 36 of Additional Protocol I of 1977*, text in *International Review of the Red Cross* 88 (2006), pp 931 ff). On weapons review and Art 36 of Additional Protocol I, see W Hays Parks, 'Conventional Weapons and Weapons Review', *Yearbook of International Humanitarian Law* 8 (2005), pp 55 ff.

[44] Duncan Blake and Joseph S Imburgia, '"Bloodless Weapons"? The Need to Conduct Legal Reviews of Certain Capabilities and the Implications of Defining Them as "Weapons"', *Air Force Law Review* 66 (2010), p 159.

[45] US Air Force Instruction 51–402, as updated in July 2011, for instance, requires a legal review of cyber capabilities used in cyber operations (Air Force Instruction 51–402 (27 July 2011)). The review includes establishing at a minimum: '3.1.1. Whether there is a specific rule of law, whether by treaty obligation of the United States or accepted by the United States as customary international law, prohibiting or restricting the use of the weapon or cyber capability in question. 3.1.2. If there is no express prohibition, the following questions are considered: 3.1.2.1. Whether the weapon or cyber capability is calculated to cause superfluous injury, in violation of Article 23(e) of the Annex to Hague Convention IV; and 3.1.2.2. Whether the weapon or cyber capability is capable of being directed against a specific military objective and, if not, is of a nature to cause an effect on military objectives and civilians or civilian objects without distinction' (p 3). On US practice in relation to weapons review, see Parks, 'Conventional Weapons', pp 113 ff.

[46] Blake and Imburgia, '"Bloodless Weapons"?', p 166.

[47] Justin McClelland, 'The Review of Weapons in Accordance With Article 36 of Additional Protocol 1', *International Review of the Red Cross* 85 (2003), p 405.

[48] Sandoz, Swinarski, and Zimmermann (eds), *Commentary*, paras 1469–70.

[49] *Nuclear Weapons*, para 51.   [50] See Chapter 1, Sections III.1 and III.2.

look at the general rules of the law of armed conflict.[51] In particular, a means or method of warfare would be unlawful if it is of a nature to cause superfluous injury or unnecessary suffering to combatants, it or its effects are indiscriminate, or is intended, or may be expected, to cause widespread, long-term, and severe damage to the natural environment.[52] It should be pointed out that a means or method of warfare is unlawful as such when it can never, under any circumstances, be used consistently with at least one of the above principles. Otherwise, it would be its use in specific situations, and not the means or method itself that is unlawful. Dinstein has effectively argued, '[t]he fact that certain weapons are used indiscriminately in a particular military engagement does not stain the weapons themselves with an indelible imprint of illegality, since in other operations the same weapons may be employed within the framework of LOIAC.[53]

According to the ICJ, the prohibition of superfluous injury or unnecessary suffering is a cardinal principle of international humanitarian law that protects combatants.[54] The prohibition first appeared in the Preamble of the 1868 St Petersburg Declaration and is contained, with slight language differences, in Article 23(e) of the Hague Regulations and Article 35(2) of Additional Protocol I. It reflects customary international law applicable to both international and non-international armed conflicts,[55] and is incorporated in Rule 42 of the Tallinn Manual with regard to means and methods of cyber warfare.[56] The fact that the suffering caused by the weapon is extensive is not sufficient to outlaw it: the suffering must be 'unnecessary', ie causing 'a harm greater than that unavoidable to achieve legitimate military objectives'.[57] However, as the ICRC concedes, 'views differ on how it can actually

---

[51] *Nuclear Weapons*, para 51. According to the Tokyo District Court, '[i]t can naturally be assumed that the use of a new weapon is legal, as long as international law does not prohibit it. However, the prohibition in this context is to be understood to include not only the case where there is an express rule of direct prohibition, but also the case where the prohibition can be implied *de plano* from the interpretation and application by analogy of existing rules of international law (customary international laws and treaties). Further, the prohibition must be understood also to include the case where, in the light of principles of international law which are at the basis of these positive rules of international law, the use of a new weapon is deemed to be contrary to these principles, for there is no reason why the interpretation of rules of international law should be limited to literal interpretation, any more than the interpretation of rules of municipal law' (*Ryuichi Shimoda et al v The State*, District Court of Tokyo, Judgment of 7 December 1963, *International Law Reports* 32 (1963), pp 628–9).

[52] See UK Ministry of Defence, *The Manual of the Law of Armed Conflict*, pp 103–4.

[53] Yoram Dinstein, *The Conduct of Hostilities under the Law of International Armed Conflict*, 2nd edn (Cambridge: Cambridge University Press, 2010), p 62.

[54] *Nuclear Weapons*, para 78.

[55] Rule 70, in Henckaerts and Doswald-Beck, *Customary International Humanitarian Law*, Vol I, p 237. The injury and suffering cover not only physical injury but also severe psychological harm (Dinstein, *The Conduct*, p 64).

[56] Tallinn Manual, p 143. It is unclear, however, whether the prohibition of unnecessary suffering applies only to means or also to methods of warfare. The extension to methods of warfare is supported by the German *Military Manual* (German Ministry of Defence, *Humanitarian Law in Armed Conflicts*, Section 402). According to Akande, the limitation to means of warfare is intentional, as in targeting 'IHL has already made the calculation as to what is necessary from the military perspective' (Dapo Akande, 'Clearing the Fog of War? The ICRC's Interpretive Guidance on Direct Participation in Hostilities', *International and Comparative Law Quarterly* 59 (2010), p 192).

[57] *Nuclear Weapons*, para 78.

be determined that a weapon causes superfluous injury or unnecessary suffering'.[58] While certain authorities require a balance between the military advantage gained from the use of the weapon in question and the suffering caused,[59] others resort to a comparative approach: '[s]uffering is unnecessary when, in the circumstances, another practicable military means, causing less suffering to the adverse combatants, could have been used to place the adversary *hors de combat*'.[60] A third approach focuses on the effects of the weapons and the principle of humanity: when the suffering they cause are 'repugnant to the public conscience', the weapon is unlawful 'whatever might be the military advantage sought to be achieved'.[61] It has also been suggested that factors like whether the weapon renders death inevitable or leads to permanent disabilities must be considered.[62] Whatever approach is preferred, only exceptionally would a weapon cause 'unnecessary' suffering in *all* circumstances.[63] The ICRC Commentary of Article 35(2) of Additional Protocol I provides some of the 'not very numerous' examples.[64]

Article 51(4)(b) and (c) of Additional Protocol I also prohibit indiscriminate means and methods of warfare, ie those 'which cannot be directed at a specific military objective; or...which employ a method or means of combat the effects of which cannot be limited as required by this Protocol; and consequently, in each such case, are of a nature to strike military objectives and civilians or civilian objects without distinction'.[65] This rule, which has been incorporated in Rule 43 of the Tallinn Manual,[66] reflects customary international law and applies to both

---

[58] Henckaerts and Doswald-Beck, *Customary International Humanitarian Law*, Vol I, p 240.

[59] See *Nuclear Weapons*, Separate Opinion of Judge Guillaume, para 5; Separate Opinion of Judge Higgins, para 14.

[60] Robert Kolb and Richard Hyde, *An Introduction to the International Law of Armed Conflicts* (Oxford and Portland: Hart, 2008), p 155. See also Dinstein, *The Conduct*, p 65.

[61] *Nuclear Weapons*, Dissenting Opinion of Judge Shahabuddeen, p 403.

[62] Henckaerts and Doswald-Beck, *Customary International Humanitarian Law*, Vol I, p 241.

[63] Giulio Bartolini, 'Armed Forces and the International Court of Justice: The Relevance of International Humanitarian Law and Human Rights Law to the Conduct of Military Operations', in *Armed Forces and International Jurisdictions*, edited by Marco Odello and Francesco Seatzu (Cambridge, Antwerp, and Portland: Intersentia, 2013), p 72.

[64] The ICRC's examples include explosive bullets and projectiles filled with glass, 'dum-dum' bullets, bullets of irregular shape or with a hollowed out nose, poison and poisoned weapons, any substance intended to aggravate a wound, asphyxiating or deleterious gases, bayonets with a serrated edge, lances with barbed heads (Sandoz, Swinarski, and Zimmermann (eds), *Commentary*, para 1419). See also Henckaerts and Doswald-Beck, *Customary International Humanitarian Law*, Vol I, pp 243–4. With regard to dum-dum bullets, the Elements of Crimes in the ICC Statute require the perpetrator to be 'aware that the nature of the bullets was such that their employment would *uselessly* aggravate suffering or the wounding effect', thus suggesting that, in certain situations, the aggravated suffering deriving from the use of these weapons would not be useless (Art 8(2)(b)(xix)(3); emphasis added). See Roy S Lee and Hakan Friman (eds), *The International Criminal Court—Elements of Crimes and Rules of Procedure* (Ardsley, NY: Transnational, 2001), p 181.

[65] While sub-para (b) refers to the case of a weapon that cannot be targeted at a specific military objective, sub-para (c) focuses on its effects, whether or not the weapon is capable to strike specific military objectives. It is not clear if the principle of proportionality also affects the evaluation of the indiscriminate nature of a weapon: see William H Boothby, *Weapons and the Law of Armed Conflict* (Oxford: Oxford University Press, 2009), p 79.

[66] Tallinn Manual, pp 144–5.

international and non-international armed conflicts.[67] The ICJ has confirmed that '[s]tates must never make civilians the object of attack and must consequently never use weapons that are incapable of distinguishing between civilian and military targets'.[68] It has been observed, however, that 'few weapons can be regarded as intrinsically "incapable of distinguishing between civilians and military targets"'.[69] Examples that have been made are long-range missiles with faulty or no guidance systems and biological weapons.[70]

According to Article 35(3) of Additional Protocol I, '[i]t is prohibited to employ methods or means of warfare which are intended, or may be expected, to cause widespread, long-term and severe damage to the natural environment'. Article 55(1) adds 'a prohibition of the use of methods or means of warfare which are intended or may be expected to cause such damage to the natural environment and thereby to prejudice the health or survival of the population'.[71] Whether or not these provisions reflect customary international law is a matter of debate.[72] While Article 35(3) protects the environment per se, the ultimate goal of Article 55(1) is to protect human beings against the effects of hostilities, as evidenced by the addition of a further element (the prejudice to the health or survival of the population).[73] Apart from this, however, the two provisions present more similarities than differences as they both prohibit 'warfare which is intended, or may be expected to have widespread, long-term and severe effects on the natural environment'.[74] If methods or means of warfare necessarily caused 'widespread, long-term and severe damage', they would then be prohibited regardless of any considerations of military necessity or proportionality.[75] There is no indication in Additional Protocol I of what 'widespread', 'long-term', or 'severe' mean, but the threshold is very high.

---

[67] Rule 71, Henckaerts and Doswald-Beck, *Customary International Humanitarian Law*, Vol I, p 244; Michael N Schmitt, Charles HB Garraway, and Yoram Dinstein, *The Manual on the Law of Non-International Armed Conflict With Commentary* (Sanremo: International Institute of Humanitarian Law, 2006), Rule 2.2.1.1, p 29, <http://www.iihl.org/iihl/Documents/The%20Manual%20on%20 the%20Law%20of%20NIAC.pdf>.

[68] *Nuclear Weapons*, para 78.

[69] Yoram Dinstein, 'The Principle of Distinction and Cyber War in International Armed Conflicts', *Journal of Conflict and Security Law* 17 (2012), p 262.

[70] Dinstein, *The Conduct*, p 62.

[71] On the protection of the environment in armed conflict, see the contributions in the special issues of the *Nordic Journal of International Law* 82 (2013), pp 1 ff. See also Karine Mollard-Bannelier, *La protection de l'environnement en temps de conflit armé* (Paris: Pedone, 2001).

[72] See Section III.3 of this Chapter, pp 229–30, n 427.

[73] 'Health' is used in a broad sense and the connection with 'survival' means that temporary, short-term and non-serious effects are not contemplated by the provision (Michael Bothe, Karl J Partsch, and Waldemar A Solf, *New Rules for Victims of Armed Conflicts: Commentary on the Two 1977 Protocols Additional to the Geneva Conventions of 1949* (The Hague: Nijhoff, 1982), pp 346–7). The fact that the term 'population' is not preceded by the adjective 'civilian' entails that the provision aims to protect the population in general, including combatants (p 346).

[74] Liesbeth Lijnzaad and Gerard J Tanja, 'Protection of the Environment in Times of Armed Conflict: The Iraq-Kuwait War', *Netherlands International Law Review* 40 (1993), p 179. The United Kingdom has declared that it 'shall interpret [Article 35(3)] in the same way as [Article 55] which in our view is a fuller and more satisfactory formulation' (quoted in Sandoz, Swinarski, and Zimmermann (eds), *Commentary*, p 420, footnote 131).

[75] Michael Bothe, 'The Protection of the Environment in Times of Armed Conflict', *German Yearbook of International Law* 34 (1991), p 56.

It appears from the *travaux préparatoires*, for instance, that 'long-term' should be interpreted as referring to a period of years or decades.[76] The 1992 German Military Manual also maintains that ' "widespread", "long-term", and "severe" damage to the natural environment is a major interference with human life or natural resources which considerably exceeds the battlefield damage to be regularly expected in a war'.[77] According to Rogers, '[a]n examination of the various commentaries on Protocol I leads one to infer that "severe" means prejudicing the continued survival of the civilian population or involving the risk of major health problems and that "widespread" means more than the standard of several hundred square kilometres considered in connection with the ENMOD [Environmental Modification] Convention'.[78]

Are cyber capabilities inherently indiscriminate, and/or of a nature to cause superfluous injury or unnecessary suffering, and/or always intended or expected to cause widespread, long-term, and severe damage to the natural environment? If so, they would be unlawful means and methods of warfare that could never be used in the conduct of hostilities. As the Commentary to Rule 48 of the Tallinn Manual explains, the information necessary to conduct such assessment 'includes a technical description of the cyber means or method, the nature of the generic targets it is to engage, its intended effect on the target, how it will achieve this effect, its precision and ability to distinguish the target system from any civilian systems with which it is networked, and the scope of intended effects'.[79] Whether or not cyber capabilities and operations are lawful means or methods of warfare, however, is a question that cannot be answered *in abstracto*. Malware may be designed to spread indiscriminately, a cyber operation might alter medical records so that combatants will receive wrong medical treatment thus causing unnecessary suffering, or cyber weapons might disable the cooling system of a nuclear reactor and cause the release of radioactive substances with consequent widespread, severe, and long-term damage to the natural environment. The Tallinn Manual also makes the example of a cyber operation that takes control of an 'Internet-addressable pacemaker device with a built-in defibrillator' of an enemy combatant in order to cause unnecessary suffering, for instance by stopping and re-starting the heart several times before finally killing the target.[80] Malware that disrupts the air traffic control system may also not be able to distinguish between civil and military aircraft. But malware may also be introduced into a closed military network,[81] or be written so as to negatively affect exclusively certain systems.[82] While Stuxnet was promiscuous, for

---

[76] Official Records of the Diplomatic Conference on the Reaffirmation and Development of International Law Applicable in Armed Conflicts, Vol 15, p 360.
[77] German Ministry of Defence, *Humanitarian Law in Armed Conflicts*, Section 403.
[78] APV Rogers, *Law on the Battlefield*, 2nd edn (Manchester: Manchester University Press, 2004), p 171.
[79] Tallinn Manual, pp 155–6.   [80] Tallinn Manual, p 144.
[81] Commentary to Rule 43, in Tallinn Manual, p 146.
[82] Owens, Dam, and Lin, *Technology, Policy, Law, and Ethics*, p 123. It is also likely that, when launching and monitoring the execution of a cyber attack, the operator will be sitting in an office surrounded by senior military officials and military lawyers and will therefore be better advised on what actions to undertake.

instance, it made itself inert if the specific Siemens software used at Iran's Natanz uranium enrichment plant was not found on infected computers, and contained safeguards to prevent each infected computer from spreading the worm to more than three others, before self-destructing on 24 June 2012.[83] In addition, it caused no more than inconvenience to infected computers other than the Natanz operating system, as the worm did not self-replicate indefinitely so as to slow down computer functions.[84] 'Flood attacks' are also a method of cyber warfare that is perfectly discriminate, as it only affects the system targeted by the multiple requests. For these reasons, one cannot but conclude that cyber means and methods of warfare are so diverse and their effects so depending on the circumstances, including the characteristics of the targeted system, that a legal review can only be conducted on each individual capability and that, in most cases, it will be *how* the means or method is used, more than the means or method itself, that may be incompatible with the law of armed conflict. It is to this aspect that the analysis now turns.

III. The Law of Targeting

Even when a cyber operation employs a lawful means or method of cyber warfare, it must still be carried out consistently with the law applicable to the conduct of hostilities: one thing is the use of unlawful means or methods of warfare, another is the unlawful use of a lawful means or method of warfare. It is, therefore, essential to determine *how* the lawful means and methods of warfare may be used consistently with the law. To this aim, the present Section analyses the rules that cyber operations must comply with when conducted by the belligerents in armed conflict, in particular the law of targeting, which identifies what and who may be attacked, and how.[85] The analysis will first determine to what cyber operations the law of targeting applies and will then investigate the issues arising from the application of the principles of distinction and proportionality, as well as from the duty to take active and passive precautions, to those operations.

1. The obligation to direct attacks exclusively against military objectives

While Article 51(1) of Protocol I Additional to the Geneva Conventions generally states that '[t]he civilian population and individual civilians shall enjoy general

---

[83] Jeremy Richmond, 'Evolving Battlefields: Does Stuxnet Demonstrate a Need for Modifications to the Law of Armed Conflict?', *Fordham International Law Journal* 35 (2011–12), p 856.

[84] Richmond, 'Evolving Battlefields', p 861.

[85] The US Joint Doctrine defines 'targeting' as 'the process of selecting and prioritizing targets and matching the appropriate response to them, considering operational requirements and capabilities' (*Joint Targeting*, Joint Publication 3-60, 31 January 2013, p I-1, <http://www.jfsc.ndu.edu/schools_programs/jc2ios/io/student_readings/1F4_jp3-60.pdf>). The rules on targeting also apply to cyber operations by which a party takes control of enemy weapons or weapon systems (Tallinn Manual, p 105).

protection against dangers arising from military operations', Article 48 more specifically provides for the 'basic' obligation of the belligerents to 'at all times distinguish between the civilian population and combatants and between civilian objects and military objectives and accordingly [to] direct their operations only against military objectives'. This obligation of distinction also applies in non-international armed conflicts, as can be inferred from Article 3(1) Common to the Geneva Conventions, which prohibits 'violence to life and person' on those 'taking no active part in hostilities',[86] and more explicitly from Article 13(2) of Additional Protocol II (even though only with regard to civilians and the civilian population, not civilian objects).[87] The ICC Statute also criminalizes '[i]ntentionally directing attacks against the civilian population as such or against individual civilians not taking direct part in hostilities' in a non-international armed conflict.[88]

The customary status of the principle of distinction is well established.[89] On 19 December 1969, the UN General Assembly adopted Resolution 2444 (XXIII) by unanimous vote, which expressly recognized the principle of civilian immunity and its corollary requiring the warring parties to distinguish civilians from combatants at all times.[90] The United States acknowledged that the Resolution, which does not distinguish between different domains of warfare, is declaratory of customary international law.[91] The customary nature of the principle of distinction has also been firmly upheld by national and international courts: in particular, according to the 1996 ICJ's Advisory Opinion on the *legality of the threat or use of nuclear weapons*, the obligation to distinguish between combatants and non-combatants is an 'intransgressible' principle of customary international law, to be observed by all states whether or not they have ratified the conventions that contain it.[92] According to the ICTY, 'it is now a universally recognised principle... that deliberate attacks on civilians or civilian objects are absolutely prohibited by international humanitarian law'.[93] The ICTY has also in several cases affirmed that, as far as customary international law is concerned, the prohibition of attacks on civilians applies both in international and non-international armed conflicts.[94]

---

[86] Rogers, *Law on the Battlefield*, p 225.

[87] Rule 10 of the ICRC Study on *Customary International Humanitarian Law* on the protection of civilian objects against attack, however, applies both to international and non-international armed conflicts (Henckaerts and Doswald-Beck, *Customary International Humanitarian Law*, Vol I, p 34).

[88] Article 8(2)(e)(i).

[89] Henckaerts and Doswald-Beck, *Customary International Humanitarian Law*, Vol I, p 3. See also Christopher Greenwood, 'Customary Law Status of the 1977 Additional Protocols', in *Humanitarian Law of Armed Conflict—Challenges Ahead*, edited by Astrid JM Delissen and Gerard J Tanja (Dordrecht, Boston, and London: Nijhoff, 1991), p 108; US Department of Defense, *Conduct of the Persian Gulf War: Final Report to Congress (1992)*, International Legal Materials 31 (1992), pp 621–2.

[90] GA Res 2444 (XXIII), 19 December 1968. The preamble to this Resolution states that these fundamental humanitarian law principles apply 'in all armed conflicts', both international and non-international.

[91] Arthur W Rovine, 'Contemporary Practice of the United States Relating to International Law', *American Journal of International Law* 67 (1973), p 122.

[92] *Nuclear Weapons*, paras 78–9.

[93] ICTY, *Prosecutor v Kupreškić*, Case No IT–96–16–T, Trial Chamber Judgment, 14 January 2000, para 521. See also EECC, *Partial Award, Western Front, Aerial Bombardment and Related Claims Eritrea's Claims 1, 3, 5, 9–13, 14, 21, 25, and 26*, 19 December 2005, RIAA, Vol XXVI, Part VIII, para 95.

[94] *Tadić*, Decision on the Defence Motion, para 127.

While both Article 48 and Article 51(1) refer to the broader notion of 'military operations', which includes 'all movements and acts related to hostilities that are undertaken by armed forces',[95] the ICRC Commentary makes clear that the expression should be interpreted 'in the context of the whole of the Section' as referring only 'to military operations during which violence is used', ie 'attacks'.[96] This is confirmed by the language of Articles 51(2) and 52(1) of Additional Protocol I (that give effect to the principle of civilian immunity), according to which the civilians, the civilian population and civilian objects 'shall not be the object of *attack*'.[97] With regard to other military operations, only a more general obligation of 'constant care...to spare the civilian population, civilians and civilian objects' applies.[98] Rule 1 of the ICRC Study on *Customary International Humanitarian Law*, which incorporates the principle of distinction, also refers to 'attacks', and not 'military operations'.[99] What is prohibited, then, is to make civilians and civilian property the object of a direct 'attack', ie to use violence against them *qua* civilians. If protected persons and objects are incidentally hit in the context of an attack on a military objective, the attack would not be inconsistent with the principle of distinction, although, to be lawful, it will have to comply with the principle of proportionality.[100]

Accordingly, only cyber operations amounting to 'attacks' as defined in Article 49(1) of Additional Protocol I are subject to the principle of distinction.[101] The problem is to determine when cyber operations can be qualified as such: this is discussed in the next sub-Section.

1.1 When does a cyber operation amount to an 'attack'?

'Attack' is defined in Article 49(1) of Additional Protocol I as 'acts of violence against the adversary, whether in offence or in defence'. In other words, attacks are only those military operations amounting to acts of hostilities that are characterized by 'violence'.[102] Unlike other acts of hostilities, non-violent military harm is not sufficient: the ICRC Commentary of Article 49 explains that 'attack' involves 'combat action'.[103] A belligerent's cyber operation that deletes, corrupts, or alters data

---

[95] Sandoz, Swinarski, and Zimmermann (eds), *Commentary*, para 1875. The *Commentary* subsequently rephrases the definition as 'any movements, manœuvres and other activities whatsoever carried out by the armed forces with a view to combat' (para 2191). Inconsistently, the Commentary of Art 13 of Additional Protocol II defines 'military operations' more narrowly as 'movements of attack or defence by the armed forces in action' (para 4769).
[96] Sandoz, Swinarski, and Zimmermann (eds), *Commentary*, para 1875. See Michael N Schmitt, 'Wired Warfare: Computer Network Attack and *Jus in Bello*', *International Law Studies* 76 (2002), pp 193–4; David Turns, 'Cyber War and the Concept of "Attack" in International Humanitarian Law', in *International Humanitarian Law and the Changing Technology of War*, edited by Dan Saxon (Leiden: Brill, 2013), p 217; Geiss and Lahmann, 'Cyber Warfare', p 2.
[97] Emphasis added.   [98] Article 57(1) of Additional Protocol I.
[99] Henckaerts and Doswald-Beck, *Customary International Humanitarian Law*, Vol I, p 3.
[100] See Section III.2.1 of this Chapter.
[101] Koh, 'International Law in Cyberspace', p 595. See also Rules 31–32, in Tallinn Manual, pp 110, 113.
[102] See Stefan Oeter, 'Methods and Means of Combat', in *The Handbook of International Humanitarian Law*, 3rd edn, edited by Dieter Fleck (Oxford: Oxford University Press, 2013), p 166.
[103] Sandoz, Swinarski, and Zimmermann (eds), *Commentary*, para 1880.

on troop displacement or military plans stored in the computers of the adversary would therefore not be an 'attack', although it is an act of hostilities. Consistently with what has been argued in Chapter 2 in relation to the use of armed force,[104] it is not the author, the target, or the intention that define an 'act of violence'. Rather, a cyber operation amounts to an 'attack' in the sense of Article 49(1) of Additional Protocol I when it employs means or methods of warfare that have or are reasonably likely to result in violent effects.[105] There is general agreement that, if a cyber operation causes or is likely to cause loss of life, injury to persons or more than minimal material damage to property, it is an 'attack' and the principle of distinction fully applies. As explained in the Commentary to Rule 1(e) of the HPCR *Manual on Air and Missile Warfare*, for instance, '[t]he definition of "attacks" also covers "non-kinetic" attacks (i.e. attacks that do not involve the physical transfer of energy, such as certain CNAs…) that result in death, injury, damage or destruction of persons or objects'.[106] Rule 30 of the Tallinn Manual also defines a 'cyber attack' as 'a cyber operation, whether offensive or defensive, that is reasonably expected to cause injury or death to persons or damage or destruction to objects'.[107] Had it been conducted in the context of an armed conflict between Iran and those responsible for the cyber operation, Stuxnet would have been an example of such 'attack' because of the damage allegedly caused to the centrifuges of the Natanz uranium enrichment facility. The relevant violent effects of a cyber attack include 'any reasonably foreseeable consequential damage, destruction, injury, or death', whether or not the computer system is damaged or data corrupted.[108] If the attack is intercepted and the reasonably expected violent effects do not occur, or occur to a lesser degree, the operation would still qualify as an 'attack' for the purposes of Article 49(1).[109]

If it is generally accepted that cyber operations that result or are reasonably likely to result in physical damage to property, loss of life or injury to persons amount to 'attack', the problem is more complicated with regard to operations that disrupt the functionality of infrastructures without causing material damage. The above-quoted Rule 30 of the Tallinn Manual appears to exclude that such operations qualify as 'attacks'.[110] Similarly, according to the Commentary to Rule 1(e) of the

---

[104] See Chapter 2, Section II.1, pp 49–50.

[105] Interpreting Art 49 of Additional Protocol I in the light of its context and purpose, Michael Schmitt arrives at the same conclusion by focusing exclusively on the consequences of the act: he suggests that violence should not be referred to the means employed, but rather to the consequences of the act, ie human physical or mental suffering, destruction of physical property or 'permanent loss of assets, for instance money, stock, etc., directly transferable into tangible property' (Schmitt, 'Wired Warfare', p 194). Turns suggests to combine the consequence-based and the intention-based approaches: 'if consequences of physical harm or damage are intended, the action amounts to an "attack"' (Turns, 'Cyber War', p 225). Establishing the 'intention' of states, however, is notoriously difficult, and is even more in the cyber context.

[106] HPCR Manual, p 12.

[107] Tallinn Manual, p 106. The Manual includes 'serious illness and severe mental suffering' in the notion of 'injury' (p 108).

[108] Tallinn Manual, p 107.   [109] Tallinn Manual, p 110.

[110] Dinstein also excludes that 'breaking through a computer's "fire wall"; planting a "worm" in digital software; extracting secret data; gaining control over codes; and disrupting communications' are as such attacks in the sense of international humanitarian law (Dinstein, 'The Principle of Distinction',

HPCR Manual on International Law Applicable to Air and Missile Warfare, 'the term "attack" does not encompass CNAs that result in an inconvenience (such as temporary denial of internet access)'.[111] Therefore, 'a CNA which interferes with air traffic control but does not cause any "death, injury, damage or destruction" does not qualify as an attack'.[112] The majority of the experts that drafted the Tallinn Manual, however, maintained that disruptive cyber operations may be 'attacks' 'if restoration of functionality requires replacement of physical components'.[113] The problem with this view, which still relies on the occurrence of physical damage, is that the attacker may not be able to know in advance whether the restoration of functionality will require replacement of physical components or mere reinstallation of the operating system: the attacker could claim, therefore, that it was not aware that it was conducting an 'attack' to which the law of targeting applied.

The limits of the doctrine of kinetic equivalence, which requires the occurrence of physical consequences, become evident if one considers that, under the Tallinn and HPCR Manuals' approach, a cyber attack that shuts down the national grid or erases the data of the entire banking system of a state would not be an 'attack', while the physical destruction of one server would.[114] Other commentators have therefore tried to extend the notion of 'attack' to include at least some disruptive cyber operations. Dörmann, for instance, recalls that the definition of 'military objective' in Article 52(2) of Additional Protocol I mentions not only destruction but also 'neutralization' of the object and concludes that, when the object (person or property) is civilian, '[i]t is irrelevant whether [it] is disabled through destruction or in any other way'.[115] Therefore, the incapacitation of an object, such as a civilian power station, without destroying it would still qualify as an 'attack'. Fleck concurs and argues that it would be 'less than convincing to insist that the term "attacks" should be limited to acts directly causing injury or physical destruction, when the same action can, eg lead to disrupt essential supplies for hospitals or other important civilian infrastructure'.[116] Melzer adopts a different approach to reach the same conclusion and argues that the principles of distinction, proportionality, and precautions apply not to 'attacks', but rather to the broader notion of 'hostilities': therefore, 'the applicability of the restraints imposed by IHL on the conduct of hostilities to cyber operations depends not on whether the operations

---

p 264). What is needed, in his view, is 'death/injury to human beings or more than nominal damage to property' (Yoram Dinstein, 'Cyber War and International Law: Concluding Remarks at the 2012 Naval War College International Law Conference', *International Law Studies* 89 (2013), p 284).

[111] HPCR Manual, p 13.   [112] HPCR Manual, p 114.
[113] Tallinn Manual, p 108.
[114] Rain Liivoja and Tim McCormack, 'Law in the Virtual Battlespace: The Tallinn Manual and the *Jus in Bello*', Melbourne Legal Studies Research Paper no 650, 23 July 2013, p 8, <http://papers.ssrn.com/sol3/papers.cfm?abstract_id=2297159>. However, as, according to the Manual, 'injury' includes severe mental suffering, if the deletion of the bank system's data causes panic among the civilian population, it could still be an 'attack' (p 9).
[115] Knut Dörmann, *Applicability of the Additional Protocols to Computer Network Attacks*, p 6, <http://www.icrc.org/eng/assets/files/other/applicabilityofihltocna.pdf>.
[116] Dieter Fleck, 'Searching for International Rules Applicable to Cyber Warfare—A Critical First Assessment of the New Tallinn Manual', *Journal of Conflict and Security Law* 18 (2013), p 341.

in question qualify as "attacks" (that is, the predominant form of conducting hostilities), but on whether they constitute part of the "hostilities" within the meaning of IHL'.[117] According to this view, cyber operations disrupting the enemy radar system would not amount to 'attack' because of the lack of violent consequences, but, as an act of hostilities, they would still be subject to the restrictions imposed by international humanitarian law on the choice and use of methods and means of warfare.[118] A similar view is put forward by Harrison Dinniss, who claims that even non-violent operations may not be conducted against civilian objects, as Article 48 refers to 'military operations' and not only to 'attacks': 'wherever an activity takes place in conjunction with hostilities it must be restricted to military objectives'.[119] As has been seen, however, it has been persuasively demonstrated that the rules contained in Part IV, Section I of Additional Protocol I essentially apply to 'attacks' and not to 'hostilities' or 'military operations'.

It is submitted that a better way of including at least certain disruptive cyber operations in the definition of 'attack' under Article 49(1) of Additional Protocol I is to interpret the provision taking into account the recent technological developments and to expand the concept of 'violence' to include not only material damage to objects, but also incapacitation of infrastructures without destruction. This is suggested by Panama in its views on cyber security submitted to the UN Secretary-General, where it qualifies cyber operations as a 'new form of violence'.[120] Indeed, as has already been observed,[121] the dependency of modern societies on computers, computer systems, and networks has made it possible to cause significant harm through non-destructive means. After all, if the use of graphite bombs, which spread a cloud of extremely fine carbon filaments over electrical components, thus causing a short-circuit and a disruption of the electrical supply, would undoubtedly be considered an 'attack' even though it does not cause more than nominal physical damage to the infrastructure, one cannot see why the same conclusion should not apply to the use of viruses and other malware that achieve the same effect. It is, however, only those cyber operations that go beyond mere inconvenience and cause at least functional harm to structures or systems[122] that can qualify as 'attacks' in the sense of Article 49(1).

According to Article 49(1) of Additional Protocol I, an attack has to be 'against the adversary'. As has been seen, most NCIs are not owned by the government, but by the private sector: the governmental or private character of the targeted infrastructure is, however, not relevant to the determination of the existence of an attack 'against the adversary'. Already the Israeli Supreme Court had established that acts which by nature and objective are intended to cause damage either to the

---

[117] Nils Melzer, *Cyberwarfare and International Law*, UNIDIR, 2011, p 27, <http://www.isn.ethz.ch/Digital-Library/Publications/Detail/?lng=en&id=134218>.
[118] Melzer, *Cyberwarfare*, pp 27–8.
[119] Heather Harrison Dinniss, *Cyber Warfare and the Laws of War* (Cambridge: Cambridge University Press, 2012), p 200.
[120] UN Doc A/57/166/Add.1, 29 August 2002, p 5.   [121] See Chapter 2, Section II.1.2.
[122] Rid and McBurney, 'Cyber-Weapons', p 7.

army or to civilians are acts of hostilities.[123] As the ICRC *Interpretive Guidance on the Notion of Direct Participation in Hostilities* correctly explains, '[t]he phrase "against the adversary" does not specify the target, but the belligerent nexus of an attack, so that even acts of violence directed specifically against civilians or civilian objects may amount to direct participation in hostilities'.[124] It is also irrelevant that the cyber attack is conducted in offence or in defence: if it qualifies as an 'act of violence' against the adversary, in both cases it will be submitted to the rules on attacks, including the principles of distinction, proportionality and precautions.[125]

The principle of distinction, as applicable to attacks, is composed of two parts: a prohibition of direct attacks against civilian objects and a prohibition of direct attacks against civilians and the civilian population. The next sub-Section will focus primarily on the former, while Section 1.3 will look at the specific aspects of targeting individuals.

1.2 The definition of 'military objective'

The principle of distinction requires that 'attacks' be directed exclusively at military objectives. The first definition of 'military objective' to appear in a legal text can be found in the 1923 Hague Rules of Aerial Warfare: 'an object of which the destruction or injury would constitute a distinct military advantage to the belligerent'.[126] To clarify the definition, the Rules provided an illustrative list of military objectives.[127] The Rules, however, have never been adopted in treaty form. No definition appears in the 1949 Geneva Conventions, although the term is employed.[128] On the other hand, 'military objectives' are expressly defined in Article 52(2) of the 1977 Additional Protocol I as

---

[123] *Public Committee Against Torture in Israel et al v The Government of Israel et al*, Israel's Supreme Court, HCJ 769/02, 11 December 2005 ('*Targeted Killings*'), para 33 (per Judge Barak).

[124] ICRC, *Interpretive Guidance*, p 49 (footnote omitted). See also Commentary to Rule 30, in Tallinn Manual, p 108.

[125] Sandoz, Swinarski, and Zimmermann (eds), *Commentary*, para 1880.

[126] Article 24(1) (text in Roberts and Guelff, *Documents*, pp 141 ff). Without referring to the notion of military objective, Art 2 of the 1907 Hague Convention IX Concerning Bombardment by Naval Forces in Time of War contains a list of objects that may be destroyed.

[127] Article 24(2). The list includes military forces; military works; military establishments or depots; factories constituting important and well-known centres engaged in the manufacture of arms, ammunition or distinctively military supplies; lines of communication or transportation used for military purposes. It is doubtful whether the list is exhaustive (Rogers, *Law on the Battlefield*, p 60).

[128] See Arts 4, 19(2) of Geneva Convention I and Arts 4, 18(5) of Geneva Convention IV. The 1956 New Delhi Draft Rules for the Limitation of the Dangers Incurred by the Civilian Population in Time of War, drafted by the ICRC, proposed a list of military objectives, to be reviewed at intervals of no more than ten years by a group of experts; however, even if an object had belonged to one of the listed categories, it would not have been a military objective if its total or partial destruction, in the circumstances ruling at the time, had offered no military advantage (Art 7; text at <http://www.icrc.org/ihl/INTRO/420?OpenDocument>). Another attempt to define the concept of 'military objective' was made by the Institute of International Law in 1969 (*Annuaire de l'Institut de droit international* (1969–II), p 359).

those objects which by their nature, location, purpose or use make an effective contribution to military action and whose total and partial destruction, capture or neutralization, in the circumstances ruling at the time, offers a definite military advantage.

This definition has been incorporated into several military manuals[129] and, in spite of the unclear position of certain states like the United States (which will be discussed below), it is largely thought to reflect customary international law.[130] The definition is also applicable in non-international armed conflicts.[131]

If one applies the above definition of 'military objective' to targeting in the cyber context, however, several problems arise. First of all, what 'objects' are relevant? Cyber operations can be directed at cyber targets, ie data, software, or networks, and/or hard targets, ie information hardware (eg computers, servers, routers, fibre-optic cables, and satellites), physical infrastructures, or persons,[132] in order to produce the primary, secondary, or tertiary effects described above. When the cyber operation aims to cause material damage to physical property or persons or incapacitation of infrastructures, or such effects are foreseeable, the attacked 'object' is not only, and not mainly, the information itself, but rather the persons, property or infrastructure attacked *through* cyberspace.[133] In the case of Stuxnet, for instance, the relevant 'object' was not the Siemens software that operated the centrifuges at the Natanz uranium enrichment facility in Iran, but the centrifuges themselves. Commentators have debated whether data are per se an 'object' for the purpose of Article 52(2) of Additional Protocol I.[134] The Experts that drafted the Tallinn Manual did not manage to achieve consensus on this point so no solution was incorporated in the black-letter rules.[135] The problem should not be overestimated. As has been said, if the cyber operation deletes, corrupts or alters data in order to cause damage to or disrupt the functioning of an infrastructure, it is such infrastructure that is the intended 'object' of the attack. Similarly, if the cyber operation deletes or alters medical records, so that patients receive the wrong treatment, it is those individuals that are (also) targeted. If, on the other hand, the cyber operation only results in the corruption, deletion, or alteration of data without

---

[129] Henckaerts and Doswald-Beck, *Customary International Humanitarian Law*, Vol II, pp 183 ff.

[130] See Rule 8 of the ICRC Study, in Henckaerts and Doswald-Beck, *Customary International Humanitarian Law*, Vol I, p 29.

[131] Rule 8 of the ICRC Study, in Henckaerts and Doswald-Beck, *Customary International Humanitarian Law*, Vol I, p 29; Art 2(6) of the 1980 Protocol on Prohibitions or Restrictions on the Use of Mines, Booby-Traps and Other Devices and Art 1(f) of the 1999 Protocol II of the Hague Convention for the Protection Cultural Property in the Event of Armed Conflict. See also Schmitt, Garraway, and Dinstein, *The Manual on the Law of Non-International Armed Conflict*, Rule 1.1.4, p 3; UK Ministry of Defence, *The Manual of the Law of Armed Conflict*, p 391; Dieter Fleck, 'The Law of Non-international Armed Conflicts', in *The Handbook*, edited by Fleck, p 591.

[132] Hard targets can be attacked both by kinetic or cyber means, while software and data can be attacked only by cyber means (Karl F Rauscher and Andrey Korotkov, 'Working Towards Rules for Governing Cyber Conflict. Rendering the Geneva and Hague Conventions in Cyberspace', EastWest Institute, January 2011, p 19).

[133] Commentary to Rule 30, in Tallinn Manual, p 108.

[134] Schmitt, 'Cyber Operations', p 96; Noam Lubell, 'Lawful Targets in Cyber Operations: Does the Principle of Distinction Apply?', *International Law Studies* 89 (2013), pp 267 ff.

[135] Commentary to Rule 38, in Tallinn Manual, p 127.

consequences in the analogue world, it will not be an 'attack' in the sense discussed above, and the law of targeting and the notion of 'military objective' will therefore not apply, whether or not the data are an 'object'.

(a) 'effective contribution to military action'
According to Article 52(2), two cumulative elements must be present for an 'object' to be a military objective and therefore targetable: it must effectively contribute to military action *and* its total or partial destruction, capture or neutralization, in the circumstances ruling at the time, must offer a definite military advantage. When infrastructures are attacked by affecting their operating system, these two elements must be tested against such infrastructures: the fact that they are attacked through cyberspace and not from land, sea, or air neither changes the requirements for their qualification as military objectives nor raises special issues, at least when civilians on land are affected.

Article 52(2) indicates the criteria to evaluate whether the object effectively contributes to military action, ie nature, location, purpose, or use.[136] Effective contribution to military action by nature characterizes those objects which are inherently military and cannot be employed but for military purposes, for instance computers designed specifically to be used as components of weapon systems or to facilitate logistic operations.[137] Other examples include military command, communication, and control networks used for the transmission of orders or tactical data and military air defence networks.[138] The premises from where the military cyber operations are conducted (such as US CYBERCOM headquarters at Fort Mead or the 12-storey building in the Pudong New Area of Shanghai which is allegedly the home of the People's Liberation Army's Unit 61398)[139] are also military objectives by nature.[140] An example of effective contribution by use would be a server normally used for civilian purposes which is taken over by the military,

---

[136] Confusingly, the Commentary to Rule 38 of the Tallinn Manual makes the example of a cyber operation against a website that inspires 'patriotic sentiments' among the population as a case of non-effective contribution to military action (Tallinn Manual, p 130): however, such an operation would not be an 'attack' in the sense of either Art 49(1) of Additional Protocol I or Rule 30 of the Manual itself.

[137] Dinstein, 'Cyber War', pp 284–5. As has been observed, however, it is normally the software rather than the hardware that turns a computer into a military objective (Dinstein, 'The Principle of Distinction', p 263).

[138] Responses to advance questions, Nomination of Lt Gen Keith Alexander for Commander, US Cyber Command, US Sen Armed Serv Committee, 15 April 2010, p 13, <http://armed-services.senate.gov/statemnt/2010/04%20April/Alexander%2004-15-10.pdf>. The three USD internal networks, for instance, would be examples of networks that are military objectives by nature. In particular, the Secret Internet Protocol Router Network (SIPRNet), which is not connected to the internet, is used for classified information and to transmit military orders, while the Joint Worldwide Intelligence Communications System (JWICS) is used to communicate intelligence information to the military. On the three DoD networks, see Richard A Clarke and Robert K Knake, *Cyber War. The Next Threat to National Security and What To Do About It* (New York: Harpercollins, 2010), pp 171–3.

[139] Mandiant, *APT1. Exposing One of China's Cyber Espionage Units*, February 2013, p 11 <http://intelreport.mandiant.com/Mandiant_APT1_Report.pdf>.

[140] Davis Turns, 'Cyber Warfare and the Notion of Direct Participation in Hostilities', *Journal of Conflict and Security Law* 17 (2012), p 297.

even if it is used for non-combat purposes.[141] If the server is about to be used by the military but this has not occurred yet, it may be a military objective by purpose.[142] As to military objectives by location, the Commentary to Rule 38 of the Tallinn Manual makes the example of a cyber attack on a water reservoir's SCADA system to cause the release of water and thus prevent the use of a certain area by the enemy.[143]

The use of an object by the military is then sufficient to make it a military objective (providing that its destruction or neutralization also offer a definite military advantage in the circumstances ruling at the time). Most cyber infrastructures, however, are dual-use, ie at the same time used by civilians and the military. It is well known, for instance, that about 98 per cent of US government communications travel through civilian-owned or civilian-operated networks.[144] Servers, fibre-optic cables, satellites, and other physical components of cyberspace are also almost entirely dual-use, as well as most technology and software used in this field: everyday applications such as web browser, e-mail client and even command line (cmd.exe) can be used as an instrument for cyber attacks. The advent of cloud computing, where military and civilian data are stored remotely side by side, is nothing but the latest manifestation of the dual-use character of information technology.[145] The fact that an object is *also* used for civilian purposes does not affect its qualification under the principle of distinction: if the two requirements provided in Article 52(2) of Additional Protocol I are present, the object is a military objective but the neutralization of its civilian component needs to be taken into account when assessing the incidental damage on civilians and civilian property under the principle of proportionality.[146] What is prohibited is to attack the dual-use cyber infrastructure *because* of its civilian function or to attack a dual-use facility where the anticipated concrete and direct military advantage of the attack is outweighed by the expected civilian damage and/or injury. It should be recalled that, under Article 52(3), '[i]n case of doubt whether an object which is normally dedicated to civilian purposes...is being used to make an effective contribution to military action, it shall be presumed not to be so used'.[147] Unlike its counterpart with

---

[141] Dinstein, 'Cyber War', p 285.

[142] According to the ICRC Commentary, purpose is 'the intended future use of an object, while... use is concerned with its present function' (Sandoz, Swinarski, and Zimmermann (eds), *Commentary*, para 2022; emphasis omitted).

[143] Tallinn Manual, p 128.

[144] Geiss and Lahmann, 'Cyber Warfare', p 386.

[145] Jensen has for instance claimed that 'Microsoft Corporation Headquarters in Washington State is a valid dual-use target, based on the support it provides to the U.S. war effort by facilitating U.S. military operations' (Eric Talbot Jensen, 'Unexpected Consequences From Knock-On Effects: A Different Standard from Computer Network Operations?', *American University International Law Review* 18 (2002–03), p 1160). However, he eventually denies that it is a lawful military objective because of doubts with regard to the military advantage that can be gained from its destruction or neutralization (pp 1167–8).

[146] See Rule 39, Tallinn Manual, p 134. As has been observed, an 'object becomes a military objective even if its military use is only marginal compared to its civilian use' (Cordula Droege, 'Get Off My Cloud: Cyber Warfare, International Humanitarian Law, and the Protection of Civilians', *International Review of the Red Cross* 94 (2012), p 563).

[147] The provision only applies when doubt concerns the use of the object, not its nature, location or purpose (William H Boothby, *The Law of Targeting* (Oxford: Oxford University Press, 2012), p 71).

regard to persons (Article 50(1) of Additional Protocol I), the customary status of this provision is, however, dubious: it is, for instance, not included in the ICRC Study on *Customary International Humanitarian Law*.[148] It has also been observed that satellites, cables, routers, and servers are not 'normally dedicated to civilian purposes', as they are widely used by the military.[149]

The effective contribution must be to 'military action'. 'Military action' has a broad meaning that corresponds to the 'general prosecution of the war'.[150] The United States' definition of 'military objective', however, is wider than that contained in Additional Protocol I, since it covers all objects which 'effectively contribute to the enemy's war-fighting or war sustaining capability'.[151] If 'war fighting' can be considered equivalent to 'military action', 'war sustaining' is much broader and includes activities not directly connected to the hostilities: it would therefore allow attacks aimed to incapacitate political and economic targets in order to 'persuade' the enemy to stop fighting.[152]

The US definition of 'military objective' is also reflected in the cyber context. According to the US Air Force's *Cornerstones of Information Warfare*, the United States 'may target any of the adversary's information functions that have a bearing on his will or capability to fight'.[153] This view would, for instance, legitimize attacks like the 2012 cyber operations against Saudi Aramco, the world's largest oil producer, which destroyed the data of about 30,000 company computers and, according to Saudi Arabia, targeted the country's economy with the purpose of preventing the pumping of oil into domestic and international markets.[154] The US CYBERCOM Head, General Keith Alexander, has also declared that power grids, banks, and other financial institutions and networks, transportation-related networks, and national telecommunication networks are 'all potential targets of military attack, both kinetic and cyber, under the right circumstances', although only when 'used solely to support enemy military operations'.[155]

This expanded notion of military objective is at odds with the definition contained in Article 52(2) of Additional Protocol I, which is largely considered to

---

[148] Rule 40 of the Tallinn Manual also does not prohibit attacks in case of doubt about the nature of the object, but limits itself to require that 'a careful assessment' is previously made (Tallinn Manual, p 137).

[149] Geiss and Lahmann, 'Cyber Warfare', p 386.

[150] Rogers, *Law on the Battlefield*, p 67.

[151] *The Commander's Handbook on the Law of Naval Operations*, July 2007, para 8.2, <http://www.usnwc.edu/getattachment/a9b8e92d-2c8d-4779-9925-0defea93325c>.

[152] For critical comments of the US position and the documents in which it appears, see Giulio Bartolini, 'Air Operations Against Iraq (1991 and 2003)', in *The Law of Air Warfare—Contemporary Issues*, edited by Natalino Ronzitti and Gabriella Venturini (Utrecht: Eleven Publishing, 2006), pp 235–6.

[153] Department of the Air Force, *Cornerstones of Information Warfare*, 17 April 1997, p 3, footnote 5, <http://www.dtic.mil/cgi-bin/GetTRDoc?AD=ADA323807>. It also appears that China sees cyber operations against financial systems, power generation, transmission facilities, and other NCIs as part of a conflict with another state (Owens, Dam, and Lin, *Technology, Policy, Law, and Ethics*, p 333).

[154] 'Saudi Aramco says cyber attack targeted kingdom's economy', *Al Arabiya News*, 9 December 2012, <http://www.alarabiya.net/articles/2012/12/09/254162.html>. Oil production, however, remained uninterrupted.

[155] Responses to advance questions, Nomination of Lt Gen Keith Alexander, p 13.

reflect customary international law. It should be noted, however, that the 1976 US Air Force Pamphlet incorporated a definition of military objective analogous to that contained in the Protocol.[156] The subsequent 1998 USAF Intelligence Targeting Guide also incorporates the Protocol's definition to the letter,[157] and the 1997 edition of the Report on US practice in international law notes that '[t]he *opinio juris* of the U.S. government recognizes the definition of military objectives in Article 52 of Additional Protocol I as customary law', although it adds that 'United States practice gives a broad reading to this definition, and would include areas of land, objects screening other military objectives, and war-supporting economic facilities as military objectives'.[158]

(b) 'definite military advantage'

Even objects that, because of their nature, use, purpose, or location, effectively contribute to military action are not, as such, military objectives unless their total or partial destruction, capture or neutralization, in the circumstances ruling at the time, are militarily necessary, ie offer a 'definite military advantage'.[159] Article 52(2) envisages not only the total or partial destruction of the attacked object, but also its capture or neutralization, which includes 'an attack for the purpose of denying the use of an object to the enemy without necessarily destroying it'.[160] These words fit cyber operations that incapacitate but do not destroy infrastructures like a glove. As the ICRC has observed, 'the fact that a cyber operation does not lead to the destruction of an attacked object is...irrelevant': the definition of military objective, which refers to neutralization, 'implies that it is immaterial whether an object is disabled through destruction or in any other way'.[161]

The advantage must be of a military nature and 'definite', ie not speculative or indirect: the Commentary explains that 'it is not legitimate to launch an attack which only offers potential or indeterminate advantages'.[162] Shutting down the computer system operating the adversary's air defences would, for instance, provide an evident 'definite' military advantage. By contrast, a cyber attack aimed at demoralizing the civilian population would be unlawful. The problem with establishing the definite military advantage requirement in the cyber context is

---

[156] *Air Force Pamphlet 110–31, International Law—The Conduct of Armed Conflict and Air Operations*, 1976, para 5–3(b)(1).
[157] US Air Force, *Intelligence Targeting Guide*, Air Force Pamphlet 14–210 Intelligence, 1 February 1998, para 1.7.1, <http://www.fas.org/irp/doddir/usaf/afpam14-210/part17.htm>.
[158] Cited in Henckaerts and Doswald-Beck, *Customary International Humanitarian Law*, Vol II, p 188.
[159] Boothby, *The Law of Targeting*, p 103. DeSaussure refers to the examples of the 1972 Christmas bombing of Hanoi or the never implemented bombing of a depot in the heart of Argentina during the Falklands war, which would have not helped the British reoccupy the islands ('The Sixth Annual American Red Cross-Washington College of Law Conference on International Humanitarian Law: A Workshop on Customary International Law and the 1977 Protocols Additional to the 1949 Geneva Conventions', *American University Journal of International Law and Policy* 2 (1987), Remarks of Professor Hamilton DeSaussure, p 513).
[160] Bothe, Partsch, and Solf, *New Rules*, p 325.
[161] ICRC, *International Humanitarian Law and the Challenges*, p 37.
[162] Sandoz, Swinarski, and Zimmermann (eds), *Commentary*, para 2024.

that measurement of effects can often be difficult: it is still not confirmed, for instance, whether Stuxnet did destroy any centrifuges at Natanz and, if so, with what consequences on the Iranian nuclear programme (while Iran denied that the incident caused significant damage, the IAEA reported that Iran stopped feeding uranium into thousands of centrifuges:[163] it is however unclear whether this was due to Stuxnet or to technical malfunctions inherent to the equipment used).[164] The fact that it might not be evident whether a definite military advantage was effectively gained does not necessarily deprive the object of its qualification as a military objective, as long that the attacker had a reasonable expectation that the intended results would occur.

As will be seen in the context of the principle of proportionality, according to the NATO states the definite military advantage is what results from the attack as a whole, and not from its isolated parts. The NATO reservation, made in relation to Articles 51 and 57 of Additional Protocol I, should also apply to Article 52(2).[165] The destruction, capture or neutralization of the object must also offer a definite military advantage 'in the circumstances ruling at the time'. This excludes any potential future advantage but also implies that an object which could not normally be considered a military objective may become one if it is used in direct support of the hostilities, and vice versa. The time between the identification of the target, the planning of the attack and the execution of the attack must therefore be reasonably short, as the circumstances could rapidly change and an object that qualified as a military objective at a certain time may subsequently turn into a civilian one.[166]

(c) Is the internet a military objective?

The internet can be disrupted by attacking its hardware or software components. The former type of attack targets servers, fibre-optic cables and other physical internet infrastructure used to ensure connectivity, while the latter affects systems like the Domain Name System (DNS), which translates domain names into IP addresses: if the DNS is compromised, the web browser would not know where to direct the visit. The China Internet Network Information Centre (CNNIC), for instance, reported that the national domain name resolution registry came under a series of a sustained DDoS attacks on 25 August 2013, which interrupted or slowed down traffic.[167] A cyber operation against either the hardware or software components of the internet would qualify as an 'attack' in the sense of Article 49(1) of Additional Protocol I if material damage or significant loss of functionality of infrastructure ensue.

---

[163] William J Broad, 'Report Suggests Problems with Iran's Nuclear Effort', *The New York Times*, 23 November 2010, <http://www.nytimes.com/2010/11/24/world/middleeast/24nuke.html>.

[164] Katharina Ziolkowski, *Stuxnet—Legal Considerations*, NATO CCDCOE, 2012, p 5.

[165] Oeter, 'Methods and Means', p 176.

[166] Luisa Vierucci, 'Sulla nozione di obiettivo militare nella guerra aerea: recenti sviluppi della giurisprudenza internazionale', *Rivista di diritto internazionale* 89 (2006), p 708.

[167] James Vincent, 'Chinese domains downed by "largest ever" cyber-attack', *The Independent*, 27 August 2013, <http://www.independent.co.uk/life-style/gadgets-and-tech/news/chinese-domains-downed-by-largest-ever-cyberattack-8786091.html>.

The internet is a computer network, which is a type of communication network: the question whether or not the internet is a military objective, then, can first be approached reasoning by analogy with more traditional means of communication.[168] The 1956 ICRC list of military objectives includes 'the lines and means of communication, installations of broadcasting and television stations, telephone and telegraph exchanges of fundamental military importance'.[169] Military manuals also include 'communication installations used for military purposes' as an example of military objective.[170] Communication nodes have been a high priority in all recent armed conflicts. Media and broadcasting systems were included in the target list both in Operation Desert Storm and in Operation Allied Force.[171] In the former case, the attacks were justified by the United States not only on the ground that the facilities were part of the military communications network, but also because they were used for Iraqi propaganda.[172] On 23 April 1999, NATO aircraft bombed the headquarters of the Radio Television of Serbia (RTS) in Belgrade.[173] According to the Organization, it was a lawful target, since the station was used for military purposes as part of the control mechanism and of the propaganda machinery of the Milošević government.[174] The *ICTY Final Report* concluded that the attack was lawful because it was aimed mainly at disabling the Serbian military command and control system and at destroying the apparatus that kept Milošević in power.[175] On 12 November 2001, the Kabul office of Al-Jazeera news television was hit by a guided bomb during Operation Enduring Freedom,[176] and other radio/television stations were attacked because they were used as means of propaganda by the Taliban.[177] In Operation Iraqi Freedom, the United States bombed the Ministry of Information, the Baghdad Television Studio and Broadcast Facility and the Abu Ghraib Television Antennae Broadcast Facility.[178] Unlike in the 1991 operation,

---

[168] Louise Doswald-Beck, 'Some Thoughts on Computer Network Attack and the International Law of Armed Conflict', *International Law Studies* 76 (2002), p 167. See also Knut Dörmann, *Computer Network Attack and International Humanitarian Law*, 19 May 2001, para 15 <http://www.icrc.org/eng/resources/documents/misc/5p2alj.htm>.

[169] The list is reprinted in *Final Report to the Prosecutor by the Committee Established to Review the NATO Bombing Campaign Against the Federal Republic of Yugoslavia*, 8 June 2000, para 39, <http://www.icty.org/x/file/Press/nato061300.pdf>.

[170] See eg UK Ministry of Defence, *The Manual of the Law of Armed Conflict*, p 57.

[171] William J Fenrick, 'Attacking the Enemy Civilian as a Punishable Offence', *Duke Journal of Comparative and International Law* 7 (1996–97), p 70.

[172] Bartolini, 'Air Operations', pp 244–5.

[173] See George H Aldrich, 'Yugoslavia's Television Studios as Military Objectives', *International Law Forum du droit international* 1 (1999), pp 149–50.

[174] Amnesty International, *'Collateral Damage' or Unlawful Killings? Violations of the Laws of War by NATO during Operation Allied Force*, June 2000, AI-Index EUR 70/18/00, p 47.

[175] *ICTY Final Report*, para 76. This conclusion is criticized by Paolo Benvenuti, 'The ICTY Prosecutor and the Review of the NATO Bombing Campaign against the Federal Republic of Yugoslavia', *European Journal of International Law* 12 (2001), pp 522–4.

[176] Marc W Herold, 'A Dossier on Civilian Victims of United States' Aerial Bombing of Afghanistan: A Comprehensive Accounting', March 2002, <http://www.cursor.org/stories/civilian_deaths.htm>.

[177] Robert Cryer 'The Fine Art of Friendship: *Jus in Bello* in Afghanistan', *Journal of Conflict and Security Law* 7 (2002), p 55.

[178] Human Rights Watch, *Off Target. The Conduct of the War and Civilian Casualties in Iraq*, 2003, pp 46–9.

however, it seems that in 2003 the emphasis in the legal justification of the attacks was more on the facilities being part of the military communications network than on their use to spread propaganda.[179] The antennas of Libya's state broadcaster were also attacked by NATO aircraft during the 2011 Operation Unified Protector.[180] The attack had the purpose 'of degrading Qadhafi's use of satellite television as a means to intimidate the Libyan people and incite acts of violence against them' and was motivated on the fact that 'TV was being used as an integral component of the regime apparatus designed to systematically oppress and threaten civilians and to incite attacks against them'.[181]

Although, as a means of communication, the internet could potentially qualify as a military objective, it would still have to meet the two requirements of Article 52(2). If internet disruption had the sole purpose of stopping propaganda, undermining civilian morale or psychologically harassing the population, its neutralization would not offer a 'definite' military advantage (even though it might weaken the political support for the enemy government).[182] On the other hand, if the internet had become part of the adversary's military communication system, it would effectively contribute to military action, but, if connection can be easily and promptly restored, it can be doubted that its neutralization or destruction would provide a definite military advantage. As has been observed, 'an attacker nowadays must probably destroy a network of telecommunication *in toto* (or at least its central connection points) in order to paralyse the command and control structures of the enemy armed forces, which in themselves clearly constitute a legitimate military objective'.[183] This may be particularly difficult to achieve in the case of the internet, which is characterized by a high level of resilience: if certain channels become unusable as a consequence of a cyber or kinetic attack on servers, the data flow will simply find another path to reach its destination and it might well be that the destruction or neutralization of certain internet infrastructure has no practical effect at all.[184] Connection would probably slow down, but it would still continue through mirror servers, mobile phones, or satellites.[185]

---

[179] Bartolini, 'Air Operations', p 245.

[180] Giulio Bartolini, 'Air Targeting in Operation *Unified Protector* in Libya. Jus ad bellum and IHL Issues: An External Perspective', in *Legal Interoperability and Ensuring Observance of the Law Applicable in Multinational Deployments*, edited by Stanislas Horvat and Marco Benatar (Brussels: International Society for Military Law and the Law of War, 2013), p 273.

[181] 'NATO strikes Libyan state TV satellite facility', Statement by the Spokesperson for NATO Operation Unified Protector, Colonel Roland Lavoie, regarding air strike in Tripoli, 30 July 2011, <http://www.nato.int/cps/en/natolive/news_76776.htm>.

[182] *ICTY Final Report*, para 55.

[183] Oeter, 'Methods and Means', p 174. The *ICTY Final Report on the NATO bombing campaign against Yugoslavia* emphasized that, even if the RTS building in Belgrade was considered a military objective (*rectius*: effectively contributed to military action), broadcasting was interrupted only for a brief period and in any case Yugoslavia's command and control network, of which the RTS building was allegedly a part, could not be disabled with a single strike (*ICTY Final Report*, para 78).

[184] Geiss and Lahmann, 'Cyber Warfare', p 388.

[185] Resilience does not, however, mean invulnerability. For instance, 88 per cent of Egyptian internet access was shut down as a consequence of the withdrawal of 3,500 Border Gateway Protocol (BGP) routes by Egyptian ISPs (Christopher Williams, 'How Egypt shut down the internet', *The Telegraph*, 28 January 2011, <http://www.telegraph.co.uk/news/worldnews/africaandindianocean/egypt/8288163/How-Egypt-shut-down-the-internet.html>).

The *ICTY Final Report on the NATO bombing campaign against Yugoslavia* suggests that a broadcasting station may also constitute a military objective when it is employed to incite the population to commit war crimes or crimes against humanity as in the case of Radio Mille Collines in Rwanda in 1994.[186] As has been seen, this argument was also used by NATO to justify the attack on Libya's state television in 2011, although the justification was offered more for the purposes of the Operation's protective mandate than from an international humanitarian law perspective.[187] Although it is doubtful that these views are consistent with the *lex lata*, as propaganda or incitement to commit crimes do not amount to 'effective contribution to military action',[188] it cannot be excluded that, in parallel with the developments in international criminal law, a customary international law rule is emerging that allows attacks against these heinous uses of means of communication. If this is the case, and should the internet be used for such purposes, connectivity could be disrupted through a kinetic or cyber operation against its components.

The internet, however, is not only a means of communication, but also an important economic resource. As the German Military Manual recalls, only economic objectives that make an effective contribution to military action may be considered lawful targets,[189] and the same view is contained in the 1998 USAF *Intelligence Targeting Guide*.[190] According to the 2002 US *Joint Doctrine for Targeting*, however, lawful targets also include economic facilities that '*indirectly but effectively* support and sustain the enemy's warfighting capability'.[191] The EECC also affirmed that '[t]he infliction of economic losses from attacks against military objectives is a lawful means of achieving a definite military advantage' and that 'there can be few military advantages more evident than effective pressure to end an armed conflict'.[192] While it is accepted that certain economic targets, ie those 'which make an effective contribution to military action (transport facilities, industrial plants, etc.)' are military objectives,[193] the Commission seems to justify attacks against *any* economic target.[194] This view goes too far and is not consistent with the definition of 'military objective' contained in Additional Protocol I, which reflects customary international law.

It is worth noting that, even if the internet qualified in certain situations as a military objective, the attacker would still have to take into account the disruption caused to its civilian function and to neutrals. The possibility of shutting down specific segments, websites, or networks should therefore be explored first.[195]

---

[186] *ICTY Final Report*, para 55. [187] Bartolini, 'Air Targeting', p 274.
[188] See Bartolini, 'Air Targeting', pp 274–5.
[189] German Ministry of Defence, *Humanitarian Law in Armed Conflicts*, Section 443.
[190] US Air Force, *Intelligence Targeting Guide*, para A4.2.2.1.
[191] *Joint Doctrine for Targeting*, Joint publication 3-60 (17 January 2002), Appendix A, p A–3 <http://www.bits.de/NRANEU/others/jp-doctrine/jp3_60(02).pdf> (emphasis added).
[192] EECC, *Partial Award, Western Front*, para 121.
[193] German Ministry of Defence, *Humanitarian Law in Armed Conflicts*, Section 443. See also UK Ministry of Defence, *The Manual of the Law of Armed Conflict*, p 57.
[194] Vierucci, 'Sulla nozione', p 707. [195] Commentary to Rule 39, in Tallinn Manual, p 136.

1.3 Targetable individuals

Article 51(2) of Additional Protocol I prescribes that '[t]he civilian population as such, as well as individual civilians, shall not be the object of attack'.[196] Combatants, on the other hand, are military objectives. In the Hague law, the expression 'combatant' in international armed conflicts is defined by what the person in question does, ie whether he is engaging in hostilities: a military staff officer based at a rear headquarters, then, would not be a 'combatant', although he would be entitled to POW status in case of capture.[197] In Article 43(2) of Additional Protocol I, 'combatant' becomes a status term based on membership of the armed forces: all members of the armed forces, as defined in Article 43(1) of the Protocol, are combatants regardless of what they do, with the exception of medical and religious personnel.[198] Combatants in international armed conflicts can be 'privileged' or not, depending on whether they are entitled to combatant immunity, ie not to be prosecuted for taking up arms against the enemy, and to POW status upon capture, which depends on the fulfilment of certain requirements.[199] Although there is no combatant status in non-international armed conflicts, the term 'combatant' is sometimes used in the generic sense of someone who is taking direct part in the hostilities. The ICRC Study on *Customary International Humanitarian Law* argues, for instance, that '[f]or purposes of the principle of distinction…, members of State armed forces may be considered combatants in both international and non-international armed conflicts',[200] and then specifies that the expression 'combatant' in non-international armed conflicts 'is only used in its generic meaning

---

[196] The provision adds that '[a]cts or threats of violence the primary purpose of which is to spread terror among the civilian populations' are also prohibited. This prohibition is also included in Art 13(3) of Additional Protocol II and, according to the United Kingdom and the United States, constitutes 'a valuable reaffirmation of existing customary rules of international law designed to protect civilians' (cited in Antonio Cassese, 'The Geneva Protocols of 1977 on the Humanitarian Law of Armed Conflict and Customary International Law', *UCLA Pacific Basin Law Journal* 3 (1984), p 87). See also Rule 36, Tallinn Manual, p 122; AIV/CAVV, *Cyber Warfare*, No 77, AIV/No 22, CAVV, December 2001, p 26 <http://www.aiv-advies.nl/ContentSuite/upload/aiv/doc/webversie__AIV77CAVV_22_ENG.pdf>. There is no prohibition to undermine military and civilian morale through non-violent means.

[197] Article 3 of the Hague Regulations Respecting the Laws and Customs of War on Land. See Charles Garraway, '"Combatants"—Substance or Semantics?', in *International Law and Armed Conflict: Exploring the Faultlines. Essays in Honour of Yoram Dinstein*, edited by Michael Schmitt and Jelena Pejic (Leiden-Boston: Nijhoff, 2007), p 321.

[198] See also Rule 3 of the ICRC Study, in Henckaerts and Doswald-Beck, *Customary International Humanitarian Law*, Vol I, p 11. The German Military Manual adds that '[p]ersons who are members of the armed forces but, by virtue of national regulations, do not have any combat mission, such as judges, government officials and blue-collar workers, are non-combatants' (German Ministry of Defence, *Humanitarian Law in Armed Conflicts*, Section 313).

[199] See Section III.1.3.b of this Chapter, pp 196–9. The expression 'privileged combatant' is employed here in a purely descriptive manner as indicating a combatant who is entitled to combatant immunity and POW status: this excludes mercenaries, combatants who do not comply with the laws and customs of war and the visibility requirements, and combatants engaging in espionage or sabotage if they are caught in the act behind enemy lines and do not wear a uniform, as well as fighters in non-international armed conflict. It does not imply that there is a third category of individuals in addition to 'civilians' and 'combatants'. See Garraway, '"Combatants"', p 327.

[200] Henckaerts and Doswald-Beck, *Customary International Humanitarian Law*, Vol I, p 11.

and indicates that these persons do not enjoy the protection against attack accorded to civilians, but does not imply a right to combatant status or prisoner-of-war status, as applicable in international armed conflicts'.[201] To avoid semantic confusion, the *Manual on the Law of Non-International Armed Conflict* uses the expression 'fighters' to refer to the 'members of armed forces and dissident armed forces or other organized armed groups, or taking an active (direct) part in hostilities'.[202] Whatever expression is preferred—fighter or combatant—to indicate the members of the armed forces of a non-state party to a non-international armed conflict, it should be borne in mind that, unlike combatants in international armed conflicts, such individuals are never entitled to combatant immunity or POW status upon capture.

The question is to determine when individuals involved in the conduct of cyber operations are combatants/fighters who may be attacked by cyber or kinetic means consistently with the principle of distinction, or civilians. Of course, if attacked exclusively by cyber means, individuals could be affected only indirectly, as in the textbook case of a cyber operation that shuts down the air traffic control system and causes airplanes to crash, with consequent human casualties. In the following pages, different categories of individuals involved in cyber operations will be considered in order to determine whether or not they are targetable by cyber or kinetic means. The analysis will first focus on members of the armed forces of states and non-state actors and other militias belonging to a party to the conflict, then will move to civilians taking direct part in hostilities. Finally some specific cases will be discussed, ie civilians accompanying the armed forces, civil defence personnel and participants in a *levée en masse*. It should be recalled that this Chapter focuses on the use of force, and not on internment and detention. Discussion of whether the above individuals have POW status if captured and what their treatment is when they fall in the hands of the enemy is beyond the scope of the present work and will be treated only incidentally.

(a) Members of a belligerent state's armed forces, including members of militias or volunteer corps forming part of such armed forces

'Armed forces' are defined in Article 43(1) of Additional Protocol I as

> all organized armed forces, groups and units, which are under a command responsible to that Party for the conduct of its subordinates, even if that Party is represented by a government or an authority not recognized by an adverse Party. Such armed forces shall be subject to an internal disciplinary system which, *inter alia*, shall enforce compliance with the rules of international law applicable in armed conflict.

More succinctly, the ICRC Study on *Customary International Humanitarian Law* defines 'armed forces' in an international armed conflict as 'all organized armed

---

[201] Henckaerts and Doswald-Beck, *Customary International Humanitarian Law*, Vol I, p 12.
[202] Schmitt, Garraway, and Dinstein, *The Manual on the Law of Non-International Armed Conflict*, Rule 1.1.2(a), p 4.

forces, groups and units which are under a command responsible to that party for the conduct of its subordinates'.[203] As implied in Article 4(A)(1) of Geneva Convention III, a state's armed forces also include 'members of militias or volunteer corps forming part of such armed forces'. This covers members of paramilitary and armed law enforcement agencies incorporated into the armed forces.[204]

As has been seen, under Article 43(2) of Additional Protocol I all members of the armed forces of a belligerent state, as defined in Article 43(1), other than medical and religious personnel are combatants by status, not conduct: their membership of the armed forces entails a presumption that they are entitled, trained and able to use methods and means of warfare whatever function they actually perform within the armed forces.[205] They are, therefore, military objectives and may be attacked by enemy forces during an armed conflict at all times.[206] Combatants who become *hors the combat*, however, may not be attacked unless they commit 'hostile acts' or try to escape.[207] Medical personnel and chaplains are not combatants: they may not be attacked unless they commit 'acts harmful to the enemy'.[208] Reservists, ie those 'who, after a period of basic training or active membership, leave the armed group and re-integrate into civilian life', are also civilians 'until and for such time as they are called back to active duty'.[209]

The above rules also apply to cyber units that form part of, or have been incorporated into, the armed forces. It has been seen that several states have established special sections of the armed forces responsible for cyber operations. The United States has famously established the CYBERCOM, a sub-unit of the US Strategic Command, while Colombia has created an Armed Forces Joint Cyber Command, tasked with preventing and countering cyber threats or attacks affecting national

---

[203] Rule 4 of the ICRC Study, in Henckaerts and Doswald-Beck, *Customary International Humanitarian Law*, Vol I, p 14. The Study specifies that, for the purposes of the principle of distinction, the definition may also apply to the state armed forces in a non-international armed conflict (Vol I, p 14).

[204] Article 43(3) of Additional Protocol I. Such incorporation must be notified to the other parties to the conflict, although lack of doing so does not prejudice combatancy.

[205] Sandoz, Swinarski, and Zimmermann (eds), *Commentary*, para 1677. See Art 43(2) of Additional Protocol I. As Corn puts it, this is 'based on the presumption that a fully functional member of an enemy belligerent group represents an ongoing threat, and attacking that individual is linked to bringing about the submission of the group writ large' (Geoffrey S Corn, 'Geography of Armed Conflict: Why it is a Mistake to Fish for the Red Herring', *International Law Studies* 89 (2013), p 97. See also Dinstein, *The Conduct*, p 33; Knut Ipsen, 'Combatants and Non-Combatants', in *The Handbook*, edited by Fleck, p 81. Membership of regular state armed forces usually depends on their 'formal integration into permanent units distinguishable by uniforms, insignia, and equipment' (ICRC, *Interpretive Guidance*, p 25).

[206] The Commentary to Art 52(2) of Additional Protocol I expressly states that members of the armed forces are military objectives (Sandoz, Swinarski, and Zimmermann (eds), *Commentary*, para 2017).

[207] Article 41(2) of Additional Protocol I.

[208] Article 13(1) of Additional Protocol I. If military personnel are performing medical duties on a temporary basis, the immunity from attack applies 'if they are carrying out these duties at the time when they come into contact with the enemy or fall into his hands' (Art 25 of Geneva Convention I).

[209] ICRC, *Interpretive Guidance*, p 34. An example would be the UK Joint Cyber Reserve established to support the GOSCC (House of Commons Defence Committee, *Defence and Cyber-Security*, Sixth Report of Session 2012–13, Vol I, 18 December 2012, p 34).

values and interests.[210] The Chinese PLA and North Korea's Army also allegedly have units entrusted with the conduct of cyber operations.[211] As combatants, the members of these military cyber units, as well as their headquarters, are military objectives and may thus be attacked at all times, unless *hors de combat* and regardless of whether the individual in question conducts cyber operations amounting to attack, cyber attacks short of 'attack' or cyber exploitation, or is merely entrusted with the defence of networks.[212]

(b) Members of other militias and volunteer corps belonging to a belligerent state

Members of militias and volunteer corps, including those of organized resistance movements, other than those integrated into a state's armed forces are combatants in an international armed conflict when they 'belong' to a party to the conflict.[213] This would also apply to those members of militias that conduct cyber operations.[214] The ordinary meaning of 'militia' and 'corps' implies that the group is armed and organized in a manner analogous to that of regular armed forces: if these elementary characteristics are lacking, there could be a collective of civilians taking direct part in hostilities, but not 'militias' or 'corps'. The Commentary to Rule 26 of the Tallinn Manual maintains that 'organization' for the purposes of establishing combatant and POW status is the same as 'organization' for the purposes of defining a non-international armed conflict:[215] as has been seen, the chances that an armed group that exists exclusively online will be sufficiently organized are slim.[216] The most likely scenario, then, is that of traditional militias or corps that *also* conduct cyber operations. 'Belonging to a Party to the conflict' means that the militia's actions are attributable to a belligerent state under the law of state responsibility.[217] It is not clear, however, what degree of control is required. The effective control standard elaborated by the ICJ in *Nicaragua* appears inappropriate to define what 'belonging to a Party to the conflict' means, as, unlike the overall control and complete dependency standards, it expresses control over the *act* and not over the actor and thus focuses on specific activities.[218] According to the ICTY Appeals Chamber,

---

[210] UN Doc A/67/167, 23 July 2012, p 5.

[211] On China's Unit 61398, see Mandiant, *APT1. Exposing One of China's Cyber Espionage Units*. On North Korea, see Arie J Schaap, 'Cyber Warfare Operations: Development and Use Under International Law', *Air Force Law Review* 64 (2009), p 133.

[212] The fact that combatants may be directly attacked under the principle of distinction does not entail that there are no legal restraints. In particular, the prohibition of inflicting unnecessary suffering and of perfidious acts constitute limits to the conduct of hostilities against enemy combatants. See *Targeted Killings*, para 23 (per Judge Barak).

[213] This is implicitly suggested by the combined effect of the *chapeau* of Art 4(A)(2) of Geneva Convention III and Art 50 of Additional Protocol I.

[214] A 'cyber militia' has been defined as 'a group of volunteers who are willing and able to use cyber attacks in order to achieve a political goal' (Rain Ottis, 'Proactive Defence Tactics Against On-Line Cyber Militia', in *Proceedings of the 9th European Conference on Information Warfare and Security*, edited by Josef Demergis (Reading: Academic Publishing Ltd, 2010), p 234).

[215] Tallinn Manual, p 98.   [216] See Chapter 3, Section IV.1.2.

[217] ICRC, *Interpretive Guidance*, p 23. See Chapter 1, Section IV.

[218] Marko Milanović, 'State Responsibility for Acts of Non-state Actors: A Comment on Griebel and Plücken', *Leiden Journal of International Law* 22 (2009), p 317. On the meaning of the effective control, overall control and complete dependency standards, see Chapter 1, Section IV, pp 37–8.

the 'ingredients' of 'belonging to a Party to the conflict' are 'control over [irregulars] by a Party to an international armed conflict and...a relationship of dependence and allegiance of these irregulars *vis-à-vis* that Party to the conflict'.[219] The necessary degree of control is, according to the ICTY, that of 'overall control'.[220] The stricter complete dependency standard developed by the ICJ in the *Genocide* judgment,[221] however, seems more consonant to the ordinary meaning of 'belonging to' than the looser 'overall control', which in fact is not an attribution standard but rather—as has been seen—an element of the primary rule providing for the definition of international(ized) armed conflict.[222] Either way, in contrast with what argued in the ICRC *Interpretive Guidance on the Notion of Direct Participation in Hostilities*,[223] a mere contractual relationship or the 'tacit agreement' of a party to the conflict are not sufficient to establish the existence of the 'belonging to a Party to the conflict' requirement and neither is the fact that the hackers were incited by a belligerent to conduct cyber operations against the adversary (as seems to be the case of the cyber operations against Georgia in 2008): such individuals would therefore be criminals or, if there is a belligerent nexus, civilians taking direct part in hostilities.

It is often suggested that, in order to have combatant status, militias and volunteer groups have to fulfil further conditions than mere belonging to a party to the conflict. Article 1 of the Regulations Respecting the Laws and Customs of War on Land attached to Hague Convention IV specifies that

[t]he laws, rights, and duties of war apply not only to armies, but also to militia and volunteer corps fulfilling the following conditions: 1. To be commanded by a person responsible for his subordinates; 2. To have a fixed distinctive emblem recognizable at a distance; 3. To carry arms openly; and 4. To conduct their operations in accordance with the laws and customs of war.

Article 4(A)(2) of Geneva Convention III focuses on POW status and accords it to:

[m]embers of other militias and members of other volunteer corps, including those of organized resistance movements, belonging to a Party to the conflict and operating in or outside their own territory, even if this territory is occupied, provided that such militias or volunteer corps, including such organized resistance movements, fulfil the following conditions:
(a) that of being commanded by a person responsible for his subordinates;
(b) that of having a fixed distinctive sign recognizable at a distance;
(c) that of carrying arms openly;
(d) that of conducting their operations in accordance with the laws and customs of war.[224]

---

[219] ICTY, *Prosecutor v Tadić*, Case No IT-94-1-A, Appeals Chamber Judgment, 15 July 1999, para 94.
[220] *Tadić*, Appeals Chamber Judgment, para 137.
[221] *Application of the Convention on the Prevention and Punishment of the Crime of Genocide (Bosnia and Herzegovina v Serbia and Montenegro)*, Merits, Judgment, 26 February 2007, ICJ Reports 2007, paras 392–3.
[222] See Chapter 3, Section II.3.2, pp 138–9.   [223] ICRC, *Interpretive Guidance*, p 23.
[224] While organization, belonging to a party to the conflict, and the presence of a responsible commander are requirements that must be possessed by the group, the others (visibility and compliance

It is also well known that Article 44 of Additional Protocol I has controversially removed some of the above requirements (in particular, the duty to comply with the laws and customs of war and the requirement of wearing a fixed distinctive sign recognizable at a distance), while that of carrying arms openly is limited to the period of active engagement and the deployment preceding the attack in 'situations in armed conflicts where, owing to the nature of the hostilities an armed combatant cannot so distinguish himself'.[225] Article 44(4) goes as far as providing that even if the individual does not carry arms openly in the situations where he is required to, 'he shall, nevertheless, be given protections equivalent in all respects to those accorded to prisoners of war by the Third Convention and by this Protocol'. This has been criticized for removing any distinction between regular and irregular forces.[226]

Much has been written on the applicability of the above requirements in the cyber context. Certain commentators have recommended the formal incorporation of cyber personnel into the armed forces so to avoid problems.[227] Taking a more technical approach, others have suggested that special visibility standards be agreed on for those conducting cyber operations: it has been argued, for instance, that cyber attacks should originate exclusively from military IP addresses,[228] or that 'universally recognized electronic identifiers' should be determined to identify the status of persons and sites.[229] Sean Watts has recommended a re-interpretation of existing criteria and suggested that the only relevant one in the cyber context is the requirement of state affiliation.[230] The requirements of command structure, visibility signs and compliance with the laws and customs of war are, in Watts's opinion, irrelevant in cyberspace because the remote character of the operations renders

---

with the laws and customs of war) must be respected both by the group and the individual members of the group: if the group does not comply with these requirements, its members are not entitled to POW status, even though the captured member complied with the visibility requirements and respected the laws of war (Natalino Ronzitti, *Diritto internazionale dei conflitti armati*, 4th edn (Torino: Giappichelli, 2011), p 172). According to the preferable view, the requirements provided in Art 4(A)(2) are implicit for members of regular armed forces (Art 4(A)(1): see Commentary to Rule 26, in Tallinn Manual, p 97).

[225] Article 44(3). The United Kingdom made a formal statement at the moment of ratification limiting the application of Art 44 to occupied territories and wars of national liberation (Roberts and Guelff, *Documents*, p 510).

[226] See eg Dinstein, *The Conduct*, pp 53–5.

[227] Doswald-Beck, 'Some Thoughts', p 172; Susan W Brenner, *Cyberthreats: The Emerging Fault Lines of the Nation States* (New York: Oxford University Press, 2009), pp 255–9.

[228] Harrison Dinniss, *Cyber Warfare*, p 146, who however eventually concludes that, if there is physical proximity between the combatant engaging in cyber operations and the opposing forces, the traditional visibility requirements apply, while, if there is no risk of deception or of mistaken status, the need for visibility requirements does not arise (p 148). IP addresses uniquely identify the computer system, but not its user.

[229] Davis Brown, 'A Proposal for an International Convention to Regulate the Use of Information Systems in Armed Conflict', *Harvard International Law Journal* 47 (2006), p 196. For example, military networks could be identified by the domain '.mil', medical networks could use '.med', and so on (Karl F Rauscher and Andrey Korotkov, *Working Towards Rules for Governing Cyber Conflict. Rendering the Geneva and Hague Conventions in Cyberspace*, EastWest Institute, January 2011, p 30; Commentary to Rule 72, in Tallinn Manual, p 208).

[230] Sean Watts, 'Combatant Status and Computer Network Attack', *Virginia Journal of International Law* 50 (2010), pp 441–2.

the possibility of capture unlikely, decreases the chances of abuses like pillage and looting on the battlefield and makes the appearance of the attackers irrelevant.[231] The presence of an internal disciplinary system, which is more easily enforceable because cyber operations are normally conducted from domestic territory, also reduces the importance of the compliance with the laws of war requirement.[232] While most of the above points are sensible ones, it seems counter-intuitive, however, to argue that the requirement of compliance with the law of armed conflict should be dropped just because it can be more easily enforced.

Be that as it may, Article 4 of Geneva Convention III and Article 44 of Additional Protocol I prescribe the requirements to be entitled to POW, not combatant status: they are therefore important for the application of the law of detention, but not for the law of targeting.[233] For the latter, the question to answer in order to establish whether an individual who conducts cyber operations is a combatant or a civilian in an international armed conflict is not whether the individual in question meets the visibility requirements and complies with the law of armed conflict, but whether he is a member of the regular armed forces, including incorporated forces, or of an irregular armed force that belongs to a party to the conflict.[234] Compliance with the visibility requirements and with the laws and customs of war, in other words, determines *privileged* combatancy and not combatancy, and would only be relevant to determine treatment in case of capture. Indeed, spies and saboteurs who are members of the armed forces but do not wear a uniform when conducting their operations behind enemy lines remain combatants, and do not become civilians because of non-compliance with the visibility requirements, although they are sanctioned with loss of entitlement to POW status. It is therefore not correct, as the UK Military Manual states, that a person that takes direct part in hostilities but does not comply with the principle of distinction 'forfeits his combatant status':[235] such person cannot claim POW status if captured, but remains a combatant and may be attacked at all times.[236] The opposite conclusion would contradict the rationale of the principle of distinction: if a combatant loses his status by not

---

[231] See similarly Schmitt, Harrison Dinniss, and Wingfield, *Computers and War*, p 12; Jenny Döge, 'Cyber Warfare—Challenges for the Applicability of the Traditional Laws of War Regime', *Archiv des Völkerrechts* 48 (2010), p 495 (as '[t]here is no physical proximity that requires distinguishing combatants from civilians... [t]he need to distinguish is obsolete').

[232] Watts, 'Combatant Status', p 440.

[233] Akande, 'Clearing the Fog of War?', p 184; Garraway, '"Combatants"', p 327.

[234] Rule 106 of the ICRC Study confirms that combatants that fail to distinguish themselves from the civilian population are not entitled to POW (not combatant) status (Henckaerts and Doswald-Beck, *Customary International Humanitarian Law*, Vol I, p 384). See also Rule 26, in Tallinn Manual, p 96 ('members of the armed forces of a party to the conflict who, in the course of cyber operations, fail to comply with the requirements of combatant status lose their entitlement to combatant immunity and prisoner of war status').

[235] UK Ministry of Defence, *The Manual of the Law of Armed Conflict*, pp 43–4.

[236] In fact, the UK Military Manual subsequently states that '[m]embers of irregular armed forces, whether they comply with the rule of distinction or not, are legitimate objects of attack when taking a direct part in hostilities' (UK Ministry of Defence, *The Manual of the Law of Armed Conflict*, p 43). This is, however, only partly correct, as, because of their combatant status, members of irregular forces are military objects for as long as their membership lasts, not only when they take direct part in hostilities.

distinguishing himself from the civilian population or by not complying with the laws and customs of war, he would enjoy the more protective regime of civilians and would be targetable not at all times but only 'for such time as' he takes direct part in hostilities.[237]

While membership of regular armed forces, including incorporated forces, is purely based on status, in the absence of formal acts of integration the ICRC *Interpretive Guidance on the Notion of Direct Participation in Hostilities* suggests that membership of an irregular armed force may be established on the basis of conduct, ie a 'continuous combat function' that the person exercises within that force in the same way as it happens for membership of the armed forces of a non-state actor.[238] The 'continuous combat function' will be discussed next.

(c) Members of the armed forces of a non-state actor

In non-international armed conflicts, there is no combatancy status in its technical meaning: one can therefore only use the expression 'combatant' in its generic sense of person engaging in hostilities, or employ alternative language like 'fighters', to refer to the armed forces of a non-state party to the conflict.[239] Whether the 'armed forces' of a non-state party to a non-international armed conflict qualify as (unprivileged) combatants/fighters or as civilians taking direct part in hostilities is controversial.[240] While the qualification would not be relevant with regard to the application of the law of detention (in neither case would the captured 'fighter' be entitled to POW status, as such status does not exist in non-international armed conflicts), it is of paramount importance in the context of the law of targeting: a 'fighter' may be attacked at all times (unless *hors de combat*), but if he remains a civilian he would be targetable only 'for such time as' he takes direct part in hostilities, ie exclusively when he performs certain conduct.[241]

---

[237] ICRC, *Interpretive Guidance*, p 22. See also Akande, 'Clearing the Fog of War?', p 184.

[238] ICRC, *Interpretive Guidance*, p 25. A distinct question is whether members of cyber firms hired by a party to the conflict to conduct cyber operations against another belligerent qualify as 'mercenaries'. According to the United States, '[w]hile existing international law has provisions governing the use of mercenaries, the use of proxies in cyberspace raises new and significant issues with wide-ranging implications' (UN Doc A/66/152, 15 July 2011, p 19). The problem, however, should not be overestimated. Article 47 of Additional Protocol I requires that, to be a mercenary, the person must 'in fact, take a direct part in the hostilities', which excludes those that conduct cyber exploitation activities. Furthermore, hackers employed by cyber crime firms are often either nationals of a party to the conflict or are not recruited for a specific conflict (Schmitt, Harrison Dinniss, and Wingfield, *Computers and War*, p 14). Finally, the qualification as mercenaries is important for the law of detention as they do not qualify for POW status, but is less relevant for the law of targeting: as they take direct part in hostilities, they may be attacked at least for such time as they do so.

[239] Schmitt, Garraway, and Dinstein, *The Manual on the Law of Non-International Armed Conflict*, Rule 1.1.2(a), p 4.

[240] Of course, when the insurgents have been recognized as belligerents by the government, they are entitled to combatant privilege and, if captured, POW status like government soldiers (Kevin Jon Heller, 'The Law of Neutrality Does Not Apply to the Conflict with Al-Qaeda, and It's a Good Thing, Too: A Response to Chang', *Texas International Law Journal* 47 (2011–12), p 124). Members of national liberation movements are also privileged combatants under certain conditions (Art 1(4) of Additional Protocol I).

[241] See Chapter 4, Section III.1.3.d, pp 209–10. The Commentary to Art 13(3) of Additional Protocol II also suggests that '[t]hose who belong to armed forces or armed groups may be attacked at any time' (Sandoz, Swinarski, and Zimmermann (eds), *Commentary*, para 4789).

According to Recommendation II of the 2009 ICRC *Interpretive Guidance on the Notion of Direct Participation in Hostilities*, 'organized armed groups constitute the armed forces of a non-State party to the conflict'. A non-state actor's armed forces must then be: (1) organized; (2) armed; and (3) belong to a non-state party to the conflict. The requirement of 'organization' entails that the group is organized in a military fashion in a way analogous, but not necessarily identical, to that of a state's armed forces. The second requirement implies that the armed forces of a non-state party to the conflict are only those that form part of the 'armed or military wing of a non-State party'.[242] Indeed, the 'organized armed group', ie the armed forces of a non-state actor, should be distinguished from the non-state party to the conflict itself, such as insurgencies or national liberation movements.[243] The third requirement should be interpreted as suggested above in relation to a state's irregular armed forces, ie as control over the actor and not over specific acts.[244]

While the above are group requirements, a fourth, individual requirement is that the person in question is a member of the non-state actor's armed forces. Within the notion of 'organized armed group', the ICRC *Interpretive Guidance* distinguishes 'dissident armed forces' and 'other organized armed groups': the former 'essentially constitute part of a State's armed forces that have turned against the government', while the latter 'recruit their members primarily from the civilian population but develop a sufficient degree of military organization to conduct hostilities on behalf of a party to the conflict, albeit not always with the same means, intensity and level of sophistication as State armed forces'.[245] If dissident armed forces might 'remain organized under the structures of the State armed forces to which they formerly belonged', which therefore determines individual membership, there is no 'act of integration' into 'other organized groups' made explicit through uniforms, fixed distinctive signs, or identification cards.[246] In this case, membership is determined not by status, but by 'whether a person assumes a *continuous function* for the group involving his or her direct participation in hostilities'.[247] In other words, membership depends 'on whether the continuous function assumed by an individual corresponds to that collectively exercised by the group as a whole, namely the conduct of hostilities on behalf of a non-State party to the conflict'.[248] Only those that have a continuous combat function are then members of 'other organized armed groups' or dissident forces not organized under their former structures, while those whose function is not a combat one, or it is but in a non-continuous fashion,

---

[242] ICRC, *Interpretive Guidance*, p 32.   [243] ICRC, *Interpretive Guidance*, p 32.

[244] On the problem of the lack of rules of attribution to non-state actors, see Liesbeth Zegveld, 'Accountability of Organized Armed Groups', in *Non-state Actors and International Humanitarian Law*, edited by Marco Odello and Gian Luca Beruto (Milano: Franco Angeli, 2010), pp 111–12.

[245] ICRC, *Interpretive Guidance*, pp 31–2. It is not explained, however, when dissident armed forces can still be considered 'organized under the structures of the State armed forces to which they formerly belonged' (Kenneth Watkin, 'Opportunity Lost: Organized Armed Groups and the ICRC "Direct Participation in Hostilities" Interpretive Guidance', *New York University Journal of International Law and Politics* 42 (2009–10), p 654).

[246] ICRC, *Interpretive Guidance*, pp 32–3.

[247] ICRC, *Interpretive Guidance*, p 33 (emphasis added).

[248] ICRC, *Interpretive Guidance*, p 33.

are not members. According to the *Interpretive Guidance*, '[a] continuous combat function may be openly expressed through the carrying of uniforms, distinctive signs, or certain weapons. Yet it may also be identified on the basis of conclusive behaviour, for example where a person has repeatedly directly participated in hostilities in support of an organized armed group in circumstances indicating that such conduct constitutes a continuous function rather than a spontaneous, sporadic, or temporary role assumed for the duration of a particular operation.'[249] The members of an organized armed group that have a continuous combat function are military objectives for the entire period where they have that function, whether they exercise it or not,[250] until they clearly disengage from the armed group by an express declaration or conclusive behaviour.[251]

The question is what activities related to cyber operations amount to a 'continuous combat function'. 'Combat' function implies conduct that amounts to direct participation in hostilities:[252] as will be seen in the following sub-Section, this rules out cyber operations for strategic intelligence gathering and propaganda purposes. The ICRC *Interpretive Guidance* confirms that 'recruiters, trainers, financiers and propagandists' and 'individuals whose function is limited to the purchasing, smuggling, manufacturing and maintaining of weapons and other equipment outside military operations or to the collection of intelligence other than of a tactical nature' do not exercise a 'combat' function and therefore are not members of the armed forces of the non-state party to the conflict.[253] Also excluded are support activities such as the design and testing of malware, as well as general technical maintenance of IT services.[254] By contrast, a combat function clearly includes the conduct of cyber attacks causing or reasonably likely to cause material damage to property, loss of life or injury to persons or disruption of infrastructures. It also includes the conduct of operations that form an integral part of an attack, eg cyber exploitation operations for target acquisition. Even when the function is a combat one, however, it has to be 'continuous': if such function is an intermittent and irregular one, the individual would remain a civilian that takes direct part in hostilities and that may be attacked only for such time as he does so. 'Continuous' does not mean 'exclusive': a continuous combat function does not 'exclude the parallel, or even predominant, exercise of non-combat functions'.[255]

While helpful, the use of the continuous combat function criterion to identify the members of the armed forces of a non-state actor is not without problems. Indeed, it creates an imbalance between the armed forces of states on the one hand and those of non-state actors (in particular dissident armed forces not organized under their former structures and 'other organized armed groups') on the other. Put differently, members of a state's armed forces may be targeted at all times, whether

---

[249] ICRC, *Interpretive Guidance*, p 35.   [250] ICRC, *Interpretive Guidance*, p 34.
[251] ICRC, *Interpretive Guidance*, p 72.   [252] ICRC, *Interpretive Guidance*, p 34.
[253] ICRC, *Interpretive Guidance*, pp 34–5.   [254] See Section III.1.3.d of this Chapter, pp 207–8.
[255] Nils Melzer, 'Keeping the Balance between Military Necessity and Humanity: A Response to Four Critiques of the ICRC's Interpretive Guidance on the Notion of Direct Participation in Hostilities', *New York University Journal of International Law and Politics* 42 (2009–10), p 848.

or not they have a continuous combat function (with the exclusion of medical personnel and chaplains), while members of the armed forces of a non-state party to the conflict may be attacked only if they have a continuous combat function: if not, they are civilians that may be attacked only for such time as they take direct part in hostilities. Hence, a computer technician providing general IT support may be attacked at all times if incorporated in a state's armed forces, while if he is a member of the armed forces of a non-state actor he may not be attacked because he does not have a continuous combat function.

(d) Civilians taking direct part in (cyber) hostilities

Thanks to the low cost of and ease of access to technology, cyber operations can be conducted not only by armed forces or militarily organized armed groups, but also by civilians: all it takes is a computer, software, and a connection to the internet.[256] This easy access adds to the general trend to outsource traditionally military functions to intelligence agencies and civilian firms that characterizes modern warfare.[257] The RBN, a cyber crime firm specializing in phishing, malicious code, botnet command-and-control, DDoS attacks, and identity theft is for instance suspected of having executed the cyber attacks against Georgia on behalf of Russia.[258] It has also been reported that Iranian hackers work for the Revolutionary Guard's paramilitary Basij group and include 'university instructors and students, as well as clerics'.[259] In the United Kingdom, the GOSCC, whose role is 'to proactively and reactively defend MoD networks 24/7 against cyber attack to enable agile exploitation of MoD information capabilities across all areas of the Department's operations', is formed not only by military but also by MoD civilian and contractor personnel from industry partners, although only military members can be sent to operational theatres.[260]

As Susan Brenner has observed, this 'integration of civilians into military efforts can create uncertainty as to whether someone is acting as a "civilian" (noncombatant) or as a military actor (combatant)'.[261] Under the principle of distinction, civilians may not be the direct object of an attack and need also to be protected as far as

---

[256] Roger W Barnett, 'A Different Kettle of Fish: Computer Network Attack', *International Law Studies* 76 (2002), p 22.

[257] The CIA's involvement in the US drone attacks and the use of private military security companies in Iraq and Afghanistan are good examples of this trend.

[258] Eneken Tikk, Kadri Kaska, Kristel Rünnimeri, Mari Kert, Anna-Maria Talihärm, and Liis Vihul, *Cyber Attacks Against Georgia: Legal Lessons Identified* (CCDCOE, November 2008), p 11, <http://www.carlisle.army.mil/DIME/documents/Georgia%201%200.pdf>; John Markoff, 'Before the Gunfire, Cyberattacks', *The New York Times*, 12 August 2008, <http://www.nytimes.com/2008/08/13/technology/13cyber.html>.

[259] Nasser Karimi, 'Iran's paramilitary launches cyber attack', *The Associated Press*, 14 March 2011, <http://www.washingtonpost.com/wp-dyn/content/article/2011/03/14/AR2011031401029.html?referrer=emailarticle>.

[260] House of Commons Defence Committee, *Defence and Cyber-Security*, p 17.

[261] Brenner, *Cyberthreats*, p 197.

possible from the effects of an attack on a military objective.[262] 'Civilians' are defined negatively in Article 50 of Additional Protocol I as

> any person who does not belong to one of the categories of the persons referred to in Article 4A(1); (2) and (6) of the Third Convention and in Article 43 of this Protocol. In case of doubt whether a person is a civilian, that person shall be considered to be a civilian.

'Civilian population' includes 'all persons who are civilians'.[263] In conventional international humanitarian law, these definitions apply only to international armed conflicts. Rule 5 of the ICRC Study on *Customary International Humanitarian Law*, however, defines civilians in both international and non-international armed conflicts as 'persons who are not members of the armed forces'.[264] Recommendation II of the ICRC *Interpretive Guidance on the Notion of Direct Participation in Hostilities* provides for a more detailed definition of 'civilians' in non-international armed conflicts, ie 'all persons who are not members of State armed forces or organized armed groups of a party to the conflict', where—as has already been seen—'organized armed groups' are 'the armed forces of a non-State party to the conflict' consisting 'of individuals whose continuous function it is to take a direct part in hostilities'.[265]

International humanitarian law contains no express prohibition on civilians to take direct part in hostilities, but it sanctions them if they do so by suspending their immunity from attacks.[266] Under Article 51(3) of Additional Protocol I, 'civilians shall enjoy the protection afforded by [Section I of Part IV of the Protocol on general protection against the effects of hostilities], unless and for such time as they take a direct part in hostilities'. According to Israel's Supreme Court, the rule reflects customary international law in its entirety.[267] An identical provision is contained in Article 13(3) of Additional Protocol II.[268] Rule 6 of the ICRC Study on *Customary International Humanitarian Law* incorporates Article 51(3) of Additional Protocol I and extends it to non-international armed conflicts.[269]

---

[262] On the other hand, civilians may be interned if they pose a security threat to the belligerent state, whether or not they take direct part in hostilities (Geneva Convention IV, Arts 5, 27, 41–3, and 78).

[263] Article 50(2) of Additional Protocol I. Recommendation I of the ICRC *Interpretive Guidance on the Notion of Direct Participation in Hostilities* defines civilians in international armed conflicts as 'all persons who are neither members of the armed forces of a party to the conflict nor participants in a *levée en masse*' (ICRC, *Interpretive Guidance*, pp 20–1). The 1956 ICRC Draft Rules for the Limitation of the Dangers Incurred by the Civilian Population in Time of War defined the civilian population as 'all persons not belonging to one or other of the following categories: (a) Members if the armed forces, or of their auxiliary or complementary organisations. (b) Persons who do not belong to the forces referred to above, but nevertheless take part in the fighting' (Art 4, text at <http://www.icrc.org/ihl/INTRO/420?OpenDocument>).

[264] Henckaerts and Doswald-Beck, *Customary International Humanitarian Law*, Vol I, p 17.

[265] Recommendation II, in ICRC, *Interpretive Guidance*, p 27.

[266] Rule 29, in Tallinn Manual, p 104.   [267] *Targeted Killings*, para 30 (per Judge Barak).

[268] Common Art 3 of the Geneva Conventions refers to '[p]ersons taking no active part in the hostilities, including members of armed forces who have laid down their arms and those placed "hors de combat" by sickness, wounds, detention, or any other cause', who must be treated humanely. 'Active' and 'direct' participation in hostilities are considered synonymous (ICTR, *Prosecutor v Akayesu*, Case No ICTR-96-4-T, Judgment, 2 September 1998, para 629).

[269] Henckaerts and Doswald-Beck, *Customary International Humanitarian Law*, Vol I, pp 19–20.

In broad strokes, three main categories of civilians involved in cyber operations can be identified: those who design and develop malicious software and programs, those who install them on computer systems, act as service administrators and provide technical assistance, and those who actually activate the software and conduct cyber operations.[270] These activities may be carried out by civilians upon instructions of a belligerent or independently in order to support a belligerent to the detriment of another, as in the case of 'patriotic hackers'. A fourth case is that of civilians involved in the conduct of cyber operations without being aware of it, when their computer is taken over by a botmaster to carry out a DDoS attack. The question is whether civilians engaging in the above acts are directly participating in hostilities under Article 51(3) of Additional Protocol I. The point is important for the application of the law of targeting, as direct participation in hostilities turns civilians into military objectives, even though, unlike combatants, only 'for such time as' the participation lasts. From the perspective of the law of detention, civilians taking direct participation in hostilities are not entitled to POW status if captured and may therefore be prosecuted under domestic law for taking up arms as well as for the violations of the laws and customs of war that they may have committed. They are, however, entitled to the minimum guarantees spelt out in Article 75 of Additional Protocol I, which reflects customary international law.[271]

In 2009, the ICRC adopted an *Interpretive Guidance on the Notion of Direct Participation in Hostilities*, which contains ten recommendations as well as commentaries.[272] Although certain rules contained in the Guidance are controversial, it is a good starting point in order to clarify one of the most obscure notions in international humanitarian law.[273] The *Interpretive Guidance* preliminarily points out that 'participation' in hostilities refers to specific hostile acts, ie 'the (individual) involvement of a person in these hostilities', as otherwise 'it would be impossible to determine with a sufficient degree of reliability whether civilians not currently preparing or executing a hostile act have previously done so on a persistently recurrent basis and whether they have the continued intent to do so again'.[274] The *Interpretive Guidance* then identifies three elements that need to be cumulatively present for a civilian to take 'direct' part in hostilities: threshold of harm, direct causation and belligerent nexus.

Threshold of harm. According to the Commentary of Article 51 of Additional Protocol I, direct participation refers to 'acts of war which by their nature or purpose are likely to cause actual harm to the personnel and equipment of the enemy armed forces'.[275] The Protocol only refers to harm to the military and do not specify whether this should be only of a physical character or not. In the *Targeted Killings*

---

[270] Turns, 'Cyber Warfare', p 289. [271] *Targeted Killings*, para 25 (Barak).
[272] ICRC, *Interpretive Guidance*. The ICRC *Interpretive Guidance* reflects the view of the ICRC and not of governments, and was not adopted by majority of the experts participating in the process.
[273] For a critical view of certain aspects of the ICRC *Interpretive Guidance*, see the contributions of Bill Boothby, Michael N Schmitt, Kenneth Watkin, and W Hays Parks in *New York Journal of International Law and Politics* 42 (2009–10).
[274] ICRC, *Interpretive Guidance*, pp 43, 45.
[275] Sandoz, Swinarski, and Zimmermann (eds), *Commentary*, para 1944.

Judgment, however, Israel's Supreme Court rightly found that 'acts which by nature and objective are intended to cause damage to civilians should be added to that definition', although it did not specify what kind of 'damage' is necessary.[276] The ICRC *Interpretive Guidance* clarifies that '[t]he act must be likely to adversely affect the military operations or military capacity of a party to an armed conflict or, *alternatively*, to inflict death, injury, or destruction on persons or objects protected against direct attack'.[277] The *Interpretive Guidance*, then, extends the notion of harm to also include damage to protected persons and objects, although only if of a physical nature: if any act that negatively affects the military operations or military capacity of the enemy is sufficient, including '[e]lectronic interference with military computer networks…, whether through computer network attacks (CNA) or computer network exploitation (CNE)',[278] acts that have non-military consequences are only pertinent if they produce material damage in the form of death, injury or destruction. Rule 29(iii) of the HPCR *Manual on Air and Missile Warfare* translates this into the cyber context and concludes that '[e]ngaging in… computer network attacks targeting military objectives, combatants or civilians directly participating in hostilities, or which is intended to cause death or injury to civilians or damage to or destruction of civilian objects' is an example of direct participation in hostilities.[279]

A cyber operation that alters data in a military database containing deployment plans of the enemy's armed forces, disrupts the command and control system of the enemy or shuts down the operating system of unmanned aerial vehicles so that they cannot be employed on the theatre of operations,[280] therefore, would reach the threshold of (military) harm necessary for direct participation in hostilities, even if no physical damage occurs. Similarly, 'operations designed to enhance one's own capabilities', such as cyber defence of military objectives, would also cause military harm to the enemy and would therefore meet this requirement.[281] On the other hand, according to the *Interpretive Guidance*, 'the interruption of electricity, water, or food supplies' as well as 'the manipulation of computer networks' that do not have 'adverse military effects' would not cause the harm required for direct participation in hostilities unless they result in death, injury or destruction of protected persons or objects, even when they have 'a serious impact on public security, health, and commerce'.[282] This seems too restrictive. If the targeted computer system or

---

[276] *Targeted Killings*, para 33 (Barak).
[277] Recommendation V(1), in ICRC, *Interpretive Guidance*, p 46 (emphasis added).
[278] ICRC, *Interpretive Guidance*, p 48.
[279] HPCR Manual, p 137.
[280] 'US drones infected by key logging virus', Al Jazeera, 8 October 2011, <http://www.aljazeera.com/news/americas/2011/10/201110816388104988.html>.
[281] Schmitt, 'Cyber Operations', p 101. See, *contra*, Harrison Dinniss, according to whom cyber defences may not even be a use of force and do not cause casualties (Harrison Dinniss, *Cyber Warfare*, p 171). Direct participation in hostilities, however, does not necessarily imply either a use of force or physical damage to property or persons. As the Israeli Supreme Court's Judgment on *Targeted Killings* found, 'it is possible to take part in hostilities without using weapons at all' (*Targeted Killings*, para 33 (per judge Barak)).
[282] ICRC, *Interpretive Guidance*, p 50. See also Turns, 'Cyber Warfare', p 284.

infrastructure are dual-use, so that their loss of functionality not only affects civilian services but *also* causes military harm, the threshold of harm is crossed with or without the concomitant occurrence of physical consequences, as the *Interpretive Guidance* does not require that military harm be the sole result of the operation. As to cyber operations that incapacitate civilian infrastructures without causing neither physical damage to civilians or civilian objects nor military harm, some commentators have suggested that the notion of 'harm' should be extended to also cover damage to information installed in computer networks because of the crucial role played by those networks.[283] For Melzer, 'any harm caused to the civilian population for reasons related to the conflict, including mere harassment or inconvenience, would have to be regarded as part of military hostilities, triggering not only the applicability of IHL on the conduct of hostilities, but also the loss of civilian protection for all those directly involved'.[284] This view goes too far: causing mere inconvenience is not sufficient to justify participation in hostilities and the loss of civilian immunity. However, if the loss of functionality of infrastructure caused by the cyber attack is sufficiently significant to be equated to (temporary) 'destruction' in the sense of Recommendation V(1) of the ICRC *Interpretive Guidance*, then the threshold of harm should be considered crossed.

In light of the above, if the cyber operations against Georgia during the 2008 conflict with Russia did affect the Georgian government's ability to communicate and the operability of the armed forces, the threshold of harm requirement would have been met, even though they did not result in physical damage on protected persons and property.[285] Had it been conducted in the context of an armed conflict with Iran, Stuxnet would have also reached the required threshold of harm for direct participation in hostilities because the worm is believed to have caused physical damage to civilian property (if the uranium enrichment facility at Natanz is exclusively for civilian purposes) and perhaps also military harm (if the facility is actually used to develop nuclear weapons).[286]

It should be pointed out that, according to the ICRC *Interpretive Guidance*, it is not necessary that the required harm materializes for the conduct to amount to direct participation in hostilities. What is relevant is the 'objective *likelihood* that the act will result in such harm', on the basis of what 'may reasonably be expected to result from an act in the prevailing circumstances'.[287]

Direct causation. According to the ICRC *Interpretive Guidance*, 'direct causation' requires that harm is produced in 'one causal step' either by the specific act itself or by the 'coordinated military operation of which that act constitutes an

---

[283] John Ricou Heaton, 'Civilians at War: Reexamining the Status of Civilians Accompanying the Armed Forces', *Air Force Law Review* 57 (2005), p 201.
[284] Melzer, *Cyberwarfare*, p 29.
[285] Report of the Independent Fact-Finding Mission on the Conflict in Georgia, September 2009, Vol II, pp 217–18, <http://www.ceiig.ch/Report.html>; Tikk, Kaska, Rünnimeri, Kert, Talihärm, and Vihul, 'Cyber Attacks', pp 7–12.
[286] Turns, 'Cyber Warfare', p 287.
[287] ICRC, *Interpretive Guidance*, p 47 (emphasis in the original).

integral part'.[288] The 'one causal step' standard should not be applied to the secondary and tertiary effects of cyber operations, but rather to their primary ones, of which the other effects are consequences. Of the categories of civilians potentially involved in cyber operations identified above, those who execute the payloads and conduct cyber operations obviously take 'direct' part in hostilities. Those who decide or plan the act are also taking direct part in the hostilities themselves.[289] Temporal and geographical proximity is not essential in this context:[290] neither the fact that the cyber operation is conducted remotely nor that the prejudicial effects occur some time after the malware has been installed preclude the 'direct' character of the participation. Israel's Supreme Court confirmed that 'a person who operates weapons which unlawful combatants use, or supervises their operation, or provides service to them, be the distance from the battlefield as it may' takes direct part in hostilities.[291]

On the other hand, civilians who recruit and train hackers or provide general technical maintenance of systems and networks only take indirect part in hostilities as the causation of the required threshold of harm requires further 'causal steps'.[292] The same considerations apply to those conducting cyber exploitation for intelligence gathering. According to the ICRC *Interpretive Guidance*, however, 'where a specific act does not on its own directly cause the required threshold of harm, the requirement of direct causation would still be fulfilled where the act constitutes an integral part of a concrete and coordinated tactical operation that directly causes such harm'.[293] In the cyber context, the direct causation requirement would then be met in case of tactical cyber exploitation operations that allow 'the identification and marking of targets, [or] the analysis and transmission of tactical intelligence to attacking forces', cyber communications providing 'the instruction and assistance given to troops for the execution of a specific military operation',[294] or in the cases of '[l]oading mission control data to military aircraft/missile software systems'.[295]

With regard to malware designers and developers, in particular, the Group of Experts on the Notion of Direct Participation in Hostilities discussed whether 'civilian scientists and weapon experts' take direct part in hostilities.[296] Some of the

---

[288] Recommendation V(2), in ICRC, *Interpretive Guidance*, p 46. The Commentary to Art 43 of Additional Protocol I also requires a 'direct causal relationship between the activity engaged in and the harm done to the enemy at the time and the place where the activity takes place' (Sandoz, Swinarski, and Zimmermann (eds), *Commentary*, para 1679), while the Commentary to Art 13(3) of Additional Protocol II refers to a 'sufficient causal relationship' (Sandoz, Swinarski, and Zimmermann (eds), *Commentary*, para 4787).

[289] *Targeted Killings*, para 37 (per Judge Barak).

[290] ICRC, *Interpretive Guidance*, p 55. See also Memorandum for Chiefs of the Military Services, Commanders of the Combatant Commands, Directors of the Joint Staff Directorates, *Joint Terminology for Cyberspace Operations*, November 2010, p 5.

[291] *Targeted Killings*, para 35 (per Judge Barak).

[292] ICRC, *Interpretive Guidance*, p 53; Sandoz, Swinarski, and Zimmermann (eds), *Commentary*, para 1806. See also ICTY, *Prosecutor v Strugar*, Case No IT-0-42-A, Appeals Chamber Judgment, para 177. If, however, the technicians maintain the network from which a cyber operation is launched, ie a weapon system, it has been suggested that they take direct part in hostilities (Schmitt, Harrison Dinniss and Wingfield, *Computers and War*, p 13).

[293] ICRC, *Interpretive Guidance*, pp 54–5.   [294] ICRC, *Interpretive Guidance*, p 55.

[295] Rule 29(xi), HPCR Manual, p 140.   [296] Turns, 'Cyber Warfare', p 291.

experts suggested that this might be so 'in extreme situations, namely where the expertise of a particular civilian was of very exceptional and potentially decisive value for the outcome of an armed conflict, such as the case of nuclear weapons experts during the Second World War'.[297] Other commentators have observed that 'cyber weapons are different from tanks or planes in that the weapon must itself be modified continuously to react to unexpected and evolving defences within a specific target', which requires 'weapons designers to work much more directly with military and intelligence counterparts during the course of the attack'.[298] All in all, what counts is that the activity aims to execute a *specific* cyber attack of which it is an integral part, not to merely establish the *capacity* to conduct attacks.[299]

Belligerent nexus. The third requirement prescribes that the activities conducted by the civilians be linked to the 'hostilities': 'an act must be specifically designed to directly cause the required threshold of harm in support of a party to the conflict and to the detriment of another'.[300] The belligerent nexus requirement has already been discussed in Chapter 3 as a precondition for the application of the *jus in bello* to cyber operations occurring in the context of a kinetic armed conflict.[301] This requirement also distinguishes civilians taking direct part in the hostilities by cyber means, to which the law on the conduct of hostilities applies, from hackers conducting cyber operations for criminal or other purposes unrelated to the hostilities, who fall under the law enforcement framework. The ICRC *Interpretive Guidance* states that the determination of whether or not a civilian's activity has a belligerent nexus

must be based on the information reasonably available to the person called on to make the determination, [and] must always be deduced from objectively verifiable factors. In practice, the decisive question should be whether the conduct of a civilian, in conjunction with the circumstances prevailing at the relevant time and place, can reasonably be perceived as an act designed to support one party to the conflict by directly causing the required threshold of harm to another party.[302]

It is worth recalling that, like direct causation, the belligerent nexus is not a geographical requirement: the remote character of the act and the distance from the battlefield would not necessarily imply that the cyber operation is not related to the hostilities.

---

[297] ICRC, *Interpretive Guidance*, p 53, footnote 122.
[298] Vijai M Padmanabhan, 'Cyber Warriors and the *Jus in Bello*', *International Law Studies* 89 (2013), p 293.
[299] ICRC, *Interpretive Guidance*, pp 53, 66. See Watts, 'Combatant Status', p 429, according to whom designing or building a weapon is 'too remote from its effects to be considered "direct" [participation]'.
[300] Recommendation V(3), in ICRC, *Interpretive Guidance*, p 58. Schmitt has criticized this formulation inasmuch as it requires the act to be both 'in support of a party to the conflict *and* to the detriment of another' (emphasis added): in his view, the presence of either element is sufficient (Michael N Schmitt, 'Deconstructing Direct Participation in Hostilities: The Constitutive Elements', *New York University Journal of International Law and Politics* 42 (2009–10), p 736). This view, however, risks extending the belligerent nexus to include any criminal act conducted in support of a party to the conflict (such as a bank robbery where the loot is donated to a belligerent).
[301] Chapter 3, Section II.2.
[302] ICRC, *Interpretive Guidance*, pp 63–4 (footnote omitted).

'For such time as'. Civilians taking direct part in hostilities may not be attacked at all times like combatants, but only 'for such time as' their participation lasts: outside that timeframe, lethal force may be used at the stricter conditions provided under the law enforcement paradigm. As hostilities cover 'not only the time that the civilian actually makes use of a weapon, but also, for example, the time that he is carrying it, as well as situations in which he undertakes hostile acts without using a weapon',[303] 'for such time as' includes preparatory measures and the deployment to and return from the place where the attack was executed.[304] The ICRC *Interpretive Guidance* specifies that, if the execution of the hostile act is conducted remotely (as is the case of cyber attacks), 'the duration of direct participation in hostilities will be restricted to the immediate execution of the act and preparatory measures forming an integral part of that act'.[305] The question is what amounts to 'execution' of a cyber operation. If it is clear that direct participation covers 'the period over which the functionality required to carry out the attack on the targeted system(s) is installed or deployed',[306] the situation is more complicated with regard to the compromise phase, ie 'the period over which the confidentiality, integrity, or availability attributes of the targeted system(s) are compromised', and to 'the time period over which the victim actually suffers the ill effects of such compromises'.[307] For instance, in case of a time or logic bomb, does the duration of the participation cover only the installation of the malware or continues until the 'bomb' actually starts to produce its effects (which may be a long time afterwards)? And when malware produces prejudicial effects that continue well after the individual has executed the installation, is the individual in question still participating in the hostilities because the effects are continuing even though he is no longer involved? While these questions do not have easy answers, it is submitted that a strict interpretation of 'for such time as', as suggested in the ICRC *Interpretive Guidance*, is preferable also in the cyber context. In particular, one fails to see the military necessity of attacking a civilian who is not playing any longer a role in the operation: the act of hostilities may well continue, but the direct participation would not. Referring to the notion of continuing act to justify an extension of the duration of the participation in hostilities so to also cover the effects of the act is not helpful.[308] Indeed, Article 14 of the ILC Articles on State Responsibility distinguishes between a wrongful act and its effects: as stated in the Commentary of the provision, '[a]n act does not have a continuing character merely because its effects or consequences extend in time. It must be the wrongful act as such which continues'.[309]

---

[303] Sandoz, Swinarski, and Zimmermann (eds), *Commentary*, para 1943.
[304] See Recommendation VI, in ICRC, *Interpretive Guidance*, p 65. See also Commentary to Rule 35, in Tallinn Manual, p 121.
[305] ICRC, *Interpretive Guidance*, p 68.
[306] Owens, Dam, and Lin, *Technology, Policy, Law, and Ethics*, p 90.
[307] Owens, Dam, and Lin, *Technology, Policy, Law, and Ethics*, p 90.
[308] See the sceptical comments on this view by Heather Harrison Dinniss, 'Participants in Conflict—Cyber Warriors, Patriotic Hackers and the Laws of War', in *International Humanitarian Law*, edited by Saxon, pp 274–6.
[309] ILC Articles on State Responsibility, *Yearbook of the International Law Commission*, 2001, Vol II, Part Two, p 60.

If the individual in question conducts repeated cyber operations amounting to direct participation, whether or not each act must be treated separately depends on the circumstances of each case. In the *Targeted Killings* Judgment, Israel's Supreme Court held that 'a civilian taking a direct part in hostilities one single time, or sporadically, who later detaches himself from that activity, is a civilian who, starting from the time he detached himself from that activity, is entitled to protection from attack. He is not to be attacked for the hostilities which he committed in the past'.[310] By contrast, 'a civilian who has joined a terrorist organization which has become his "home", and in the framework of his role in that organization he commits a chain of hostilities, with short periods of rest between them, loses his immunity from attack "for such time" as he is committing the chain of acts. Indeed, regarding such a civilian, the rest between hostilities is nothing other than preparation for the next hostility'.[311] It has been suggested that, applied in the cyber context, this entails that 'the only reasonable interpretation of "for such time" is that it encompasses the entire period during which the direct cyber participant is engaging in repeated cyber operations'.[312]

Botnets. In some cases, civilians may be taking direct part in hostilities without being aware of it. 'Botnets' (short for 'robot networks'), which are the source of most spam, are networks of infected computers ('zombies') hijacked from their unaware owners by external users: linked together, such networks can be used to mount massive DDoS attacks.[313] The Mariposa botnet was one of the world's biggest with up to 12.7 million computers controlled.[314]

In the context of an armed conflict, a belligerent's cyber operation aimed at hijacking enemy civilian computers in order to conduct a DDoS attack might be a violation of the prohibition to compel civilians of enemy nationality to do work 'directly related to the conduct of military operations' or to compel protected persons in occupied territory 'to serve in its armed or auxiliary forces'.[315] The cyber operation aimed at taking control of computers to create a botnet, however, is not, as such, an 'attack' in the sense of Article 49(1) of Additional Protocol I, as it does not per se have violent consequences, and is not even an act of 'direct' participation in hostilities, as it normally necessitates further steps to cause the required threshold of military or civilian harm. On the other hand, the DDoS attack conducted through the botnet by the botmaster might well be an 'attack' if it causes violent

---

[310] *Targeted Killings*, para 39 (per Judge Barak).

[311] *Targeted Killings*, para 39 (per Judge Barak). The 'chain of hostilities' resembles the continuous combat function identified by the ICRC *Interpretive Guidance* and discussed in Section III.1.3.c, pp 200–2, of this Chapter.

[312] Schmitt, 'Cyber Operations', p 102.

[313] Stewart Baker, Shaun Waterman, and George Ivanov, *In the Crossfire, Critical Infrastructure in the Age of Cyber War* 2009, p 6, <http://www.mcafee.com/us/resources/reports/rp-in-crossfire-critical-infrastructure-cyber-war.pdf>.

[314] Charles Arthur, 'Alleged controllers of "Mariposa" botnet arrested in Spain', *The Guardian*, 3 March 2010, <http://www.guardian.co.uk/technology/2010/mar/03/mariposa-botnet-spain>.

[315] Articles 40(2) and 51(1) of Geneva Convention IV, respectively. Article 147 of Geneva Convention IV also provides that it is a grave breach of the Convention to compel 'a protected person to serve in the forces of a hostile Power'.

consequences, in which case it is submitted to the law of targeting, including the principles of distinction, proportionality, and precautions. If the botmaster that operates the botnet and mounts the DDoS operation causing the required threshold of harm to an adverse party is a civilian, he is taking direct part in hostilities and may thus be targeted for such time as he is doing so. As to the hijacked computers, the situation is analogous to that of civilians that unknowingly transport weapons. As the ICRC *Interpretive Guidance* states, 'when civilians are totally unaware of the role they are playing in the conduct of hostilities...or when they are completely deprived of their physical freedom of action', they 'remain protected against direct attack despite the belligerent nexus of the military operation in which they are being instrumentalized. As a result, these civilians would have to be taken into account in the proportionality assessment during any military operation likely to inflict incidental harm on them'.[316]

(e) Civilians accompanying the armed forces

Under Article 4(A)(4) of Geneva Convention III, civilians 'who accompany the armed forces without actually being members thereof, such as civilian members of military aircraft crews, war correspondents, supply contractors, members of labour units or of services responsible for the welfare of the armed forces, provided that they have received authorization from the armed forces which they accompany', are entitled to POW status.[317] In general, this category includes civilians that do not perform functions directly related to combat, for instance technicians tasked with the general maintenance of computers, networks, and websites, as long as they do not prepare or execute cyber operations or are involved in the defence of military objectives.[318] The ordinary meaning of 'to accompany' suggests that the provision applies only to those that physically follow the armed forces.

As civilians accompanying the armed forces do not have combat or combat-related functions, they retain their civilian status and may not be directly attacked. They become military objectives only if they take direct part in hostilities for such time as they do so, and can also be incidentally hit in an attack aimed at a military objective, in which case they fall in the proportionality calculation.

(f) Civil defence personnel

Civilians that exercise humanitarian functions 'intended to protect the civilian population against the dangers, and to help it to recover from the immediate

---

[316] ICRC, *Interpretive Guidance*, p 60. *Contra*, see Geiss and Lahmann, 'Cyber Warfare', p 385, according to whom botnets are military objectives. It should be recalled that the presence of combatants within the population does not deprive the population of its civilian character (Art 50(3) of Additional Protocol I).

[317] See also Recommendation III, in ICRC, *Interpretive Guidance*, p 37. Civilian employees are different from contractors in that while the former are in an employment relationship with the armed forces, that can therefore supervise them, the latter work for a firm that has a contractual relationship with the armed forces and are supervised by the firm itself (Ricou Heaton, 'Civilians', p 174).

[318] Owens, Dam, and Lin, *Technology, Policy, Law, and Ethics*, p 266; Turns, 'Cyber Warfare', p 292; Doswald-Beck, 'Some Thoughts', p 172.

effects, of hostilities or disasters and also provide the conditions necessary for its survival', as well as members of the armed forces and military units assigned to civil defence organizations, must be respected and protected from attack.[319] Civilian civil defence organizations, their personnel, buildings, shelter, and *matériel* should be properly identified by an international distinctive sign[320] and their protection ceases if they 'commit or are used to commit, outside their proper tasks, acts harmful to the enemy'.[321]

Article 61(a) of Additional Protocol I includes as examples of civil defence tasks 'emergency repair of indispensable public utilities' and 'assistance in the preservation of objects essential for survival'. Individuals assigned to the cyber defence of exclusively civilian NCIs or to the resumption of disrupted critical civilian services may fall into this category. On the other hand, those entrusted with the defence of military objectives are excluded: they would be civilians taking direct part in the hostilities or, if incorporated in the armed forces, combatants at all effects and purposes.

(g) *Levée en masse*

The expression 'cyber *levée en masse*' is sometimes used to describe 'a mass networked mobilization that emerges from cyber-space with a direct impact on physical reality'.[322] The notion of *levée en masse* only applies in international armed conflicts. According to Article 2 of the 1907 Hague Regulations, '[t]he inhabitants of a territory which has not been occupied, who, on the approach of the enemy, spontaneously take up arms to resist the invading troops without having had time to organize themselves in accordance with Article 1, shall be regarded as belligerents if they carry arms openly and if they respect the laws and customs of war'. Even though not members of a belligerent's regular or irregular armed forces or of organized armed groups, then, participants in a *levée en masse* do not qualify as civilians,[323] and may therefore be the object of attack.[324] Article 4(A)(6) of Geneva Convention III grants POW status to persons taking part in a *levée en masse* in case of capture 'provided they carry arms openly and respect the laws and customs of war'.

It is difficult to see how participants in a cyber *levée en masse* can carry arms 'openly'.[325] Of course, no problem would exist in case of a traditional *levée en masse* where the participants, in addition to using traditional weapons, *also* conduct cyber operations against the approaching enemy.[326] If, however, the individuals in

---

[319] Articles 61, 62 and 67 of Additional Protocol I.   [320] Article 66 of Additional Protocol I.
[321] Article 65 of Additional Protocol I.
[322] See for instance Audrey Kurth Cronin, 'Cyber-Mobilization: The New *Levée en Masse*', *Parameters*, Summer 2006, p 77.
[323] See Recommendation I, in ICRC, *Interpretive Guidance*, p 20.
[324] Conventional law does not specify how many inhabitants must take up arms against the invading army (the ICRC Commentary only explains that the situation implies 'mass rising': Jean S Pictet (ed), *Commentary on the Geneva Conventions of 12 August 1949* (Geneva: International Committee of the Red Cross, 1952–60), Vol 3, p 48).
[325] Turns, 'Cyber Warfare', p 293.
[326] The ordinary meaning of 'invasion' excludes that a *levée en masse* could take place against a massive cyber attack not accompanied by the use of ground forces (Commentary to Rule 27, in Tallinn Manual, p 103).

question conduct exclusively cyber operations, Melzer suggests that the requirement to carry arms openly is met 'when cyber operations are not conducted by feigning protected, non-combatant status within the meaning of the prohibition of perfidy'.[327] Be that as it may, as has been previously observed, the visibility requirements are only relevant for the application of the law of detention, ie to determine whether or not the captured individual is entitled to POW status.[328]

Accordingly, if a *levée en masse* is occurring, an individual who is an 'inhabitant' of the territory being invaded but not yet occupied and who spontaneously conducts cyber operations amounting to 'attack' ('take up arms') to resist the invading army is a combatant and may thus be attacked for as long as he participates in the *levée*.[329] If the individual also carries arms openly and respects the laws and customs of war, he will be entitled to POW status in case he falls in the hands of the enemy.

It is doubtful whether the *levée en masse* can be conducted against the territory of the attacker, and not only against the invading army: '[t]here is no legal precedent for a *levée en masse* bringing the fight to the attacker's homeland', as would likely be the case of cyber operations.[330] The ICRC Commentary also explains that Article 4(A)(6) of Geneva Convention III 'is not applicable to inhabitants of a territory who take to the "maquis", but only to mass movements which *face* the invading forces',[331] which might be difficult to reconcile with the covert and anonymous character of cyber operations.

1.4 Are there geographical limitations to attacks on combatants and on civilians conducting cyber operations that amount to direct participation in hostilities?

The question whether combatants and civilians taking direct part in hostilities may be attacked anywhere or only in certain areas is particularly relevant in the cyber context: given the remote character of cyber operations and the interconnectivity of networks, it might well be that the operation is carried out from computers located on the territory of a non-belligerent state or even from within the territory of the attacked state. Nowhere does conventional international humanitarian law expressly provide for geographical limitations with regard to attacks on military

---

[327] Melzer, *Cyberwarfare*, p 34. Other commentators suggest that the notion of cyber *levée en masse* is 'an unworkable anachronism' and that '[w]hether due to the irresolvable distinction problem, or because of a complete dissimilarity between a traditional *levée en masse* and the cyber variant, it is untenable to maintain this combatant category in cyber warfare' (David Wallace and Shane R Reeves, 'The Law of Armed Conflict's 'Wicked' Problem: *Levée en Masse* in Cyber Warfare', *International Law Studies* 89 (2013), pp 664, 661). These commentators recommend instead that 'all assemblages of cyber participants, either ad hoc or pre-existing,... generally comply with the criteria that define militias, volunteer corps or organized resistance movements' (p 664).

[328] See Chapter 4, Section III.1.3.b, pp 198–9.

[329] The conduct of cyber operations such as the defacement of websites would thus not be sufficient for participation in a *levée en masse*.

[330] Brown, 'A Proposal', p 192. A *levée en masse* is also not possible against a retreating army (Ronzitti, *Diritto internazionale*, p 173).

[331] Pictet (ed), *Commentary*, Vol 3, p 67 (emphasis added).

objectives: on the contrary, Article 49(2) states that '[t]he provisions of [Additional Protocol I] with respect to attacks apply to all attacks *in whatever territory conducted*'.[332] If an individual is a combatant or a civilian taking direct part in hostilities in an international armed conflict, then, he may be attacked anywhere, in the former case at all times (unless he is *hors de combat*), in the latter for such time as he takes direct part in the hostilities, so long as the attack does not amount to a perfidious act and consistently with the principle of proportionality and the duty to take precautions in attack. In other words, it is the status (combatant, civilian, or civilian taking direct part in hostilities) of the person and not his location that makes him targetable or not under the *jus in bello*.[333] The case of the Japanese General Yamamoto, shot down by US military aircraft when overflying Bougainville on 18 April 1943, is exemplary in this sense.

If attacks against combatants or civilians taking direct part in hostilities on the territory of a non-belligerent state are not specifically prohibited by the *jus in bello*, however, they may amount to an unlawful use of force against that state under the *jus ad bellum* and a violation of its sovereignty, unless the consent of the territorial state to the operation has been previously obtained, self-defence may be invoked or the UN Security Council has authorized the operation.[334] In case the combatant or civilian taking direct part in hostilities is located on the territory of the attacked state itself, it has been observed that 'it would hardly seem justifiable to apply a conduct of hostilities paradigm in such a context, thereby rendering not only the individual hacker prone to a lethal attack, but also subjecting bystanding civilians to the humanitarian proportionality framework'.[335] This conclusion finds support in the Israeli Supreme Court's Judgment on *Targeted Killings*, where it states that 'a civilian taking direct part in hostilities cannot be attacked at such time as he is doing so, if a less harmful means can be employed.... Thus, if a terrorist taking a direct part in hostilities can be arrested, interrogated, and tried, those are the means which should be employed'.[336] The Court considers this solution 'particularly practical' in case of occupied territory, 'in which the army controls the area in which the operation takes place, and in which arrest, investigation, and trial are at times realizable possibilities':[337] this conclusion would seem to apply even more strongly to the state's own territory. While the Court's argument is limited

---

[332] Emphasis added. According to Hays Parks, Art 23(b) of the Hague Regulations, that prohibits the treacherous killing or wounding of enemy combatants, 'does not...preclude attacks on individual soldiers or officers of the enemy whether in the zone of hostilities, occupied territory, or elsewhere' (W Hays Parks, 'Memorandum on the Law: Executive Order 12333 and Assassination', November 1989, p 6, <http://www.hks.harvard.edu/cchrp/Use%20of%20Force/October%202002/Parks_final.pdf>).

[333] As Geoffrey Corn puts it, 'once the armed conflict door is open, threat-based strategy—focusing military action in response to threat dynamics in order to destroy or disable threat capabilities—is essentially opportunity driven: the conflict follows the belligerent target' (Corn, 'Geography', p 89).

[334] The problem has famously arisen with regard to the use of drones by the United States in Pakistan and other countries. If the conflict is of an international character, issues related to the law of neutrality would also arise: see Chapter 5.

[335] Robin Geiss, 'The Conduct of Hostilities in and via Cyberspace', *American Society of International Law Proceedings* 104 (2010), pp 373–4.

[336] *Targeted Killings*, para 40 (per Judge Barak).   [337] *Targeted Killings*, para 40 (per Judge Barak).

to civilians taking direct part in hostilities, Recommendation IX of the ICRC *Interpretive Guidance* controversially extends the capture-rather-than-kill standard to all 'persons not entitled to protection against direct attack', including combatants, in situations analogous to policing in peacetime.[338] A capture-rather-than-kill obligation may well arise from the application of international human rights law in internal armed conflicts.[339] According to the *Report of the International Commission of Inquiry on Libya*, for instance,

> international human rights law obligations remain in effect and operate to limit the circumstances when a state actor—even a soldier during internal armed conflict—can employ lethal force. This is particularly the case where the circumstances on the ground are more akin to policing than combat. For example, in encountering a member of the opposing forces in an area far removed from combat, or in situations where that enemy can be arrested easily and without risk to one's own forces, it may well be that the international humanitarian law regime is not determinative. In such situations, combatants/fighters should ensure their use of lethal force conforms to the parameters of international human rights law.[340]

It remains to be convincingly demonstrated, however, that this standard is also a specific obligation under international humanitarian law.

1.5 Ruses of war and the prohibition of perfidy

The fact that combatants and civilians taking direct part in hostilities may be attacked does not mean that this can occur without limits. The prohibition of superfluous injury and unnecessary suffering has already been examined.[341] What is left to discuss is the prohibition of perfidious acts.

Article 37(2) of Additional Protocol I explains that the 'use of camouflage, decoys, mock operations and misinformation' are lawful ruses of war: this is particularly relevant in the cyber context. Indeed, anyone launching cyber attacks can disguise their origin thanks to anonymizers like IP spoofing or the use of botnets.

---

[338] Recommendation IX of the ICRC *Interpretive Guidance* provides that '[i]n addition to the restraints imposed by international humanitarian law on specific means and methods of warfare, and without prejudice to further restrictions that may arise under other applicable branches of international law, the kind and degree of force which is permissible against persons not entitled to protection against direct attack must not exceed what is actually necessary to accomplish a legitimate military purpose in the prevailing circumstances' (ICRC, *Interpretive Guidance*, p 77). The Commentary to the Guidance specifies that Recommendation IX becomes relevant 'where armed forces operate against selected individuals in situations comparable to peacetime policing' (p 80). See also Larry May, 'Targeted Killings and Proportionality in Law', *Journal of International Criminal Justice* 11 (2013), pp 51–3. *Contra*, see W Hays Parks, 'Part IX of the ICRC "Direct Participation in Hostilities" Study: No Mandate, No Expertise, and Legally Incorrect', *New York Journal of International Law and Politics* 42 (2009–10), pp 769 ff; Amichai Cohen and Yuval Shany, 'A Development of Modest Proportions. The Application of the Principle of Proportionality in the *Targeted Killings* Case', *Journal of International Criminal Justice* 5 (2007), p 314.

[339] It seems that, in the above-mentioned judgment, the Israeli Supreme Court is also taking an international human rights law perspective. This is suggested by the reference to the European Court of Human Rights' case-law (*Targeted Killings*, para 40 (per Judge Barak)).

[340] *Report of the International Commission of Inquiry on Libya*, UN Doc A/HRC/19/68, 2 March 2012, para 145 (footnotes omitted).

[341] See Chapter 4, Section II, pp 172–3.

Using Trojan horses, ie apparently innocuous code fragments that actually conceal a harmful program or allow remote access to the computer by an external user, belligerents can also 'alter data in the enemy's computer databases, and... transmit to enemy subordinate units false message that appear to come from their headquarters'.[342] Such operations would be lawful ruses of war.[343] Other examples of cyber ruses of war are corrupting a link so that it redirects the users to a fictitious webpage where false military data are given, onion routing, ie 'establishing a path through a maze of multiple onion routers, each of which accepts a packet from a previous router and forwards it on to another onion router',[344] polymorphic malware that alters its signature every time it replicates and spreads to another computer to avoid detection, rootkits designed to hide the existence of certain processes or programs from normal methods of detection so to allow unauthorized access to the system, and honeynets, ie systems without security safeguards designed to attract intrusions and containing false information to mislead the intruder.[345] Stuxnet itself had two components: one designed to force a change in the centrifuges' rotor speed, inducing excessive vibrations or distortions, and one that recorded the normal operations of the plant and sent the recording back to plant operators so to make it look as everything was functioning normally.[346] Had it occurred in armed conflict, the second component would have been a lawful *ruse de guerre* functional to the successful execution of the attack.

The resort to ruses of war finds, however, a limit in the prohibition of perfidious acts, which protects both combatants and civilians taking direct part in hostilities. According to Article 37(1) of Additional Protocol I

[i]t is prohibited to kill, injure or capture an adversary by resort to perfidy. Acts inviting the confidence of an adversary to lead him to believe that he is entitled to, or is obliged to accord, protection under the rules of international law applicable in armed conflict, with intent to betray that confidence, shall constitute perfidy. The following acts are examples of perfidy:

(a) the feigning of an intent to negotiate under a flag of truce or of a surrender;
(b) the feigning of an incapacitation by wounds or sickness;

---

[342] Dinstein, *The Conduct*, p 240.
[343] The following examples of ruses of war, given in the UK Military Manual, can also be easily transposed into the cyber context: 'transmitting bogus signal messages and sending bogus dispatches and newspapers with a view to their being intercepted by the enemy; making use of the enemy's signals, passwords, radio code signs, and words of command; conducting a false military exercise on the radio while substantial troop movements are taking place on the ground; pretending to communicate with troops or reinforcements which do not exist; [...] giving false ground signals to enable airborne personnel or supplies to be dropped in a hostile area, or to induce aircraft to land in a hostile area; and feint attacks to mislead the enemy as to the point of the main attack' (UK Ministry of Defence, *The Manual of the Law of Armed Conflict*, p 64).
[344] Owens, Dam, Lin, *Technology, Policy, Law, and Ethics*, p 98.
[345] Katharina Ziolkowski, 'Computer Network Operations and the Law of Armed Conflict', *Military Law and Law of War Review* 49 (2010), pp 77–8. See also the examples provided in the Commentary to Rule 61, in Tallinn Manual, p 184.
[346] William J Broad, John Markoff, and David E Sanger, 'Israeli Test on Worm Called Crucial in Iran Nuclear Delay', *The New York Times*, 15 January 2011, <http://www.cfr.org/iran/nyt-israeli-test-worm-called-crucial-iran-nuclear-delay/p23850>.

(c) the feigning of civilian, non-combatant status; and
(d) the feigning of protected status by the use of signs, emblems or uniforms of the United Nations or of neutral or other States not Parties to the conflict.[347]

For a cyber operation to be perfidious, then, deception or anonymity are not sufficient: the operation has to 'invite the confidence of an adversary with respect to protection' under the law of armed conflict and betray it, and also result in the death, injury, or capture of the adversary.[348] Such consequences must occur, as a mere intention of causing them is not sufficient.[349] Sending malware attached to an email appearing to be from the United Nations, a neutral state or states not parties to the conflict which, once executed, results in infrastructure malfunction causing loss of life or injury would then be a perfidious act. As the prohibition of perfidy is not limited to 'attack', a cyber operation that defaces a news website and makes enemy combatants believe that a truce has been signed, so that they depose the arms and are killed, injured, or captured in a surprise attack, would also be a perfidious act. If the cyber operation leads to destruction or loss of functionality of infrastructure but not to death, injury, or capture of the adversary, it would not amount to a violation of Article 37 of Additional Protocol I, although it may fall under the prohibitions contained in Articles 38 and 39 of Additional Protocol I and Article 12 of Additional Protocol II on the misuse of certain emblems.[350]

2. The prohibition of indiscriminate attacks

The law of targeting prohibits not only direct attacks on civilians, the civilian population and civilian objects, but also 'indiscriminate' attacks: in this case, the attacker does not intend to directly harm protected persons or objects, but does not care about possible incidental civilian harm.[351] Under Article 51 of Additional Protocol I, indiscriminate attacks are:

4. ... (a) those which are not directed at a specific military objective;
    (b) those which employ a method or means of combat which cannot be directed at a specific military objective; or

---

[347] Rule 65 of the ICRC Study on *Customary International Humanitarian Law* extends the prohibition of perfidious acts to non-international armed conflicts (Henckaerts and Doswald-Beck, *Customary International Humanitarian Law*, Vol I, p 221). On the applicability of the prohibition of perfidy in non-international armed conflicts, see Richard B Jackson, 'Perfidy in Non-International Armed Conflicts', *International Law Studies* 88 (2012), pp 237 ff.

[348] As has been observed, in ruses of war '[t]he deceit takes place on a point of fact (on the military operations) and not on a point of law (the protections under the LOAC)' (Kolb and Hyde, *An Introduction*, p 164). It is unclear whether the inclusion of 'capture' also reflects customary international law (Dinstein, *The Conduct*, p 231).

[349] Dinstein, *The Conduct*, pp 231–2. But see the ICRC Commentary, according to which 'it seems evident that the attempted or unsuccessful act also falls under the scope of this prohibition' (Sandoz, Swinarski, and Zimmermann (eds), *Commentary*, para 1493).

[350] See also Rules 57–64 of the ICRC Study (Henckaerts and Doswald-Beck, *Customary International Humanitarian Law*, Vol I, pp 203 ff) and Rules 62–5 of the Tallinn Manual, pp 185–92. Article 39(2) of Additional Protocol I, in particular, prohibits 'to make use of the flags or military emblems, insignia or uniforms of adverse Parties while engaging in attacks or in order to shield, favour, protect or impede military operations'.

[351] Dinstein, *The Conduct*, p 127.

(c) those which employ a method or means of combat the effects of which cannot be limited as required by this Protocol;

and consequently, in each such case, are of a nature to strike military objectives and civilians or civilian objects without distinction.

5. Among others, the following types of attacks are to be considered as indiscriminate:

(a) an attack by bombardment by any methods or means which treats as a single military objective a number of clearly separated and distinct military objectives located in a city, town, village or other area containing a similar concentration of civilians or civilian objects;

and

(b) an attack which may be expected to cause incidental loss of civilian life, injury to civilians, damage to civilian objects, or a combination thereof, which would be excessive in relation to the concrete and direct military advantage anticipated.[352]

The above prohibitions also apply to cyber operations amounting to 'attack' in the sense of Article 49(1) of Additional Protocol I.[353] A cyber operation would for instance be indiscriminate if it is 'launched against all enemy computers—without any effort being made to differentiate between them on the basis of military or civilian nature, use, purpose or location'.[354]

Article 51(4)(a) concerns all attacks not specifically directed at a military objective: as the Commentary to Rule 49 of the Tallinn Manual specifies, whether an attack is indiscriminate in this sense is a situational evaluation based on factors such as the nature of the attacked system and the means or method employed, the accuracy of planning and any evidence of indifference on the part of those involved in the attack.[355] Article 51(4)(b) and (c) have been the object of Section II of this Chapter. The difference between them and the case covered in Article 51(4)(a) is that in the latter the means or method of warfare are capable of being directed at a military objective but the attacker fails to use them discriminately, while in (b) and (c) the means or method are inherently incapable of being used discriminately.

With regard to Article 51(5)(a), the Commentary to Rule 50 of the Tallinn Manual, which incorporates it, recommends that the rule should not be interpreted territorially.[356] A computer network composed of military and civilian computers or a server farm hosting both military and civilian servers, therefore, are not military objectives in their entirety if the military computers or servers can be singled out and attacked discretely.[357] If this is not possible, the attack against the network or server farm would be lawful under the principle of distinction but would still have to take into account the incidental damage caused to the civilian computers and servers, as provided by Article 51(5)(b), which incorporates, without expressly

---

[352] These prohibitions reflect customary international law and also apply in non-international armed conflicts (Rules 8, 11–13 of the ICRC Study, in Henckaerts and Doswald-Beck, *Customary International Humanitarian Law*, Vol I, pp 29, 37, 40, 43; *Tadić*, Decision on the Defence Motion, para 127).

[353] See the US comments submitted to the UN Secretary-General, UN Doc A/66/152, 15 July 2011, p 19.

[354] Dinstein, 'The Principle of Distinction', p 267.   [355] Tallinn Manual, p 157.

[356] Tallinn Manual, p 158.   [357] Tallinn Manual, p 158.

naming it, the principle of proportionality in attack. The application of this provision to cyber operations will be examined in the next sub-Section.

2.1 The principle of proportionality

Civilians and civilian objects may be incidentally hit as the collateral result of an attack directed against military objectives. According to Article 51(5)(b) of Additional Protocol I, an attack would be indiscriminate, and thus prohibited, if it 'may be expected to cause incidental loss of civilian life, injury to civilians, damage to civilian objects, or a combination thereof, which would be excessive in relation to the concrete and direct military advantage anticipated'.[358] This provision, commonly referred to as the rule of proportionality,[359] has been incorporated in several military manuals and doctrines, including the German Military Manual,[360] the UK Military Manual,[361] the French *Manuel de droit des conflits armés*,[362] and the US *Joint Doctrine for Targeting*.[363] It reflects customary international law and, although it does not appear expressly in Additional Protocol II, it is generally accepted that it also applies to attacks in non-international armed conflicts.[364]

The *Final Report by the Committee established to Review the NATO Bombing Campaign Against the Federal Republic of Yugoslavia* noted that '[t]he main problem with the principle of proportionality is not whether or not it exists but what it means and how it is to be applied'.[365] Indeed, '[t]he intellectual process of balancing the various elements is so complicated, needs to take into account such a huge amount of data and so many factors, that any attempt to design a

---

[358] The same language appears in Art 57(2)(a)(iii) and Art 57(2)(b) of Additional Protocol I in relation to the duty to take precautions in attack on which (see Section III.4.1 in this Chapter.

[359] The expression is explicitly used in the UK Military Manual (UK Ministry of Defence, *The Manual of the Law of Armed Conflict*, p 68). Proportionality operates differently in *jus in bello* and *jus ad bellum*: while in the latter case it is a requirement for the legality of a self-defence reaction, applies to the operation as a whole and balances the armed reaction against the purpose of defeating an armed attack, in the former it is a limitation that applies to each individual attack and balances the concrete and direct military advantage anticipated from the attack against the expected incidental damage to protected property and persons. Furthermore, unlike in the *jus ad bellum* notion of proportionality, where the interest of the attacked state is given superior standing with respect to that of the attacker, in the *jus in bello* proportionality is a normative technique that aims to reconcile two values of equal rank (Enzo Cannizzaro, 'Contextualizing proportionality: *jus ad bellum* and *jus in bello* in the Lebanese war', *International Review of the Red Cross* 88 (2006), pp 786–7).

[360] German Ministry of Defence, *Humanitarian Law in Armed Conflicts*, Section 456.

[361] UK Ministry of Defence, *The Manual of the Law of Armed Conflict*, p 86.

[362] Ministère de la Défense, *Manuel de droit des conflits armés*, Definition of 'Proportionnalité (principe de)', <http://www.defense.gouv.fr/content/download/77498/693317/file/Manuel_de_droit_des_conflits_armes.pdf>.

[363] *US Joint Doctrine for Targeting*, p A–1.

[364] See Rule 14 of the ICRC Study, in Henckaerts and Doswald-Beck, *Customary International Humanitarian Law*, Vol I, p 46. Indeed, it follows from Art 13(1) of Additional Protocol II, according to which civilians 'enjoy general protection against the dangers arising from military operations', that incidental civilian losses must be avoided or at least minimized (Boothby, *The Law of Targeting*, p 436). See also *Targeted Killings*, para 42 (Barak); Schmitt, Garraway and Dinstein, *The Manual on the Law of Non-International Armed Conflict*, Rule 2.1.1.4, p 22.

[365] *ICTY Final Report*, para 48.

formula which is both comprehensive and precise would be ridiculous'.[366] These complexities are even more evident in the cyber context, where 'uncertainties in outcome...are significantly greater than those usually associated with kinetic attacks in the sense that there may not be an analytic or experiential basis for estimating uncertainties at all'.[367]

It is, however, not disputed that the principle of proportionality extends to cyber operations amounting to 'attacks' under the law of armed conflict.[368] In fact, if one considers that most information infrastructures are dual-use and therefore potentially military objectives, the principle of proportionality may play in the cyber context a more significant role in the protection of civilians than the principle of distinction.[369] The question, then, is not if, but *how* the principle of proportionality applies to cyber operations. The principle entails balancing two parameters (incidental damage and military advantage) of different nature but of equal standing in specific attacks.[370] But what do incidental 'damage' and 'military advantage' mean in the cyber context? The analysis first addresses the former and then moves to the latter, before finally discussing their relationship to each other.

As with kinetic attacks, it is only the incidental damage on civilians and civilian property that is relevant in the application of the principle of proportionality: damage on military objectives and injury/death of combatants and civilians taking direct part in hostilities do not count in the equation.[371] What damage to civilians and civilian objects could then be 'incidentally' caused by a cyber operation? As has been already seen, the effects of a cyber operation can be distinguished in primary effects, ie those on the attacked data and software, secondary effects, ie those on the infrastructure operated by the attacked system (if any), and tertiary effects, ie those on the persons affected by the destruction or incapacitation of the attacked system or infrastructure. Primary, secondary, and tertiary effects all fall in the notion of incidental damage on civilians and civilian objects when they are an expected consequence of the attack.[372]

---

[366] Oeter, 'Methods and Means' p 191. In certain countries, such as the United Kingdom, targeteers are given Collateral Damage Estimate parameters in 'tiers' corresponding to different amounts of casualties: higher 'tiers' require higher levels of authorization.

[367] Owens, Dam, and Lin, *Technology, Policy, Law, and Ethics*, p 262.

[368] Koh, 'International Law in Cyberspace', pp 595–6. See Rule 51 of the Tallinn Manual, p 159. The recently published French White Paper on defence and national security also states that, with regard to hostile cyber attacks, '[l]es capacités d'identification et d'action offensive sont essentielles pour une riposte éventuelle et proportionnée à l'attaque' (*Livre blanc, Défense et sécurité nationale*, 2013, p 73, <http://www.elysee.fr/assets/pdf/Livre-Blanc.pdf>).

[369] Droege, 'Get Off My Cloud', p 566.

[370] According to the *ICTY Final Report*, '[i]t is much easier to formulate the principle of proportionality in general terms than it is to apply it to a particular set of circumstances because the comparison is often between unlike quantities and values' (*ICTY Final Report*, para 48).

[371] In the *Nuclear Weapons* Advisory Opinion, the ICJ also found that 'States must take environmental considerations into account when assessing what is necessary and proportionate in the pursuit of legitimate military objectives' (*Nuclear Weapons*, para 30).

[372] For instance, if a power station used by both the military and civilians is destroyed or incapacitated, hospitals and civilian infrastructures like water purification plants might be deprived of electricity for a certain amount of time (Henry Shue and David Wippman, 'Limiting Attacks on Dual-Use Facilities Performing Indispensable Civilian Functions', *Cornell International Law Journal* 35 (2001–02), p 564). See also Boothby, *The Law of Targeting*, pp 384–5; James W Crawford, III, 'The Law of

The fact that secondary and tertiary effects, at least those that are 'expected', should also be included in the proportionality calculation results clearly from the US Counterinsurgency Field Manual, which explicitly states that, in order to establish whether an attack is discriminate, '[l]eaders must consider not only the first-order, desired effects of a munition or action but also possible second- and third-order effects—including undesired ones'.[373] According to the US *Commander's Handbook on the Law of Naval Operations*, in case of a non-kinetic computer network attack 'factors involved in weighing anticipated incidental injury/death to protected persons can include, depending on the target, indirect effects (for example, the anticipated incidental injury/death that may occur from disrupting an electricity-generating plant that supplies power to a military headquarters and to a hospital)'.[374] In his speech at CYBERCOM, the US Department of State's Legal Advisor confirmed that proportionality 'requires parties to a conflict to assess: (1) the effects of cyber weapons on both military and civilian infrastructure and users, including shared physical infrastructure (such as a dam or a power grid) that would affect civilians; (2) the potential physical damage that a cyber attack may cause, such as death or injury that may result from effects on critical infrastructure; and (3) the potential effects of a cyber attack on civilian objects that are not military objectives, such as private, civilian computers that hold no military significance, but may be networked to computers that are military objectives'.[375] The Commentary to Rule 51 of the Tallinn Manual endorses the inclusion of both direct and indirect effects that 'should be expected by those individuals planning, approving, or executing a cyber attack'.[376] Some commentators have suggested that knock-on effects should be included in the proportionality equation when such effects 'would not have occurred "but for" the attack'.[377] Others have suggested that the fact that malware, once it becomes available, could be misused by malicious third parties should also be factored in the proportionality calculation.[378] This, however, goes too far. As the use of the adjective 'expected' in Article 51(5)(b) in relation to incidental damage suggests,[379] the crux of the matter is whether the effect is a reasonably likely or foreseeable consequence of the operation on the basis of the information available at the time of the attack: 'remote effects will generally be beyond the attacking commander's ability to reliably predict and are probably within the defenders' control'.[380]

Noncombatant Immunity and the Targeting of National Electrical Power Systems', *Fletcher Forum of World Affairs* 21 (1997), p 114; Owens, Dam, and Lin, *Technology, Policy, Law, and Ethics*, p 264.

[373] Department of the Army, *Counterinsurgency*, FM No 3-24 (December 2006), para 7-36, <http://www.fas.org/irp/doddir/army/fm3-24.pdf>.

[374] *The Commander's Handbook*, para 8.11.4.   [375] Koh, 'International Law in Cyberspace', p 596.

[376] Tallinn Manual, p 160. The relevance of indirect prejudicial effects in the law of targeting is confirmed by provisions like Arts 54 and 56 of Additional Protocol I, which prohibit attacks on certain objects because they are indispensable to the survival of the civilian population or because they may cause the release of certain dangerous forces.

[377] Schmitt, Harrison Dinniss, and Wingfield, *Computers and War*, p 9.

[378] Richmond, 'Evolving Battlefields', p 893.   [379] Harrison Dinniss, *Cyber Warfare*, p 208.

[380] Joseph Holland, 'Military Objectives and Collateral Damage: Their Relationship and Dynamics', *Yearbook of International Humanitarian Law* 7 (2004), p 62. See also ICTY, *Prosecutor v Galić*, Case No IT-98-29-T, Trial Chamber I Judgment, 5 December 2003, para 58.

With regard to the incidental primary effects of cyber operations, mere deletion or corruption of software or data may also amount to damage to civilian 'objects' for the purposes of the proportionality calculation. It is true that the ordinary meaning of the word 'object' seems to suggest a different conclusion: the ICRC Commentary describes an object as 'something placed before the eyes, or presented to the sight or other sense, an individual thing seen, or perceived, or that may be seen or perceived; a material thing'.[381] As has been observed, however, the Commentary was issued 'at a particular point in time and in a specific context', where the drafters did not envisage the destruction of data.[382] It should be recalled, however, that, if a cyber operation exclusively corrupts, alters or deletes information data without violent consequences in the analogue world, it would not be an 'attack' in the sense of Article 49(1) of Additional Protocol I and therefore issues of proportionality under Article 51(5)(b) would not arise.[383]

As to the secondary effects of cyber operations, the question arises whether incidental 'damage' on protected objects includes not only physical damage, but also loss of functionality.[384] It has been noted that, while Article 52(2) distinguishes between destruction and neutralization, Article 51(5)(b) only refers to 'damage', which is broad enough to include the loss of functionality without physical destruction.[385] In fact, '[i]t would appear counter-intuitive that only the physical destruction of a civilian object should be taken into consideration, whereas functionality loss—even if it affects the civilian population much more severely—should be irrelevant'.[386] This view is correct. Indeed, whereas incidental damage for the purposes of proportionality clearly does not include 'inconvenience, irritation, stress, or fear' as they cannot be compared to 'loss of civilian life, injury to civilians, damage to civilian objects',[387] the incapacitation of networked infrastructure has potentially as severe tertiary effects on protected persons as its destruction. All in all, what matters is that the infrastructure is rendered inoperable, whether by destroying or incapacitating it. The principle of proportionality, and in particular the notion of 'incidental damage', should then be interpreted so to keep pace with the digitalization of essential services in today's society.[388] Targeteers will need to take into account the expected consequences arising from the loss of functionality of dual-use infrastructures caused by a cyber operation amounting

---

[381] Sandoz, Swinarski, and Zimmermann (eds), *Commentary*, para 2007.
[382] Lubell, 'Lawful Targets', p 267. See also Melzer, *Cyberwarfare*, p 31.
[383] See Section III.1.1 in this Chapter.
[384] Eric Talbot Jensen, 'Cyber Attacks: Proportionality and Precautions in Attack', *International Law Studies* 89 (2013), pp 206–7.
[385] Geiss and Lahmann, 'Cyber Warfare', p 397. See also the Commentary to Rule 51 of the Tallinn Manual, p 160.
[386] Geiss and Lahmann, 'Cyber Warfare', p 397.
[387] Tallinn Manual, p 160. Similarly, mere 'passing through' civilian computers without causing damage would not count as incidental damage (John Richardson, 'Stuxnet as Cyberwarfare: Applying the Law of War to the Virtual Battlefield', *Journal of Computer and Information Law* 29 (2011), p 26).
[388] See also Droege, according to whom 'disrupting the functioning of certain systems by interfering with their underlying computer systems can amount to damage insofar as it impairs their usefulness' (Droege, 'Get Off My Cloud', p 559).

to 'attack' in their proportionality calculation.[389] This is spelt out in the 2010 US *Joint Terminology for Cyberspace Operations*: according to this document, 'collateral effect' of cyber operations in the context of targeting includes the 'unintentional or incidental...effects on civilian or dual-use computers, networks, information, or infrastructure' when 'there is a reasonable probability of loss of life, serious injury, *or* serious adverse effect on the affected nation's national security, economic security, public safety, or an combination of such effects'.[390]

It has been claimed that, as cyber operations can be used to incapacitate, instead of destroying, an object, they might expand the scope of what is targetable. Kelsey has, for instance, maintained that '[t]he potentially nonlethal nature of cyber weapons may cloud the assessment of an attack's legality, leading to more frequent violations of the principle of distinction in this new form of warfare than in conventional warfare'.[391] According to this view, international humanitarian law protects civilian objects because of the severe effects that conventional attacks have on them:[392] as disruptive cyber operations leave the object intact, they could be carried out against objects that only indirectly contribute to military action or whose neutralization offers a non-definite military advantage.[393] This view cannot be accepted, as it does not take into account that, in today's information societies, the neutralization of computer systems controlling NCIs could have far more serious effects on protected persons than certain kinetic attacks. Objects not fulfilling the definition of military objective, therefore, are and remain civilian objects and may not be attacked. However, certain attacks *against military objectives*, which would be unlawful if executed with kinetic weapons because they are expected to cause excessive incidental civilian damage, may be lawful if conducted by way of disruptive cyber operations.[394]

Unlike in kinetic attacks and similarly to biological weapons, the incidental damage caused by a cyber attack is not only that on objects and persons located within or near the attacked military objective or on the civilian function performed by the attacked dual-use installation, but also that caused to computer systems (and the infrastructure they might operate) to which the malware may spread as a consequence of its characteristics and the interconnectivity of networks. The proportionality calculation in a cyber operation that shuts down a dual-use power station, for instance, will have to factor in both the loss of the civilian function

---

[389] Tallinn Manual, p 160; Shue and Wippman, 'Limiting Attacks', p 570.
[390] *US Joint Terminology for Cyberspace Operations*, p 3 (emphasis added).
[391] Jeffrey TG Kelsey, 'Hacking into International Humanitarian Law: The Principles of Distinction and Neutrality in the Age of Cyber Warfare', *Michigan Law Review* 106 (2008), p 1439.
[392] Kelsey, 'Hacking', p 1440; Mark R Shulman, 'Discrimination in the Laws of Information Warfare', *Columbia Journal of Transnational Law* 37 (1998–99), p 964.
[393] Kelsey, 'Hacking', p 1448. See also Lawrence T Greenberg, Seymour E Goodman, and Kevin J Soo Hoo, *Information Warfare and International Law* (Washington: National Defense University Press, 1998), p 12. A similar argument seems to be implicit in Wegdwood's suggestion that proportionality could be conceived dynamically so to tolerate greater collateral damage to civilian objects to eliminate a security threats 'so long as the damage is reversible or, indeed, aid is given in its restoration' (Ruth G Wedgwood, 'Proportionality, Cyberwar, and the Law of War', *International Law Studies* 76 (2002), p 228).
[394] Dörmann, *Applicability*, p 6.

performed by the installation with consequent negative repercussions on its users *and* the fact that the malware might infect other computer systems, as long as it is an expected consequence of the operation on the basis of the information available at the time. The only inherently discriminate cyber operations from this perspective are those on systems that are part of a closed military network or 'flood' attacks, which only affect the system overloaded by the multiple requests.

In the *Kupreškić* judgment, the ICTY found that 'in case of repeated attacks, all or most of them falling within the grey area between indisputable legality and unlawfulness, it might be warranted to conclude that the cumulative effect of such acts entails that they may not be in keeping with international law. Indeed, this pattern of military conduct may turn out to jeopardise excessively the lives and assets of civilians, contrary to the demands of humanity'.[395] This is particularly relevant for multiple low-intensity cyber attacks (the 'death by a thousand cuts' scenario). According to the *ICTY Final Report*, however, the *Kupreškić* statement must be interpreted as referring to 'an *overall* assessment of the totality of civilian victims as against the goals of the military campaign', since 'the mere *cumulation* of such instances, all of which are deemed to have been lawful, cannot *ipso facto* be said to amount to a crime'.[396]

Let us now move to the other element of the proportionality equation, against which the incidental damage on civilians and civilian objects must be balanced, namely the 'concrete and direct military advantage' anticipated from the attack. If, in the context of the definition of 'military objective' (Article 52(2)), the military advantage has to be 'definite', Article 51(5)(b) of Additional Protocol I requires it to be 'concrete and direct', which is a stronger standard and imposes stricter limits on the attacker when incidental damage is expected. As a consequence, 'the advantage concerned should be substantial and relatively close, and...advantages which are hardly perceptible and those which would only appear in the long term should be disregarded'.[397] If 'concrete and direct' means 'a real and quantifiable benefit',[398] however, the problem with cyber operations is that measurement of their effects can be difficult: as already observed, for instance, it has not yet been confirmed whether Stuxnet did damage any centrifuges at Natanz and, if so, how many and how seriously.

In any case, according to the ICRC, in the context of the principle of proportionality '[a] military advantage can only consist in ground gained and in annihilating or weakening the enemy armed forces'.[399] In contrast with this narrow

---

[395] *Kupreškić*, para 526. According to the Tribunal, this interpretation follows from the application of the Martens clause codified in Art 1(2) of Additional Protocol I.
[396] *ICTY Final Report*, para 52 (emphasis added). See comments by Natalino Ronzitti, 'Is the *non liquet* of the Final Report by the Committee Established to Review the NATO Bombing Campaign Against the Federal Republic of Yugoslavia Acceptable?', *International Review of the Red Cross* 82 (2000), p 1017.
[397] Sandoz, Swinarski, and Zimmermann (eds), *Commentary*, para 2209.
[398] Commentary to Rule 51, in Tallinn Manual, p 161.
[399] Sandoz, Swinarski, and Zimmermann (eds), *Commentary*, para 2218. This interpretation has been criticized for being too narrow. According to the Commentary to the HPCR Manual, '[a] better approach is to understand military advantage as any consequence of an attack which directly enhances

interpretation, some states have claimed that military advantage should also include the protection of the attacking forces.[400] Canada's *Joint Doctrine Manual*, for instance, recalls that '[m]ilitary advantage may include a variety of considerations including the security of the attacking forces'.[401] Australia and New Zealand also issued declarations at the ratification of Additional Protocol I emphasizing that 'military advantage' includes the 'security of attacking forces'.[402] Israel's document on the 2009 Operation in Gaza claims that military advantage 'may legitimately include not only the need to neutralise the adversary's weapons and ammunition and dismantle military or terrorist infrastructure, but also—as a relevant but not overriding consideration—protecting the security of the commander's own forces'.[403] The United States' view is that '[t]he foreseeable military advantage from an attack includes increasing the security of the attacking force'.[404] In the *Targeted Killings* Judgment, Israel's Supreme Court supported the inclusion of force protection in the proportionality calculation where it held that '[t]he state's duty to protect the lives of its soldiers and civilians must be balanced against its duty to protect the lives of innocent civilians harmed during attacks on terrorists'.[405] If these views are correct, the remote character of cyber operations, and thus the enhanced security for the attacker, would increase the military advantage they provide and would therefore justify a higher amount of incidental damage on civilians and civilian objects.

The number of states that support a broad notion of 'military advantage' that includes force protection, however, is relatively limited and different positions have also been adopted.[406] The broad interpretation has the major disadvantage of introducing a further subjective element in the calculation of proportionality: for

---

friendly military operations or hinders those of the enemy. This could, e.g., be an attack that reduces the mobility of the enemy forces without actually weakening them, such as the blocking of an important line of communication' (HPCR Manual, p 36). See similarly Schmitt, Garraway, and Dinstein, *The Manual on the Law on Non-International Armed Conflict*, Rule 2.1.1.4, p 24.

[400] In literature, see Robin Geiss, 'The Principle of Proportionality: "Force Protection" as a Military Advantage', *Israel Law Review* 45 (2012), p 77. See also Dinstein, *The Conduct*, pp 141–2; Ian Henderson, *The Contemporary Law of Targeting* (Leiden: Nijhoff, 2009), p 205. *Contra*, see Fenrick, 'Attacking', p 549; Gary D Solis, *The Law of Armed Conflict* (Cambridge: Cambridge University Press, 2010), p 285.

[401] Office of the Judge Advocate General, Law of Armed Conflict Manual: At the Operational and Tactical Levels, Joint Doctrine Manual, B-GJ-005-104/FP-02, 2001, para 415, p 4–4, <http://www.fichl.org/uploads/media/Canadian_LOAC_Manual_2001_English.pdf>. Note, however, that, in the sentence, 'military advantage' is not preceded by 'concrete and direct'.

[402] Roberts and Guelff, *Documents*, pp 500, 508. Force protection is also mentioned in France's declaration upon ratification of Additional Protocol I, but in relation to the second sentence of Art 50(1) (declaration of 11 April 2011, para 9, <http://www.icrc.org/applic/ihl/ihl.nsf/Notification.xsp?action=openDocument&documentId=D8041036B40EBC44C1256A34004897B2>).

[403] State of Israel, *The Operation in Gaza (27 December 2008–18 January 2009). Factual and Legal Aspects*, July 2009, para 126, <http://www.mfa.gov.il/MFA_Graphics/MFA%20Gallery/Documents/GazaOperation%20w%20Links.pdf>.

[404] *Report on United States Practice in International Law*, 1997, cited in Henckaerts and Doswald-Beck, *Customary International Humanitarian Law*, Vol II, p 188.

[405] *Targeted Killings*, para 46 (per Judge Barak).

[406] The US Counterinsurgency Field Manual, for instance, states that the principles of proportionality and discrimination require combatants to '[a]ssume additional risk to minimize potential harm [to non-combatants]' (Department of the Army, *Counterinsurgency*, para 7–30).

instance, what is the value of the safety of military personnel with respect to civilian lives? Does it depend on the rank or the specialization of the member of the armed forces in question? Is the evaluation different when it is the safety of military *matériel* that is at stake? It should not be forgotten that only a 'concrete and direct' military advantage is relevant in the calculation of the proportionality of the attack, and not 'abstract protectiveness of the means and methods used to attack':[407] a belligerent 'cannot justify higher numbers of civilian casualties for the sole reason that it has opted for a more secure...operation instead of a less secure...operation'.[408] If force protection cannot be a determinant factor in the calculation of proportionality, however, it could be relevant in the context of the duty to take precautions.[409] As will be seen, Article 57(2)(a)(ii) of Additional Protocol I requires the belligerents to 'take all feasible precautions in the choice of means and methods of attack with a view to avoiding, and in any event to minimizing, incidental loss of civilian life, injury to civilians and damage to civilian objects'. It is clear that the determination of what is a 'feasible' precaution will consider the additional risks encountered by the attacking forces, which are under no obligation to sacrifice themselves. That this is the correct relevance of force protection in the framework of the law of targeting is confirmed in the UK Military Manual, which includes the 'risks to his own troops' among the factors that a commander needs to evaluate in the context of the duty to take precautions when choosing means or methods of attack.[410] A targeteer, then, *may* go for a certain means or method of warfare, including cyber operations, that minimizes the risk for the attacking forces (even if it is at the cost of reducing the military advantage he anticipates) but providing that this does not increase the expected incidental damage on civilians or civilian objects.

As with incidental damage, when the attack is composed of multiple hostile acts the military advantage is what results from the attack considered as a whole. Several NATO states added interpretive declarations in relation to Article 51(5)(b), stating that the attack has to be considered in its totality, and not in its specific parts.[411] Read in the light of these declarations, the concrete and direct military

---

[407] Geiss, 'The Principle of Proportionality', p 85.
[408] Geiss, 'The Principle of Proportionality', p 87.
[409] Oeter, 'Methods and Means', p 201; Tallinn Manual, p 164.
[410] UK Ministry of Defence, *The Manual of the Law of Armed Conflict*, p 84. The *Final Report to Congress on the Conduct of the Persian Gulf War* also recalls that coalition forces took the 'risk to aircraft and aircrews' into account when choosing means of warfare for attacks on targets in populated areas (US DoD, *Conduct of the Persian Gulf War*, p 622). According to the US Commander's *Handbook on the Law of Naval Operations*, 'the commander must decide, in light of all the facts known or reasonably available to him, *including the need to conserve resources and complete the mission successfully*, whether to adopt an alternative method of attack, if reasonably available, to reduce civilian casualties and damage' (*The Commander's Handbook*, para 8.3.1; emphasis added). It is not clear, however, whether this point is made in relation to the principle of proportionality or the duty to take precautions in attack. The Handbook also states that '[m]ilitary advantage may involve a variety of considerations, including the security of the attacking force', but in the context of the definition of 'military objective', not the principle of proportionality (para 8.2).
[411] Italy declared, for instance, that 'the military advantage anticipated from an attack is intended to refer to the advantage anticipated from the attack considered as a whole and not only from isolated or

advantage of the Israeli cyber operation that allegedly switched off the Syrian radar system with the aim to facilitate the bombing of a nuclear reactor in 2007 has to be evaluated jointly with the airstrike that followed it. This is also relevant for several coordinated low-intensity cyber attacks. In such cases, 'if a CNA is mounted systematically against a whole array of enemy computers, the military advantage accruing from the destruction of—or intrusion into—any particular target computer may be of little consequence by itself. Only an examination of the larger picture would divulge what is at stake'.[412] The EECC, however, has gone further than the NATO reservation and opined that 'the term "military advantage" can only properly be understood in the context of the military operations between the Parties taken as a whole, not simply in the context of specific attack' and that 'a definite military advantage must be considered in the context of its relation to the armed conflict as a whole at the time of the attack'.[413] This view has been rightly criticized by the Commission's President van Houtte in his Separate Opinion[414] and by several commentators[415] for excessively conflating the notion of 'military advantage': if military advantage is defined in relation to the armed conflict as a whole and the ultimate objective of defeating the adversary, it would justify virtually any level of incidental damage on protected persons and objects.

Having established what 'incidental damage' and 'concrete and direct military advantage' mean in the cyber context, the two parameters must now be balanced against each other. In particular, the expected incidental damage must not be 'excessive' with respect to the anticipated military advantage. 'Excessive' should not be confused with 'extensive':[416] the principle of proportionality permits even massive incidental damage on civilians and civilian objects if this is matched by a correspondently significant military advantage. On the other hand, 'if—while disrupting some military electronic systems in a minor way—[a cyber operation] causes irreparable damage to the civilian infrastructure (eg water management, research centres, banking systems, stock exchanges), this should be adjudged "excessive"'.[417] It may be recalled that, under Article 51(5)(b), it is not the incidental damage that actually occurs or the military advantage that is effectively gained

---

particular parts of the attack' (Roberts and Guelff, *Documents*, p 507). The ICRC Commentary considers such statements 'redundant' as 'it goes without saying that an attack carried out in a concerted manner in numerous places can only be judged in its entirety' (Sandoz, Swinarski, and Zimmermann (eds), *Commentary*, para 2218).

[412] Dinstein, 'The Principle of Distinction', p 271.

[413] EECC, *Partial Award, Western Front*, para 113. The same questionable position appears in certain US documents, including the *Final Report to Congress on the Conduct of the Persian Gulf War*, p 622. See also Iran's position in *Yearbook of International Humanitarian Law* 6 (2003), p 496, according to which '"military advantage" will be the advantage expected from an invasion in its entirety and not part of it'.

[414] EECC, *Partial Award, Western Front*, Separate Opinion of President van Houtte, paras 8, 10.

[415] Gabriella Venturini, 'International Law and the Conduct of Military Operations', in *The 1998–2000 War between Eritrea and Ethiopia*, edited by Andrea De Guttry, Harry HG Post, and Gabriella Venturini (The Hague: Asser Press, 2009), p 301; Vierucci, 'Sulla nozione', pp 704–5; Yoram Dinstein, 'Air Warfare', in *Max Planck Encyclopedia of Public International Law* (2012), Vol I, p 255.

[416] Dinstein, 'The Principle of Distinction', p 272.

[417] Dinstein, 'The Principle of Distinction', p 272.

from the attack that count, but rather the 'expected' damage and the 'anticipated' military advantage. The difficulty of calculating the 'expected' incidental damage and the 'anticipated' military advantage are already well-known in relation to traditional warfare, but the problems are exacerbated in the cyber context, where the interconnectivity of networks and the reverberating effects of cyber operations often make the *ex ante* evaluation an esoteric prediction. The difficulties of measuring each term of the equation must be added to those of balancing the terms against each other, which is necessarily a subjective process, depending on social and historical factors as well as on the background of the specific targeteers involved.[418] In an attempt to objectivize the test, the ICTY Trial Chamber held that, to assess the proportionality of an attack, it is necessary to determine 'whether a reasonably well-informed person in the circumstances of the actual perpetrator, making reasonable use of the information available to him or her, could have expected excessive civilian casualties to result from the attack'.[419] The ICTY Final Report refers to this person as a 'reasonable military commander'.[420] Because of the technicalities of cyber warfare, the 'reasonable military commander' will almost inevitably have to be assisted by cyber engineers in order to determine the incidental civilian damage of the cyber attack, unless he is a trained cyber expert himself.[421] Collecting information about the architecture of the attacked network (network mapping) or operating system (footprinting) through cyber exploitation will also be of decisive importance, as the damaging effects of a cyber operation also depend on the characteristics of the targeted systems. All in all, the issue is one of degree: the more the effects of a cyber operation are unclear and unforeseeable, the more indiscriminate the attack is likely to be.

To conclude on this point, cyber operations present both opportunities and dangers for the principle of proportionality. On the one hand, their potentially less damaging character may offer a better means to minimize incidental damage on civilians and civilian property, which is particularly important given the current trend towards effects-based warfare.[422] Cyber operations also present advantages for the attacking state, as they virtually entail no risk for its forces thanks to their

---

[418] As the ICTY Report on the NATO operations in Kosovo states, '[i]t is unlikely that a human rights lawyer and an experienced combat commander would assign the same relative values to military advantage and to injury to non-combatants. Further, it is unlikely that military commanders with different doctrinal backgrounds and differing degrees of combat experience or national military histories would always agree in close cases' (*ICTY Final Report*, para 50).

[419] *Galić*, para 58 (footnote omitted). See also the ICRC Commentary to Art 57(2)(a)(iii), which recommends that the evaluation made by military commanders 'must above all be a question of common sense and good faith' where the humanitarian and military interests at stake are balanced (Sandoz, Swinarski, and Zimmermann (eds), *Commentary*, para 2208).

[420] *ICTY Final Report*, para 50. See also Israel's Supreme Court, *Beit Sourik Village v The Government of Israel*, HCJ 2056/04, 30 June 2004, para 46 (per Judge Barak), and the declarations issued by Germany, Belgium, Italy, the Netherlands, and Spain in relation to Art 51 of Additional Protocol I, in Roberts and Guelff, *Documents*, pp 505, 501, 507, 508, 509, respectively.

[421] Harrison Dinniss, *Cyber Warfare*, pp 206–7.

[422] The EECC has emphasized the 'increased emphasis on avoiding unnecessary injury and suffering by civilians resulting from armed conflict' that characterizes modern effect-based warfare (EECC, *Partial Award, Western Front*, para 104). For a comparison of the campaigns against Iraq in 1991 and 2003, see Bartolini, 'Air Operations', pp 251–4.

remote character and the difficulties with regard to identification and attribution. On the other hand, the interconnectivity of military and civilian networks raises the question of the spread of malware to other computers and networks, which might be difficult to predict and therefore to avoid or minimize.[423] As required by Article 57(2), then, all feasible precautions must be adopted to ensure that the attack is consistent with the principle of proportionality.[424] It appears for instance that, even though the law of armed conflict was not applicable, those that developed and deployed Stuxnet went a long way to prevent or at least minimize incidental damage on targets other than the Natanz uranium enrichment facility: (1) the worm was activated only by the presence of the specific Siemens software used at Natanz; (2) each infected computer could spread the worm to three other computers only; (3) even when a computer was infected, the worm did not sufficiently self-replicate to inhibit computer functions and therefore only caused annoyance; and (4) it contained a command that deactivated the worm on 24 June 2012.[425]

3. Objects and persons specially protected from attack

In addition to that deriving from the general rules discussed in the previous Sections, certain objects and persons enjoy special protection: cultural objects and places of worship,[426] the natural environment,[427] medical units, personnel and means

---

[423] Schmitt, 'Wired Warfare', p 204.   [424] Article 57(2)(a).
[425] Richmond, 'Evolving Battlefields', p 856.
[426] Article 53(a) of Additional Protocol I. The provision does not refer to 'attacks' but 'acts of hostility', which is a broader term. The protection of cultural property during armed conflict also forms the object of the 1954 Hague Convention for the Protection of Cultural Property in the Event of Armed Conflict and its two Protocols, which also apply to non-international armed conflicts. See also Rules 38–41 of the ICRC Study, in Henckaerts and Doswald-Beck, *Customary International Humanitarian Law*, pp 127 ff. The case of a cyber operation that disrupts the system controlling air humidity in a museum so to cause damage to the paintings exhibited there has been suggested as an example of a cyber operation that would affect cultural property (Boothby, *The Law of Targeting*, p 397). Some digital art could also be considered 'cultural property': see Rule 82 of the Tallinn Manual and its Commentary, pp 228–9.
[427] Articles 35(3) and 55(1) of Additional Protocol I. The two provisions do not refer to 'attack', but more broadly to the use of methods and means of warfare. An example of a cyber operation negatively affecting the natural environment would be that disrupting the operating system of a chemical or nuclear plant and causing the release of chemicals or radioactive materials in the natural environment. The EECC appears to have implicitly regarded the environmental provisions contained in Additional Protocol I as customary, as it rejected Ethiopia's claims of unlawful damage to environmental resources committed by Eritrea because the destruction fell 'below the standard of widespread and long-lasting environmental damage required for liability under international humanitarian law' (the Commission did not hold relevant the fact that Eritrea had not ratified Additional Protocol I: EECC, *Partial Award, Central Front, Ethiopia's Claim 2*, 28 April 2004, RIAA, Vol XXVI, Part V, paras 53, 100). According to the ICTY Final Report, Art 55 of Protocol I 'may... reflect current customary law' (*ICTY Final Report*, para 15). The ICJ's view is not clear: in its 1996 Advisory Opinion on the *legality of the threat or use of nuclear weapons*, the Court first states that Arts 35 and 55 'embody a general obligation to protect the natural environment against widespread, long-term and severe environmental damage', and then concedes that these provisions are 'powerful constraints' only for the states having subscribed to them (*Nuclear Weapons*, para 31). The United States objects to the provisions in question, because they are 'too broad and ambiguous' ('Session One: The United States Position on

of transportation,[428] religious personnel,[429] objects indispensable to the survival of the civilian population,[430] works and installations containing dangerous forces (dams, dykes and nuclear electrical generating stations).[431] Attacks 'by any means

---

the Relation of Customary International Law to the 1977 Protocols Additional to the 1949 Geneva Conventions', Remarks of Michael J Matheson, *American University Journal of International Law and Policy* 2 (1987), p 424). In the US government's view, only the destruction of the environment not necessitated by military necessity and carried out wantonly is prohibited: damage to the environment is thus only limited by the principles of distinction and proportionality and must be balanced against the military advantage expected from the operation (*US Joint Doctrine for Targeting*, p A–6; *The Commander's Handbook*, para 8.4). According to the ICRC, '[t]he general principles on the conduct of hostilities apply to the natural environment' (Rule 43, in Henckaerts and Doswald-Beck, *Customary International Humanitarian Law*, Vol I, p 143). The special protection provided in Additional Protocol I also 'arguably' reflects customary international law and is applicable to non-international armed conflicts as well (Rule 45, Vol I, p 151). In its official response to the publication of the ICRC Study, the United States claimed that the Study's conclusions are flawed, as they do not accurately assess the practice of specially affected states, examine 'only limited operational practice' and offer no evidence of *opinio juris* (Letter from John Bellinger III, Legal Adviser, US Department of State, and William J Haynes, General Counsel, US Department of Defense, to Dr Jakob Kellenberger, President, ICRC, regarding Customary International Law Study (3 November 2006), *International Legal Materials* 46 (2007), pp 520–2). Rule 83 of the Tallinn Manual only accords the natural environment the protection arising from the principle of distinction, although it acknowledges the special protection regime for states parties to Additional Protocol I (Tallinn Manual, p 231).

[428] Articles 19(1), 24, 25, 35, and 36 of Geneva Convention I; Arts 22(1), 36, and 39 of Geneva Convention II; Arts 18(1), 20(1), 21, and 22(1) of Geneva Convention IV; Arts 12(1), 15(1), and 21 ff of Additional Protocol I; Art 9(1) of Additional Protocol II. The immunity lasts only to the extent that the objects are used exclusively for medical purposes. The provisions reflect customary international law and also apply in non-international armed conflicts (see Rules 25, 28–30 of the ICRC Study, in Henckaerts and Doswald-Beck, *Customary International Humanitarian Law*, pp 79, 91 ff). See also Rule 70 of the Tallinn Manual, p 204.

[429] Article 24 of Geneva Convention I; Art 36 of Geneva Convention II; Art 15(5) of Additional Protocol I, Art 9(1) of Additional Protocol II. See also Rules 27 and 30 of the ICRC Study, in Henckaerts and Doswald-Beck, *Customary International Humanitarian Law*, Vol I, pp 88, 102; and Rule 70 of the Tallinn Manual, p 204.

[430] Article 54(2) of Additional Protocol I. The attack, destruction, removal or 'rendering useless' of the object must be 'for the specific purpose of denying them for their sustenance value to the civilian population or to the adverse Party, whatever the motive'. The prohibition does not prevent a belligerent from conducting a scorched land strategy on the national territory under its control against an invading force if this is 'required by imperative military necessity' (Art 54(5)). The prohibition also applies to non-international armed conflicts, although the scorched land strategy is not mentioned (see Art 14 of Additional Protocol II). Rule 54 of the ICRC Study removes the limitation to acts conducted 'for the specific purpose of denying them for their sustenance value to the civilian population or to the adverse Party'. Paragraphs 3–5 of Art 54 are also omitted (Henckaerts and Doswald-Beck, *Customary International Humanitarian Law*, Vol I, p 189). See also Rule 81, Tallinn Manual, p 225.

[431] Article 56(1) of Additional Protocol I, that refers to 'attack'. The installations could be attacked only if they are used in regular, significant, and direct support of military operations and the attack is the only feasible way to terminate such support, or if the action does not cause the release of dangerous forces and consequent severe losses among the civilian population (Art 56(2)). It is not clear whether the rule reflects customary international law and, if so, if it does in its entirety (Dinstein, *The Conduct*, p 194). A more succinct version of the provision appears in Art 15 of Additional Protocol II. Rule 42 of the ICRC Study on *Customary International Humanitarian Law*, applicable to both international and non-international armed conflicts, reproduces only part of Art 56 of Additional Protocol I and provides that '[p]articular care must be taken if works and installations containing dangerous forces, namely dams, dykes and nuclear electrical generating stations, and other installations located at or in their vicinity are attacked, in order to avoid the release of dangerous forces and consequent severe losses among the civilian population' (Henckaerts and Doswald-Beck, *Customary International Humanitarian Law*, Vol I, p 139). The Rule forms the basis of Rule 80 of the Tallinn Manual, p 223.

whatsoever' are also prohibited against non-defended localities,[432] and military operations may not be extended to 'hospital and safety zones'[433] and demilitarized and neutralized zones.[434] Finally, 'personnel and objects involved in a peacekeeping mission in accordance with the Charter of the United Nations' are immune from attacks 'as long as they are entitled to the protection given to civilians and civilian objects under international humanitarian law'.[435] These special limitations apply not only to kinetic attacks, but also to cyber operations whose intended or foreseeable secondary or tertiary effects are the causation of physical damage to the above objects or persons: the rules would apply much in the same way as if the damage had been produced by traditional weapons.

As to whether the rules under examination also apply to cyber operations merely entailing loss of functionality but not physical damage, the problem would be relevant only in relation to certain of the above objects, ie medical units and means of transportation, objects indispensable to the survival of the civilian population, UN installations, materiel, units and vehicles, and works and installations containing dangerous forces. Article 54(2) of Additional Protocol I expressly protects objects indispensable to the survival of the civilian population not only from attack, destruction or removal, but also from being 'rendered useless'.[436] The expression is broad enough to include all cyber attacks aimed to disrupt the functioning of such objects 'for the specific purpose of denying them for their sustenance value to the civilian population or to the adverse Party'. Medical units and means of transportation must be 'respected and protected', which may cover all sorts of cyber operations.[437] The same duty to respect and protect is contained in Rule 74 of the Tallinn Manual with regard to UN installations, materiel, units and vehicles, although Rule 33 of the ICRC Study on Customary International Humanitarian Law only refers to 'attacks':[438] in any case, it has been seen that 'attacks' also include

---

[432] Article 59(1) of Additional Protocol I. See also Rule 37 of the ICRC Study, in Henckaerts and Doswald-Beck, *Customary International Humanitarian Law*, Vol I, p 122.

[433] Article 14 of Geneva Convention IV. See also Rule 35 of the ICRC Study, in Henckaerts and Doswald-Beck, *Customary International Humanitarian Law*, Vol I, p 119.

[434] Article 60(1) of Additional Protocol I and Art 15 of Geneva Convention IV, respectively. See also Rule 36 of the ICRC Study, in Henckaerts and Doswald-Beck, *Customary International Humanitarian Law*, Vol I, p 120.

[435] Rule 33 of the ICRC Study, in Henckaerts and Doswald-Beck, *Customary International Humanitarian Law*, Vol I, p 112. See also Art 7(1) of the 1994 Convention on the Safety of UN and Associated Personnel, Arts 8(2)(b)(iii) and 8(2)(e)(iii) of the ICC Statute, and Rule 74 of the Tallinn Manual, p 210. The immunity, for instance, would cease in case of an enforcement action under Chapter VII 'in which any of the personnel are engaged as combatants against organized armed forces' (Art 2(2) of the UN Safety Convention), using kinetic or cyber means.

[436] See also Rule 81 of the Tallinn Manual, p 225. An example might be a cyber attack that shuts down a water purification system.

[437] See also Rule 70 and its Commentary, in Tallinn Manual, pp 204–5. Rule 71 of the Tallinn Manual contains a specific obligation to respect and protect computers, computer networks, and data 'that form an integral part of the operations or administration of medical units and transports' (p 206). The protection ceases when the computers, computer networks, and data are 'used to commit, outside their humanitarian function, acts harmful to the enemy', but only after a warning has been given (Rule 73, p 208).

[438] See Rule 74, Tallinn Manual, p 210.

cyber operations that cause malfunction of infrastructures.[439] With regard to cyber operations against installations containing dangerous forces, they would fall under the stricter regulation of Article 56 only 'if such attack may cause the release of dangerous forces and consequent severe losses among the civilian population':[440] the use of 'may' entails that it is not required that the dangerous forces are actually released, but that there is the likelihood that this might be the case.[441] Be that as it may, it is not necessary that a cyber attack causes physical damage to the dam to cause flooding: the same result could be reached by inputting the wrong data into its operating system or by taking control of it and opening the gates so to release the dangerous forces. Analogous considerations can be made in relation to dykes and nuclear electrical generating stations.[442]

4. The duty to take precautions

The obligation to take active and passive precautions is contained in Articles 57 and 58 of Additional Protocol I. They have both matured into customary international law[443] and, according to the majority view, are also applicable in armed conflicts of a non-international character.[444] Whereas Article 57 deals with precautions to be taken in attack, Article 58 provides for precautions that the belligerents have to adopt to protect civilians and civilian property *from* attacks.

4.1 Precautions in attack

While there is 'nothing inherently unlawful' in conducting remote attacks where 'the target could not be verified with the naked eye',[445] commanders have a duty to take 'practicable measures to distinguish military objectives from civilians and civilian objectives'.[446] Indeed, the remote character of cyber operations and the

---

[439] See Section III.1.1, pp 179–82, in this Chapter.   [440] Article 56(1) of Additional Protocol I.
[441] Frits Kalshoven, *Reflections on the Law of War: Collected Essays* (Leiden: Nijhoff, 2007), p 235.
[442] It is worth noting that the Natanz facility in Iran, allegedly targeted by Stuxnet, is not a 'nuclear electricity generating station', but a uranium enrichment site: Art 56, therefore, does not apply to it.
[443] *Kupreškić*, para 524. According to the ICTY, this is so 'not only because they specify and flesh out general pre-existing norms, but also because they do not appear to be contested by any State, including those which have not ratified the Protocol' (para 524).
[444] Rules 15–24 of the ICRC Study, in Henckaerts and Doswald-Beck, *Customary International Humanitarian Law*, Vol I, pp 51ff. The Commentary to Additional Protocol II states that Art 13 'requires that precautions are taken both by the party launching the attack during the planning, decision and action stages of the attack, and by the party that is attacked' (Sandoz, Swinarski, and Zimmermann (eds), *Commentary*, para 4772). See also the UK Military Manual, according to which the duty to take precautions in attack in internal armed conflicts can be 'inferred from the principle of proportionality and the principle of distinction' (UK Ministry of Defence, *The Manual of the Law of Armed Conflict*, p 393). The Manual is more cautious with regard to Art 58 and limits itself to state that, while customary international law is silent, precautions against the effects of attacks in non-international armed conflicts are 'a matter of common humanity' (p 393). *The Manual on the Law of Non-international Armed Conflict* identifies a more limited set of precautions that need to be taken in planning and carrying out an attack in a non-international armed conflict (Schmitt, Garraway, and Dinstein, *The Manual on the Law of Non-International Armed Conflict*, Rule 2.1.2, p 25).
[445] *ICTY Final Report*, para 56.   [446] *ICTY Final Report*, para 56.

interconnectivity of networks make the duty to take precautions in attack even more relevant in this context than in domains of warfare allowing visual confirmation of the target.[447]

Article 57(1) of Additional Protocol I prescribes that '[i]n the conduct of military operations, constant care shall be taken to spare the civilian population, civilians and civilian objects'.[448] Note that the reference here is not to 'attacks' but to the broader notion of 'military operations': 'constant care' must therefore be exercised also when conducting cyber operations not amounting to 'attack' (as long as they are military operations). The 'constant care' standard is specified with regard to 'attacks' in the subsequent paragraphs of the provision. According to Article 57(2)(a), those who plan or decide upon an attack have to adopt the following precautions:

(i) do everything feasible to verify that the objectives to be attacked are neither civilians nor civilian objects and are not subject to special protection but are military objectives within the meaning of paragraph 2 of Article 52 and that it is not prohibited by the provisions of this Protocol to attack them;

(ii) take all feasible precautions in the choice of means and methods of attack with a view to avoiding, and in any event to minimizing, incidental loss of civilian life, injury to civilians and damage to civilian objects;

(iii) refrain from deciding to launch any attack which may be expected to cause incidental loss of civilian life, injury to civilians, damage to civilian objects, or a combination thereof, which would be excessive in relation to the concrete and direct military advantage anticipated.[449]

An attack must also 'be cancelled or suspended if it becomes apparent that the objective is not a military one or is subject to special protection or that the attack may be expected to cause incidental loss of civilian life, injury to civilians, damage to civilian objects, or a combination thereof, which would be excessive in relation to the concrete and direct military advantage anticipated' (Article 57(2)(b)).[450] It was for instance reported that, although NATO intended to upload incorrect messages and targets in Yugoslavia's air-defence command network, which would

---

[447] As the ICRC Commentary explains, when 'the attacker has no direct view of the objective...even greater caution is required' (Sandoz, Swinarski, and Zimmermann (eds), *Commentary*, para 2221). The German Military Manual confirms that the duty to take precautions in attack 'shall also apply to attacks by...remotely controlled weapons' (German Ministry of Defence, *Humanitarian Law in Armed Conflicts*, Section 447).

[448] See also Rule 52, in Tallinn Manual, p 165.

[449] Article 27 of the Hague Regulations already provided that '[i]n sieges and bombardments all necessary steps must be taken to spare, as far as possible, buildings dedicated to religion, art, science, or charitable purposes, historic monuments, hospitals, and places where the sick and wounded are collected, provided they are not being used at the time for military purposes'.

[450] See Rule 57, in Tallinn Manual, p 172. At the moment of ratification, Switzerland issued a reservation in relation to Art 57(2) stating that the provision only applies to 'commanding officers at the battalion or group level and above' (Roberts and Guelff, *Documents*, p 509). According to the UK Military Manual, the level of responsibility depends on whether the person 'has any discretion in the way the attack is carried out' (UK Ministry of Defence, *The Manual of the Law of Armed Conflict*, p 85).

have affected Yugoslavia's capacity to attack NATO airplanes during the 1999 Operation Allied Force, the operation was eventually abandoned because of the risks for civilian aviation.[451] Similarly, in the 2011 operations against Libya, a plan to use cyber operations to disrupt Libya's air defence system was set aside.[452]

The obligation to take precautions in attack is not one of result, but of conduct: the keyword is 'feasible' ('tout ce qui est pratiquement possible' in the French text).[453] Feasibility is a flexible notion that necessarily depends on the circumstances of each case and the means at the disposal of the attacker.[454] Article 3(4) of Protocol II annexed to the 1980 Convention on Prohibitions or Restrictions on the Use of Certain Conventional Weapons Which May Be Deemed to Be Excessively Injurious or to Have Indiscriminate Effects clarifies that '[f]easible precautions are those precautions which are practicable or practically possible taking into account all circumstances ruling at the time, including humanitarian and military considerations'.[455] The question is what 'feasible precautions' can be adopted when conducting a cyber operation amounting to attack. Hans Blix has helpfully identified the key components of the duty to take precautions in: (1) identifying the target with some certainty; (2) directing the attack against that target; and (3) using methods and means that will hit the target with some degree of likelihood and—one could add—that will avoid or minimize incidental damage to civilians and civilian objects.[456] With regard to the first element, the belligerent will have to map the enemy's network with sufficient accuracy through network mapping, footprinting, and other cyber exploitation operations.[457] The evaluation of the information so obtained 'must include a serious check of its accuracy, particularly as there is nothing to prevent the enemy from setting up fake military objectives or camouflaging the true ones'.[458] The US Presidential Policy Directive 20, signed in October 2012, incapsulates this first precautionary element and states that the United States government 'shall make all reasonable

---

[451] Kelsey, 'Hacking', pp 1434–5.   [452] Rid and McBurney, 'Cyber-Weapons', p 6.

[453] It should be noted that, with regard to military operations at sea or in the air, Art 57(4) does not refer to 'feasibility' but rather requires states parties to adopt 'all *reasonable* precautions to avoid losses of civilian lives and damage to civilian objects' (emphasis added). According to the ICRC Commentary of the provision, this is 'a little less far-reaching' than 'feasible' and is due to the peculiar characteristics of naval and air warfare (Sandoz, Swinarski, and Zimmermann (eds), *Commentary*, para 2230). See also Tallinn Manual, pp 164–5.

[454] The *ICTY Final Report on the NATO Bombing Campaign* also states that '[t]he obligation to do everything feasible is high but not absolute' (*ICTY Final Report*, para 29).

[455] See also Art 1(5) of Protocol III on Prohibitions or Restrictions on the Use of Incendiary Weapons and Art 3(10) of the amended Protocol II (1996).

[456] Hans Blix, 'Area Bombardment: Rules and Reasons', *British Year Book of International Law* 49 (1978).

[457] See Rule 53, in Tallinn Manual, pp 167–8.

[458] Sandoz, Swinarski, and Zimmermann (eds), *Commentary*, para 2195. The *ICTY Final Report on the NATO Bombing Campaign against Yugoslavia*, for instance, requires a military commander to 'set up an effective intelligence gathering system to collect and evaluate information concerning potential targets' and 'also direct his forces to use available technical means to properly identify targets during operations' (*ICTY Final Report*, para 29). Such considerations can be easily extended to the conduct of cyber operations.

efforts, under circumstances prevailing at the time, to identify the adversary and the ownership and geographic location of the targets and related infrastructure where [cyber effects operations] will be conducted or cyber effects are expected to occur, and to identify the people and entities, including U.S. persons, that could be affected by proposed [operations]'.[459] The second element requires, inter alia, that appropriately trained personnel are employed to execute the attack: if no relevant expertise is available, the attack may have to be cancelled.[460] One could go as far as to say that the use of inexperienced personnel might amount to an indiscriminate attack under Article 51(5)(b) of Additional Protocol I, at least when the choice of the inexperienced personnel is intentional.[461] Interestingly, the issue of computer technology training arose before the EECC. The Commission noted that the aircraft used by Eritrea were provided with 'computerized aiming systems that are designed to release bombs at the proper time to hit a target when the pilot sees it aligned with a "heads up" display in the cockpit and pushes a bomb release switch'.[462] The EECC acknowledged that 'Eritrea had little experience with these weapons and that the individual programmers and pilots were utterly inexperienced, and it recognize[d] the possibility that, in the confusion and excitement of June 5, both computers could have been loaded with the same inaccurate targeting data'.[463] The Commission concluded that the use of inexperienced pilots in the bombing of the Mekele airport was a violation of Article 57(2)(a)(ii) of Additional Protocol I.

As to the third element, the belligerents are under an obligation to take all feasible precautions in view 'to select the means (that is, weapons) or methods of attack (that is, tactics) which will cause the least incidental damage commensurate with military success'.[464] If, for instance, the military advantage sought is a temporary interruption in the provision of electricity to an enemy military base, a cyber operation that incapacitates a power station serving the base but that does not destroy it is to be preferred to a kinetic attack, as it reduces the risks of collateral damage and of a long-term impact on the civilian population. Similarly, if the military advantage is analogous (all things being equal), a 'flood' attack which causes the system to shut down by overloading it with requests should be preferred to a cyber operation employing a weaponized program that corrupts data and that could spread to other networks and systems. As has been seen, the protection of the attacking forces may also be taken into account in this context.[465]

Article 57(3) further requires that '[w]hen a choice is possible between several military objectives for obtaining a similar military advantage, the objective to be selected shall be that the attack on which may be expected to cause the least danger

---

[459] US Presidential Policy Directive/PPD–20, October 2012, p 7, <http://www.guardian.co.uk/world/interactive/2013/jun/07/obama-cyber-directive-full-text>.
[460] Droege, 'Get Off My Cloud', p 574.   [461] Vierucci, 'Sulla nozione', p 722.
[462] EECC, *Central Front*, Claim 2, para 103.   [463] EECC, *Central Front*, Claim 2, para 109.
[464] UK Ministry of Defence, *The Manual of the Law of Armed Conflict*, pp 82–3. See Rule 54 of the Tallinn Manual, p 168.
[465] See Section III.2.1 in this Chapter, p 226.

to civilian lives and to civilian objects'.[466] 'Danger' includes material damage to property, loss of life or injury to individuals, as well as loss of functionality of infrastructures, but not mere inconvenience.[467] 'Similar' military advantage means that the attack 'would achieve comparable military effects', but need not be identical.[468] If it is not possible to assess with sufficient certainty the incidental damage likely to be caused by the operation, the attacker must abstain from conducting it.[469]

The EECC has also added another element to the duty to take precautions in attack by extending it to the aftermath of the attack. The Commission noted that Eritrea failed to take 'appropriate actions' after the attacks 'to prevent future recurrence', including investigations and 'changes in Eritrean training or doctrine aimed at avoiding possible recurrence of what happened'.[470] This *post facto* obligation is not expressly contained in Article 57: the EECC has therefore contributed to the interpretation and development of the duty to take precautions.[471]

According to Article 57(2)(c), 'effective advance warning shall be given of attacks which may affect the civilian population, unless circumstances do not permit'.[472] This provision is considered customary by the ICRC in its Study on Customary International Humanitarian Law.[473] It also applies in non-international armed conflicts under customary international law.[474] It is, however, unlikely that advance warning will be given in case of cyber operations, as such warning would deprive them of two of their main advantages, ie their covert character and their surprise effect. In any case, the obligation only applies to cyber operations that 'may affect the civilian population' (which does not include mere inconvenience)[475] and only 'unless circumstances do not permit'. The warning of an incumbent attack, be it kinetic or cyber, may be given by cyber means, for instance by communicating the relative information by email, social networks or websites. According to the 2009 Goldstone Report, to be 'effective' the warning 'must reach those who are likely to be in danger from the planned attack, it must give them sufficient time to react to the warning, it must clearly explain what they should do to avoid harm and it must be a credible warning'.[476] Whether a warning delivered by cyber means would meet these standards is necessarily a situational assessment.

---

[466] See also Rule 56 of the Tallinn Manual, pp 170–1.
[467] Commentary to Rule 56, in Tallinn Manual, p 171.
[468] Commentary to Rule 56, in Tallinn Manual, p 171.
[469] As has been observed, this advantages more technologically developed belligerents, that may be more able than others to assess the incidental damage and use means to minimize it. Less developed states that do not have the same means will have to abstain from conducting the operation (Cannizzaro, 'Contextualizing Proportionality', p 788).
[470] EECC, *Central Front*, Claim 2, paras 110–11. See also *Targeted Killings*, para 40 (per Judge Barak).
[471] Venturini, 'International Law', p 303.
[472] See Rule 58, in Tallinn Manual, p 173. On warnings, see Pnina Sharvit Baruch and Noam Newman, 'Warning Civilians Prior to Attack under International Law—Theory and Practice', *Israel Yearbook on Human Rights* 41 (2011), pp 359 ff.
[473] Henckaerts and Doswald-Beck, *Customary International Humanitarian Law*, Vol I, pp 62–64.
[474] Ibid.   [475] Commentary to Rule 58, in Tallinn Manual, p 174.
[476] *Human Rights in Palestine and Other Occupied Arab Territories*, Report of the United Nations Fact-Finding Mission on the Gaza Conflict, UN Doc A/HRC/12/48, 25 September 2009 ('the Goldstone Report'), para 530.

Finally, it should be noted that, in automated active defences, the duty to take precautions assumes an even greater importance.[477] As the ICRC Commentary prophetically warned,

[t]he use of long distance, remote control weapons, or weapons connected to sensors positioned in the field, leads to the automation of the battlefield in which the soldier plays an increasingly less important role. The counter-measures developed as a result of this evolution, in particular electronic jamming (or interference), exacerbates the indiscriminate character of combat. In short, all predictions agree that if man does not master technology, but allows it to master him, he will be destroyed by technology.[478]

Automated processes may be militarily necessary when there is no time for humans to make decisions on the response 'because of the fleeting nature of the opportunity to strike back or because of the harm that rapidly accrues if the attack is not stopped'.[479] Military necessity, however, does not justify deviations from the law unless the law expressly provides so, and automated responses do increase the chance of errors and therefore of violations of the principles of distinction and proportionality.[480] In automated processes, then, the duty to take precautions will necessarily have to be applied before the attack, ie at the time that the targeting software is programmed and the data designed and uploaded, or up to any time the attack can be called off.

4.2 Precautions against the effects of an attack

Article 58 of Additional Protocol I concerns passive precautions against the effects of an attack and reads as follows:

The Parties to the conflict shall, to the maximum extent feasible:
(a) without prejudice to Article 49 of the Fourth Convention, endeavour to remove the civilian population, individual civilians and civilian objects under their control from the vicinity of military objectives;
(b) avoid locating military objectives within or near densely populated areas;
(c) take the other necessary precautions to protect the civilian population, individual civilians and civilian objects under their control against the dangers resulting from military operations.

Article 58(a) and (b) require that the belligerents segregate military objectives from civilians and civilian objects. As the 1999 DoD *Assessment of International*

---

[477] Droege, 'Get Off My Coud', p 574. Target selection in cyber operations, for instance, is often an automated process, which relies on 'mapping and filtering IP addresses and/or DNS names, for example through programmed pattern matching, network mapping, or querying databases (either public ones, or ones accessible through close-access attacks)' (Owens, Dam, and Lin, *Technology, Policy, Law, and Ethics*, p 117).
[478] Sandoz, Swinarski, and Zimmermann (eds), *Commentary*, para 1476 (footnote omitted).
[479] Owens, Dam, and Lin, *Technology, Policy, Law, and Ethics*, p 230.
[480] Matthew J Sklerov, 'Solving the Dilemma of State Responses to Cyberattacks: A Justification for the Use of Active Defenses Against States Who Neglect Their Duty to Prevent', *Military Law Review* 201 (2009), p 82.

Legal Issues in Information Operations suggests, then, '[w]here there is a choice…, military systems should be kept separate from infrastructures used for essential civilian purposes'.[481] In relation to cyber operations, segregation could apply to computer data, networks, and information infrastructure, or to the physical installations operated by the computer systems. With regard to the latter, the obligation would apply in the same way as in any other physical installations. The former case is more complicated. Military information travels through civilian networks and information infrastructures used for connection, such as satellites, fibre-optic cables and servers, are privately owned and used both by the military and civilians.[482] The separation of civilian and military networks and of information infrastructure is therefore at present not possible, either technically or financially, as 'such segregation would require the government to establish its own lines of communication throughout the world, connecting its dispersed military installations'.[483] Article 58, however, only requires the belligerents to adopt precautions 'to the maximum extent feasible' in order to protect civilians and civilian objects 'under their control'.[484] This allows a dynamic interpretation of the provision: should future developments permit the separation of civilian and military information infrastructures, Article 58 would require states to 'endeavour' to do so.[485]

Article 58(c) also requires the belligerents to take 'other necessary precautions' to protect civilians and civilian objects under their control from the dangers of military operations (not only attacks): this is broad enough to include precautions in advance of the attack.[486] This is the only paragraph of Article 58 that has been incorporated in the Tallinn Manual, although only in relation to cyber 'attacks' and not other cyber operations.[487] 'Other necessary precautions' include, for instance, the use of cyber defences and standard measures of cyber hygiene such as the use of anti-viruses, the execution of regular back-ups to facilitate data recovery and resumption of service,[488] warnings of impending or ongoing attacks and the provision of technical assistance to repair networks or reroute them to alternative

---

[481] US Department of Defense, *An Assessment of International Legal Issues in Information Operations*, May 1999, p 9, <http://www.au.af.mil/au/awc/awcgate/dod-io-legal/dod-io-legal.pdf>.

[482] In most countries, only a minority of defence information infrastructure is owned, controlled, managed, and administered by the defence (Fred Schreier, *On Cyber Warfare*, DCAF Horizon 2015 Working Paper no 7, p 39, <http://www.dcaf.ch/Publications/On-Cyberwarfare>). See also Eric Talbot Jensen, 'Cyber Warfare and Precautions Against the Effects of Attacks', *Texas Law Review* 88 (2010), p 1535.

[483] Jensen, 'Cyber Warfare', p 1551 (footnote omitted). See also Geiss and Lahmann, 'Cyber Warfare', p 393.

[484] Jensen, 'Cyber Warfare', p 1553.

[485] Note, however, that Rule 24 of the ICRC Study omits 'endeavour', although the obligation of a belligerent to remove civilian persons and objects under its control from the vicinity of military objects still applies 'to the extent feasible'.

[486] Jensen, 'Cyber Warfare', p 1554.

[487] Rule 59, in Tallinn Manual, pp 176–7. According to the majority of the Experts that drafted the Manual, 'all civilian cyber infrastructure and activities located in territory under the control of a party to the conflict are subject to this Rule' (Commentary to Rule 59, in Tallinn Manual, p 178). See also Eneken Tikk, 'Ten Rules for Cyber Security', *Survival* 53(3) (2011), p 121.

[488] Geiss and Lahmann, 'Cyber Warfare', p 395.

systems,[489] the distribution of protective software,[490] the monitoring of networks and systems,[491] 'air gapping' NCIs from the internet, in particular in time of armed conflict, so to minimize the chances of a cyber attack against networked critical infrastructures. A belligerent state might also block traffic from specific IP addresses or shield a certain network from all international traffic.[492] In extreme scenarios, a more drastic measure would be to disconnect the whole country from the internet to prevent the continuation of cyber attacks.[493]

The preservation of cyber and physical infrastructures identified as essential will necessarily require cooperation with ISPs and the private sector, which creates delicate problems of business confidentiality and right to privacy.[494] Each state will also have to clearly establish who is competent domestically to ensure protection against the effects of attacks. In the United States, for instance, CYBERCOM only protects the DoD's computers, networks, and infrastructure as the military is prohibited to execute missions domestically, while the Department of Homeland Security is responsible for protecting the private sector's system.

IV. Cyber Operations Short of 'Attack'

Cyber operations not resulting in loss of life or injury to persons, more than minimal material damage to property or loss of functionality of infrastructures are not 'attacks' in the sense of Article 49(1) of Additional Protocol I and, therefore, the rules applicable to 'attacks' are not relevant to them, unless they are an integral part of an operation qualifying as an attack. Other rules on the conduct of hostilities, however, may apply when the operations are conducted by a belligerent to the detriment of another. In particular, a general obligation of 'constant care... to spare the civilian population, civilians and civilian objects' applies to all military operations.[495] Such 'general obligation' applies, in particular, to cyber attacks that merely delete, corrupt, or alter data without consequences in the analogue world.

In the context of an armed conflict, cyber operations short of 'attack' can, for instance, aim at disseminating information for deception or propaganda purposes, or can be used to gather intelligence about the enemy. Operations of the former type were conducted before and during the 2008 armed conflict between Russia

---

[489] Jensen, 'Cyber Attacks', p 215; Tikk, 'Ten Rules', pp 126–7.
[490] Commentary to Rule 59, in Tallinn Manual, p 179.
[491] Commentary to Rule 59, in Tallinn Manual, p 179. In this perspective, a role could be played by national CERTs. A CERT is '[a] team that provides initial emergency response aid and triage services to the victims or potential victims of cyber operations or cyber crimes, usually in a manner that involves coordination between the private sector and government entities' (Tallinn Manual, p 258).
[492] Giancarlo A Barletta, William A Barletta, Vitali N Tsygichko, 'Cyber Conflict & Geo-Cyber Stability', in *The Quest for Cyber Peace*, edited by Hamadoun I Touré et al, ITU, January 2011, p 57, <http://www.itu.int/dms_pub/itu-s/opb/gen/S-GEN-WFS.01-1-2011-PDF-E.pdf>.
[493] The ICRC Commentary rules out precautions that 'go beyond the point where the life of the population would become difficult or even impossible' (Sandoz, Swinarski, and Zimmermann (eds), *Commentary*, para 2245). If only mere discontinuation of email services or internet connection is involved, however, it is unlikely that this threshold would be reached.
[494] Tikk, 'Ten Rules', p 125.   [495] Article 57(1) of Additional Protocol I.

and Georgia: the websites of Georgia's President, Minister of Foreign Affairs and National Bank were defaced and replaced with a series of pictures of Mikhail Saakashvili and Adolf Hitler.[496] Before the start of Operation Iraqi Freedom, thousands of emails were also sent by the US Central Command to Iraqi military officers warning of the imminent invasion and asking them to abandon their positions and vehicles so not to suffer harm.[497] Although not in the context of an armed conflict, examples of cyber exploitation operations of the second type are DuQu and Flame: the former had striking similarities with Stuxnet although its payload was not designed to cause physical damage but to steal information that could be used to attack industrial control systems.[498] The latter was found in May 2012 to have penetrated the computers of senior Iranian officials with the alleged purpose of stealing sensitive data.[499]

Article 24 of the Hague Regulations states that 'the employment of measures necessary for obtaining information about the enemy and the country are considered permissible' and are therefore a lawful method of warfare. These operations are not 'espionage' unless conducted under false pretences or clandestinely from within the territory controlled by the adverse party in an international armed conflict,[500] which, thanks to the interconnectivity of computer networks, is not usually the case in cyber operations.[501]

As has been seen, cyber operations for deception purposes qualify as lawful ruses of war as long as they do not become perfidious acts or entail the misuse of certain emblems.[502] As to cyber operations amounting to psychological warfare,[503] already

---

[496] Tikk, Kaska, Rünnimeri, Kert, Talihärm, and Vihul, 'Cyber Attacks', pp 7–8.

[497] Clarke and Knake, *Cyber War*, pp 9–10.

[498] Symantec, *W32.DuQu—The Precursor to the Next Stuxnet*, 23 November 2011, <http://www.symantec.com/content/en/us/enterprise/media/security_response/whitepapers/w32_duqu_the_precursor_to_the_next_stuxnet.pdf>.

[499] Ellen Nakashima, Greg Miller, and Julie Tate, 'U.S., Israel developed Flame computer virus to slow down Iranian nuclear efforts, officials say', *The Washington Post*, 19 June 2012, <http://articles.washingtonpost.com/2012-06-19/world/35460741_1_stuxnet-computer-virus-malware>.

[500] Article 46 of Additional Protocol I. The ICRC Commentary of Art 46 distinguishes between technical instruments to remotely acquire intelligence on the one hand, and espionage on the other (Sandoz, Swinarski, and Zimmermann (eds), *Commentary*, para 1765). See also Rule 66 of the Tallinn Manual, pp 192–3. According to the majority of the Experts that drafted the Manual, the nature of the information gathered is not relevant for the determination of 'espionage' in international humanitarian law (p 194). On the contrary, Rule 118 of the HPCR requires the information to be 'of military value' (HPCR Manual, p 319).

[501] HPCR Manual, p 321. Doswald-Beck argues that, when the individual conducts intelligence gathering from outside the adversary's territory through cyber exploitation, 'the situation should be no different from someone gathering data from a spy satellite' (Doswald-Beck, 'Some Thoughts', p 172). Cyber espionage would then likely occur only 'as a close access cyber operation, such as when a flash drive is used to gain access to a computer system' (Commentary to Rule 66, in Tallinn Manual, p 194). As in all other cases those conducting cyber exploitation operations remotely do not qualify as 'spies' in the sense of Art 46 of Additional Protocol I, they would not be subject to the unfavourable treatment provided for them in case of capture.

[502] See Section III.1.5 in this Chapter.

[503] Psychological operations are defined in the US Doctrine for Joint Psychological Operations as '[p]lanned operations to convey selected information and indicators to foreign audiences to influence their emotions, motives, objective reasoning, and ultimately the behavior of foreign governments, organizations, groups, and individuals. The purpose of psychological operations is to induce or reinforce

the 1923 Hague Rules of Aerial Warfare provided that '[t]he use of aircraft for the purpose of disseminating propaganda shall not be treated as an illegitimate means of warfare'.[504] The 1992 German Military Manual confirms that '[i]t is permissible to engage in political and military propaganda by spreading even false information to undermine the adversary's will to resist and to influence the military discipline of the adversary (e.g. instigation to defect).'[505] As they do not amount to 'attacks' under Article 49(1) of Additional Protocol I, the principle of distinction does not apply to cyber operations amounting to psychological warfare, which may therefore be directed not only at combatants, but also at civilians. It is, however, prohibited to incite the population to commit crimes.[506]

If conducted by civilians, cyber operations short of attack can potentially amount to direct participation in hostilities if the three requirements examined above are present.[507] The threshold of harm requires that, lacking damaging physical effects, the conduct at least causes military harm, which excludes industrial espionage to gain commercial advantages.[508] As to direct causation, a distinction must be made between those engaging in cyber operations amounting to psychological warfare on the one hand, and those conducting operations aimed at deception and information gathering on the other. With regard to the former, the Israeli Supreme Court pointed out that those who engage in propaganda are only taking indirect part in hostilities.[509] The ICRC *Interpretive Guidance* also includes propaganda as an example of 'war sustaining activities' amounting to indirect participation in hostilities.[510] As to cyber operations for the purpose of collecting intelligence, the 'gathering and transmission of military information' is provided in the Commentary to Additional Protocol I as an example of 'indirect' participation in hostilities.[511] According to the Supreme Court of Israel, however, persons collecting intelligence on the army, whether or not on issues regarding the hostilities, take direct part in hostilities.[512] In *Strugar*, the ICTY Appeals Chamber included 'intelligence agents' and those 'transmitting military information for the immediate use of a belligerent' as examples of 'active or direct participation in hostilities', while 'gathering and transmitting military information' in general was considered indirect participation.[513]

---

foreign attitudes and behavior favourable to the originator's objectives' (*Psychological Operations*, Joint Publication 3–13.2, 7 January 2010, p GL–8, <http://www.fas.org/irp/doddir/dod/jp3-13-2.pdf>).

[504] Article 21(1). The Rules, however, were never adopted in treaty form.
[505] German Ministry of Defence, *Humanitarian Law in Armed Conflicts*, Section 474. See Rogers, *Law on the Battlefield*, pp 47–8. Of course, psychological warfare could also be conducted through violent acts by kinetic or cyber means. In this case, the law of targeting would apply, including Art 51(2) of Additional Protocol I, which prohibits '[a]cts or threats of violence the primary purpose of which is to spread terror among the civilian population'.
[506] German Ministry of Defence, *Humanitarian Law in Armed Conflicts*, Section 475.
[507] See Section III.1.3.d in this Chapter.
[508] Jeremy A Rabkin and Ariel Rabkin, 'To Confront Cyber Threats, We Must Rethink the Law of Armed Conflict', Hoover Institution, 2012, p 11, <http://media.hoover.org/sites/default/files/documents/EmergingThreats_Rabkin.pdf>. Economic espionage might however breach intellectual property law and trade law.
[509] *Targeted Killings*, para 35 (per Judge Barak).
[510] ICRC, *Interpretive Guidance*, p 51. See also Bothe, Partsch, and Solf, *New Rules*, p 672.
[511] Sandoz, Swinarski, and Zimmermann (eds), *Commentary*, para 3187.
[512] *Targeted Killings*, para 35 (per Judge Barak).   [513] *Strugar*, para 177.

This seems to be the correct approach: it is only the collection of tactical (eg target acquisition), and not also strategic intelligence that amounts to 'direct' participation in hostilities.[514] The ICRC *Interpretive Guidance on the Notion of Direct Participation in Hostilities* confirms that 'where a specific act does not on its own directly cause the required threshold of harm, the requirement of direct causation would still be fulfilled where the act constitutes an integral part of a concrete and coordinated tactical operation that directly causes such harm'.[515] The gathering of intelligence amounts to direct participation in hostilities, then, only 'if carried out with a view to the execution of a specific hostile act'.[516] On the other hand, if the cyber operation only intends to create a general capability to conduct attacks, it would not entail the loss of civilian immunity. The same considerations apply to cyber operations for deception purposes.

V. Cyber Operations as Remedies Against Violations of the Law of Armed Conflict

This Section is concerned with cyber operations as means to enforce the law of armed conflict, where enforcement is intended as 'exercising coercive pressure on the deviant subject to realign his conduct to the prescriptions of the rules'.[517] States not parties to the armed conflict must be distinguished from belligerent states.[518] As to the former, it does not seem that Common Article 1 of the Geneva Conventions, which provides that '[t]he High Contracting Parties undertake to respect *and to ensure respect* for the present Convention[s] in all circumstances',[519] is, on its own, sufficient to qualify a state not party to a conflict as an 'injured state' under Article 42 of the ILC Articles on the Responsibility of States for Internationally Wrongful Acts and thus to entitle it to adopt countermeasures (including cyber operations below

---

[514] See Section III.1.3.d, p 207, in this Chapter.   [515] ICRC, *Interpretive Guidance*, pp 54–5.

[516] ICRC, *Interpretive Guidance*, pp 55, 66.

[517] Georges Abi-Saab, 'Conclusions', in *Les Nations Unies et le droit international humanitaire: Actes du Colloque international à l'occasion du cinquantième anniversaire de l'ONU*, edited by Luigi Condorelli, Anne-Marie La Rosa, and Sylvie Scherrer (Paris: Pedone, 1996), p 307. Implementation, which is a broader concept, includes 'direct application by the subjects of the legal system, or the addressees of its rules' and 'determinations by third parties—ideally judicial, but could be quasi-judicial instances as well,—in case of dispute as to the proper application by the subjects' (p 307). Implementation mechanisms include resort to human rights bodies, Protecting Powers and the International Fact-Finding Commission. See Silja Vöneky, *Implementation and Enforcement of International Humanitarian Law*, in *The Handbook*, edited by Fleck, pp 647 ff.

[518] The more specific situation of neutral states is discussed in Chapter 5, Section VI.

[519] Emphasis added. See also Additional Protocol I, Art 1(1). An analogous provision does not appear in Additional Protocol II, but it has been argued that, as the situations covered by this Protocol also fall within the scope of application of Common Art 3 of the Geneva Conventions, the obligation to respect and ensure respect applies to non-international armed conflicts as well (Luigi Condorelli and Laurence Boisson de Chazournes, 'Quelques remarques à propos de l'obligation des Etats de "respecter et faire respecter" le droit international humanitaire "en toute circonstances"', in *Studies and Essays on International Humanitarian Law and Red Cross Principles in Honour of Jean Pictet*, edited by Christophe Swinarski (1984), p 17).

the level of the use of force) in reaction to violations of international humanitarian law committed by the belligerents.[520] Under Article 48 of the ILC Articles, in case of violation of obligations *erga omnes* states other than the injured states may only make certain claims from the responsible state: 'a) cessation of the internationally wrongful act, and assurances and guarantees of non-repetition...and b) performance of the obligation of reparation...in the interest of the injured State or of the beneficiaries of the obligation breached'.[521] A state not party to a conflict, however, would be 'injured' by a violation of international humanitarian law if the violation in question occurs on its territory or the victims are its nationals:[522] such a state may adopt countermeasures against the wrongdoing belligerent under the conditions already examined in Chapter 2.[523]

As to cyber operations above the use of force level, it is the *jus ad bellum* that determines the legality of *any* initial use of armed force between belligerent and non-belligerent states, regardless of its purpose: whether or not cyber operations above the use of force level conducted by states not parties to the armed conflict against belligerents responsible for violations of the *jus in bello* are lawful, then, depends on their consistency with the UN Charter and the relevant customary provisions. Indeed, neither Common Article 1 of the Geneva Conventions nor Article 89 of Additional Protocol I constitute exceptions to Article 2(4) of the Charter.[524] The UN Security Council could play an important role in this context by authorizing individual states or international peace enforcement operations to use cyber force against states responsible for violations of the *jus in bello* and to protect civilians.[525]

On the other hand, belligerent states may adopt not only countermeasures but also belligerent reprisals to induce an adversary to comply with its obligations under international humanitarian law: once an armed conflict has broken out, it is only the *jus in bello*, and not the *jus ad bellum*, that governs the conduct of hostilities between belligerents. While armed reprisals are prohibited in peacetime,

---

[520] Marco Sassòli, 'State Responsibility for Violations of International Humanitarian Law', *International Review of the Red Cross* 84 (2002), p 424; Bartolini, 'Armed Forces', pp 79–80.

[521] Article 54 of the ILC Articles provides that, in case of obligations of a collective character, any states 'other than the injured States', to which the collective obligation is owed, may take 'lawful measures' against the wrongdoing state 'to ensure cessation of the breach and reparation in the interest of the injured State or of the beneficiaries of the obligation breached', but without specifying whether 'lawful measures' include countermeasures.

[522] Sassòli, 'State Responsibility', p 423.   [523] See Chapter 2, Section IV, pp 105–6.

[524] International Committee of the Red Cross, *Report on the Protection of War Victims, International Review of the Red Cross* 33 (1993), pp 427–8; Commentaries on Art 1(1) and Art 89 of Additional Protocol I, in Sandoz, Swinarski, and Zimmermann (eds), *Commentary*, pp 36–7, 1035; Henckaerts and Doswald-Beck, *Customary International Humanitarian Law*, Vol I, pp 512–13. See also *Legal consequences of the construction of a wall*, para 159. Article 89 of Additional Protocol I provides that '[i]n situations of serious violations of the Conventions or of this Protocol, the High Contracting Parties undertake to act, jointly or individually, in co-operation with the United Nations and in conformity with the United Nations Charter'.

[525] Chapter 2, Section V. See also Marco Roscini, 'The UN Security Council and the Enforcement of International Humanitarian Law', *Israel Law Review* 43 (2010), pp 330 ff; Jann K Kleffner and Heather A Harrison Dinniss, 'Keeping the Cyber Peace: International Legal Aspects of Cyber Activities in Peace Operations', *International Law Studies* 89 (2013), pp 512 ff; Vöneky, *Implementation*, pp 694 ff.

belligerent reprisals are lawful under the law of armed conflict, if only within certain stringent limits.[526] Belligerent reprisals are different from countermeasures as they may be adopted exclusively in the context of an international armed conflict and in reaction to an ongoing violation of the *jus in bello*, not other international law. They may be adopted by the belligerent that suffered the damage as a consequence of the original breach or, in coalition warfare, by one of its allies, but not by a state not party to the conflict.[527] Unlike countermeasures, they may involve a use of force, but their only purpose is to coerce the wrongdoing belligerent into complying with its *jus in bello* obligations and to provide reparation, including the provision of compensation and the punishment of responsible individuals, if need be.[528] As Rule 46 of the Tallinn Manual recalls, belligerent reprisals, including by way of cyber operations, are prohibited against:

(a) prisoners of war;
(b) interned civilians, civilians in occupied territory or otherwise in the hands of an adverse party to the conflict, and their property;
(c) those *hors de combat*; and
(d) medical personnel, facilities, vehicles, and equipment.[529]

Additional Protocol I also prohibits belligerent reprisals against 'civilian population, individual civilians, civilian objects, cultural objects and places of worship, objects indispensable to the survival of the civilian population, the natural environment, and dams, dykes, and nuclear electrical generating stations'.[530] In addition, cyber operations amounting to belligerent reprisals may only be undertaken as a measure of last resort and after a warning has been issued, they must be proportionate to the violation they react against, cease as soon as the wrongdoing belligerent complies with the law, and the decision to adopt them must be made at the highest level of government.[531] The same considerations with regard to the difficulties of calculating the proportionality of the effects of cyber operations that have been made earlier also apply here.[532] Note, however, that, unlike in the principle of proportionality in attacks, the comparison in the present context is between the

---

[526] Dinstein, *The Conduct*, p 254.   [527] Dinstein, *The Conduct*, p 256.
[528] Tallinn Manual, p 150. Article 3 of the 1907 Hague Convention IV on the Laws and Customs of War on Land and Art 91 of the 1977 Protocol I Additional to the Geneva Conventions on the Protection of Victims of War make clear that a belligerent state, be it the aggressor or the state acting in self-defence, 'shall be responsible for all acts committed by persons forming part of its armed forces'. See above, Chapter 1, Section IV, p 36.
[529] Tallinn Manual, p 149.
[530] Rule 47, in Tallinn Manual, p 152 (doubting the customary character of these prohibitions). See Arts 20, 51(6), 52(1), 53(c), 54(4), 55(2), and 56(4) of Additional Protocol I. When signing/ratifying Additional Protocol I, certain states appended declarations reserving the right to adopt reprisals against these objects in order to induce an adversary to terminate the violation of these provisions, after a formal warning has been issued and decision at the highest level is taken (see eg the UK statement, in Roberts and Guelff, *Documents*, p 511).
[531] Henckaerts and Doswald Beck, *Customary International Humanitarian Law*, Vol I, pp 516–18.
[532] See Chapter 2, Section III.4, pp 90–1.

damage caused by the original breach and that caused by the belligerent reprisal, and does not include the military advantage anticipated from the operation.[533]

VI. Conclusions

The above analysis has demonstrated that, if a cyber operation with a belligerent nexus to an armed conflict amounts to 'attack' in the sense of Article 49(1) of Additional Protocol I, ie it produces or is reasonably likely to produce 'violent' consequences in the form of loss of life or injury of persons, more than minimal material damage to property, or loss of functionality of infrastructures, the law of targeting will apply to it, including the principles of distinction and proportionality and the duty to take precautions.

In such case, to be lawful a cyber operation amounting to 'attack', whatever its purpose, will have to meet the following conditions:

(1) it must not employ unlawful means or methods of cyber warfare;

(2) it must be directed against a person or object that qualifies as a military objective;

(3) it must not be indiscriminate; and, in particular, it must not be expected to cause incidental damage on civilians or civilian property which is excessive with respect to the concrete and direct military advantage anticipated;

(4) it must comply with rules providing for special protection from attack, if applicable;

(5) all feasible precautions must have been taken to avoid or at least minimize incidental damage on civilians and civilian objects;

(6) it must not be contrary to other applicable rules of international humanitarian law, in particular the prohibition of perfidy and the principle of unnecessary suffering;

(7) if undertaken as a belligerent reprisal, it must comply with the stringent conditions for its adoption;

(8) it must not breach international human rights law and other peacetime international law when applicable to the cyber operation as *leges generales*.

In addition, if the target is located on the territory of a non-belligerent state, the consent of that state to the operation must be previously obtained, unless other justifications for the extraterritorial use of force under the *jus ad bellum* (self-defence, authorization by the UN Security Council) may be invoked. Finally, the law of neutrality must also be taken into account as a possible limit in international armed conflicts. This will be discussed in the next Chapter.

---

[533] Dinstein, *The Conduct*, p 255.

5
Cyber Operations and the Law of Neutrality

I. Introduction

Neutrality is 'the status of a state which is not participating in an armed conflict between other states'.[1] In its Advisory Opinion on the *legality of the threat or use of nuclear weapons*, the ICJ found that the 'principle' of neutrality is 'an established part of the customary international law', 'is of a fundamental character similar to that of the humanitarian principles and rules', and 'is applicable (subject to the relevant provisions of the United Nations Charter), to all international armed conflict'.[2]

The law of neutrality does not entirely displace the application of the law of peace between belligerent and neutral states, but partly supersedes it in order to prevent the escalation of the conflict and protect neutral states and nationals from the harmful effects of the hostilities.[3] The relevant *jus ad bellum* and the *jus in bello* rules also continue to regulate the use of armed force between the neutral and belligerent states. The rules on neutrality were codified at the beginning of

---

[1] The Federal Ministry of Defence of the Federal Republic of Germany, *Humanitarian Law in Armed Conflict*, ZDv 15/2, 1992, Section 1101, <http://www.humanitaeres-voelkerrecht.de/ManualZDv15.2.pdf>. See also Rule 13(d) of the *San Remo Manual on International Law Applicable to Armed Conflicts at Sea* (text in Adam Roberts and Richard Guelff, *Documents on the Laws of War* (Oxford: Oxford University Press, 2000), p 577); Rule 1(aa) of the HPCR *Manual on International Law Applicable to Air and Missile Warfare*, (Cambridge: Cambridge University press, 2013), p 43.

[2] *Legality of the Threat or Use of Nuclear Weapons*, Advisory Opinion, 8 July 1996, ICJ Reports 1996 ('*Nuclear Weapons*'), paras 88–9. See also *Legal Consequences for States of the Continued Presence of South Africa in Namibia (South West Africa) notwithstanding Security Council Resolution 276 (1970)*, Advisory Opinion, 21 June 1971, ICJ Reports 1971 ('*Namibia*'), Separate Opinion of Vice-President Ammoun, p 92.

[3] Wolff Heintschel von Heinegg, '"Benevolent" Third States in International Armed Conflicts: The Myth of the Irrelevance of the Law of Neutrality', in *International Law and Armed Conflict: Exploring the Faultlines*, edited by Michael N Schmitt and Jelena Pejić (Leiden and Boston: Nijhoff, 2007), p 560; Dietrich Schindler, 'Transformations in the Law of Neutrality Since 1945', in *Humanitarian Law of Armed Conflict: Challenges Ahead*, edited by Astrid JM Delissen and Gerard J Tanja (Dordrecht and Boston: Nijhoff, 1991), p 368. As Gioia argues, 'the law of neutrality essentially consists of a series of limitations of the rights and freedoms which third States would otherwise enjoy in times of peace; these limitations are accepted on the assumption that neutral States can thus reduce the damage they might otherwise suffer as a result of an armed conflict to which they are not a party' (Andrea Gioia, 'Neutrality in Air Warfare', in *The Law of Air Warfare—Contemporary Issues*, edited by Natalino Ronzitti and Gabriella Venturini, (Utrecht: Eleven International, 2006), p 184).

the twentieth century.[4] Neutrality in land warfare is regulated by the 1907 Hague Convention V Respecting the Rights and Duties of Neutral Powers and Persons in Case of War on Land, while Hague Convention XIII, opened for signature in the same year, concerns the Rights and Duties of Neutral Powers in Naval War.[5] There is general agreement that these conventions have matured into customary international law.[6] Neutrality in air warfare is addressed in the 1923 Hague Rules of Aerial Warfare: the Rules, drafted by a Commission of experts on behalf of the Washington Conference on the limitation of armament (1921–22), have never been converted into a treaty but are considered declarative of customary international law.[7] Private codification attempts like the 1994 *San Remo Manual on International Law Applicable to Armed Conflict at Sea*, the 2009 *HPCR Manual on Air and Missile Warfare* and the 2013 *Tallinn Manual on the International Law Applicable to Cyber Warfare* also contain a limited number of rules on neutrality applicable in their respective domains of warfare.[8] These codifications, as such, are of course not binding on states, but at least some of their rules are a restatement of customary international law.

In the *Nuclear Weapons* Advisory Opinion, the ICJ found that the principle of neutrality applies 'whatever type of weapons might be used'.[9] Indeed, the same exigencies that determined the gradual emergence of the law of neutrality in the eighteenth and nineteenth centuries with regard to maritime warfare, ie the 'rapid growth and increasing importance of international trade..., which led maritime states to seek a means of resisting belligerent interference with neutral trade',[10] now justify its extension to cyberspace, where neutral states have an interest in the continued operation of cyber infrastructure for communication, commercial and social purposes. That the law of neutrality also applies in the cyber context is confirmed by the US DoD *Cyberspace Policy Report*, which refers to 'actions taking place on or through computers or other infrastructure located in a neutral third country'.[11] It is also implied in the speech given by the then US DoD Legal Advisor

---

[4] Previous attempts to codify the law of neutrality included the 1856 Declaration of Paris Respecting Maritime Law and the 1871 Washington Rules of Neutral Duty, as well as Art 54 and 57–60 of the Regulations on the War on Land annexed to the 1899 Hague Convention II.

[5] Provisions on neutrality are also contained in Hague Convention VIII Relating to the Laying of Automatic Submarine Contact Mines, Hague Convention XI Relative to Certain Restrictions With Regard to the Exercise of the Right of Capture in Naval War, Hague Convention XII Relative to the Creation of the International Prize Court, and the 1928 Havana Convention on Maritime Neutrality, as well as in the 1909 Declaration of London on Naval Warfare and the 1938 Stockholm Declaration Regarding Similar Rules of Neutrality. A limited number of provisions referring to the rights and duties of neutrals is contained in the four Geneva Conventions of 1949 and their Additional Protocol I.

[6] See *Namibia*, Separate Opinion of Vice-President Ammoun, p 93.

[7] Remigiusz Bierzanek 'Commentary to the 1923 Hague Rules for Aerial Warfare', in *The Law of Naval Warfare. A Collection of Agreements and Documents with Commentaries*, edited by Natalino Ronzitti (Dordrecht, Boston, and London: Nijhoff, 1988), pp 396 ff. The text of the Rules is in Roberts and Guelff, *Documents*, pp 141 ff.

[8] Neutrality issues also arise in outer space, although there are no codified rules in this domain (see George K Walker, 'Information Warfare and Neutrality', *Vanderbilt Journal of Transnational Law* 33 (2000), pp 1170–2).

[9] *Nuclear Weapons*, para 89.   [10] Roberts and Guelff, *Documents*, p 85.

[11] US DoD, *Cyberspace Policy Report. A Report to Congress Pursuant to the National Defense Authorization Act for Fiscal Year 2011*, Section 934, November 2011, p 8, <http://www.defense.gov/

at CYBERCOM in 2012, where he indicated the 'implications of sovereignty and neutrality law' among the problems arising from the application of international law in the cyber context.[12] Furthermore, the US Presidential Policy Directive 20 states that the US government will conduct cyber operations 'consistent with its obligations under international law, including with regard to matters of sovereignty and neutrality'.[13]

It is, however, undeniable that '[t]he fact that cyberspace involves worldwide connectivity irrespective of geopolitical borders challenges certain assumptions upon which the law of neutrality is based'.[14] The present Chapter, therefore, will explore how the law of neutrality affects the conduct of cyber operations by neutrals and belligerents. It will first establish in what situations the law of neutrality applies and will subsequently explore what cyber activities carried out by the belligerent and neutral states breach their obligations under the traditional law of neutrality by distinguishing operations from, through, and with effects on neutral territory, as well as other cyber and cyber-related activities, including the use of cyber infrastructures on neutral territory for communication purposes. Section V will discuss what impact the entry into force of the UN Charter has had on the law of neutrality, while the last Section will look at the remedies against cyber activities that breach the law of neutrality, in particular the resort to forcible and non-forcible countermeasures. Not all the provisions of the law of neutrality will be discussed in this Chapter, but only those that fall within the scope of the present book, ie those that affect the conduct of cyber operations. Note also that, while there are specific provisions for land, sea, and air warfare, the fundamental principles of the law of neutrality largely apply irrespective of the domain concerned. In any case, as has been seen in Chapter 4, whenever a cyber operation affects 'civilian population, individual civilians or civilian objects on land', the law of land warfare becomes relevant.[15] It is with this understanding that the following Sections will refer to provisions of Hague Convention V in relation to cyber operations.

II. When Does the Law of Neutrality Apply?

Neutrality may be a permanent status not dependent upon the occurrence of an armed conflict.[16] Such status could derive from a treaty, as in the case of Switzerland,

---

home/features/2011/0411_cyberstrategy/docs/NDAA%20Section%20934%20Report_For%20webpage.pdf>.

[12] Harold Koh, 'International Law in Cyberspace', Speech at the US CYBERCOM Inter-Agency Legal Conference, 18 September 2012, in CarrieLyn D Guymon (ed), *Digest of United States Practice in International Law*, 2012, p 598, <http://www.state.gov/documents/organization/211955.pdf>.

[13] US Presidential Policy Directive/PPD–20, October 2012, p 4, <http://www.guardian.co.uk/world/interactive/2013/jun/07/obama-cyber-directive-full-text>.

[14] *Tallinn Manual on the International Law Applicable to Cyber Warfare* (Cambridge: Cambridge University Press, 2013), p 249.

[15] Article 49(3) of Additional Protocol I. See Chapter 4, Section I, p 166.

[16] Robert Kolb and Richard Hyde, *An Introduction to the International Law of Armed Conflicts* (Oxford and Portland: Hart, 2008), p 278.

or from a unilateral act, whether or not in execution of a previous international agreement, as in the cases of Malta, Costa Rica, and Turkmenistan.[17] Permanent neutrality entails duties that apply before the commencement of and independently from an armed conflict: in particular, a permanently neutral state is not only under an obligation not to participate in hostilities when an armed conflict breaks out, but also 'not to accept any military obligations and to abstain from acts which would render the fulfilment of its obligations of neutrality impossible should the armed conflict occur',[18] such as becoming a member of military alliances.

Apart from the case of permanent neutrality, the application of the law of neutrality is triggered by the insurgence of an armed conflict. States do not necessarily have to declare their neutrality upon the breaking out of hostilities for the law of neutrality to apply: any state that complies with the obligations provided for neutrals enjoys the corresponding rights, unless otherwise provided by a resolution of the UN Security Council acting under Chapter VII.[19]

Although some state practice seems to suggest a growing support towards its possible extension to non-international armed conflicts,[20] the conventional law of neutrality only applies to international ones, unless the insurgents have been recognized as belligerents or the conflict is a war of national liberation according to Article 1(4) of Protocol I additional to the 1949 Geneva Conventions on the Protection of Victims of War.[21] In 'ordinary' armed conflicts of a non-international character, third states are precluded from intervening in any form on the side of the insurgents by the principle of non-intervention in the internal affairs of other states, while support for the legitimate government, ie that in effective control of a sufficiently representative part of the national territory,[22] is generally considered lawful if such government requests the external intervention to quell the insurrection.[23] Of course, states may decide to remain impartial in a non-international armed conflict and not support a government despite its request, as—lacking a

---

[17] Natalino Ronzitti, *Diritto internazionale dei conflitti armati* 4th edn (Torino: Giappichelli, 2011), pp 114–15.

[18] Michael Bothe, 'The Law of Neutrality', in *The Handbook of International Humanitarian Law*, 3rd edn, edited by Dieter Fleck (Oxford: Oxford University Press, 2013), p 554.

[19] See Section V of this Chapter. See also Yoram Dinstein, *War, Aggression and Self-Defence*, 5th edn (Cambridge: Cambridge University Press, 2011), p 25; Leslie C Green, *The Contemporary Law of Armed Conflict*, 3rd edn (Manchester: Manchester University Press, 2008), p 298. See also HPCR Manual p 381.

[20] Nils Melzer, *Cyberwarfare and International Law*, UNIDIR, 2011, p 21, <http://www.isn.ethz.ch/Digital-Library/Publications/Detail/?lng=en&id=134218>.

[21] Karl S Chang, 'Enemy Status and Military Detention in the War Against Al-Qaeda', *Texas International Law Journal* 47 (2011–12), p 37; Gioia, 'Neutrality', p 211. See also *Namibia*, Separate Opinion of Vice-President Ammoun, p 92.

[22] *Tinoco Claims Arbitration (Great Britain v Costa Rica)*, Arbitral Award, 18 October 1923, RIAA, Vol I, p 381.

[23] Detlev F Vagts, 'The Traditional Legal Concept of Neutrality in a Changing Environment', *American University International Law Review* 14 (1998–99), pp 90–1. According to certain commentators, however, '[t]here is…a growing tendency to consider the assistance given to parties in a civil war, even in the form of an "intervention by invitation", as being generally inadmissible' (Kevin Jon Heller, 'The Law of Neutrality Does Not Apply to the Conflict with Al-Qaeda, and It's a Good Thing, Too: A Response to Chang', *Texas International Law Journal* 47 (2011–12), p 119). A controversial

Security Council resolution providing otherwise—intervention is the object of a political decision and not of a legal obligation.[24] As conflicts where armed groups operate against a government from the territory of another state remain non-international,[25] the law of neutrality does not apply to them. The state where the armed group is located, then, is under an obligation to terminate their actions not on the basis of the law of neutrality, but because of the general duty of states to prevent acts by individuals under their jurisdiction that are harmful to other states and to punish those that engage in such activities.[26] It goes without saying that, if the state where the insurgents are located actively participates in the hostilities on the side of the insurgents against another state, or if the insurgents are under its 'overall control',[27] there is an international armed conflict and the law of neutrality becomes potentially applicable.

The law of neutrality, then, implies the existence of an international armed conflict. The question, however, is whether it applies to *all* international armed conflicts. If neutrality traditionally became relevant only in case of war in the legal sense, ie accompanied by *animus bellandi* as expressed in a declaration of war,[28] this is no longer the case: the shift from the notion of 'war' to that of 'armed conflict' in the 1949 Geneva Conventions reflects customary international law and therefore fixes the threshold of application not only of the Geneva Conventions but also of the customary provisions contained in the Hague Conventions.[29] The extension

---

case is that of wars of national liberation. During the Algerian civil war, in particular, some states argued that in such wars the right of self-determination imposed on third states the obligation not to support the government and even a right to support the insurgents. This position, however, was rejected by France and the NATO states (Patrick M Norton, 'Between the Ideology and the Reality: The Shadow of the Law of Neutrality', *Harvard International Law Journal* 17 (1976), p 274). In any case, support to the national liberation movement may not consist of forcible measures. According to the General Assembly's Declaration on Friendly Relations (GA Res 2625 (XXV), 24 October 1970), in resisting actions depriving them of their right to self-determination, 'peoples are entitled to seek and to receive support in accordance with the purposes and principles of the Charter', therefore including Art 2(4). Article 7 of the Resolution on the Definition of Aggression (GA Res 3314 (XXIX), 14 December 1974) also requires that the support sought and received by peoples struggling to exercise their right to self-determination must be 'in accordance with the principles of the Charter and in conformity with the above-mentioned Declaration [on Friendly Relations]'. Conversely, a colonial, racist or occupying power in a war of national liberation may not use force on the territory of a neutral state that provides logistical support to a national liberation movement (Ronzitti, *Diritto internazionale*, pp 319–20). As to national liberation movements, they would have to respect the rights of third states (Schindler, 'Transformations', p 377).

[24] Heller, 'The Law of Neutrality', p 121.   [25] See Chapter 3, Section II.3.2, pp 139–40.
[26] Eric Talbot Jensen, 'Sovereignty and Neutrality in Cyber Conflict', *Fordham International Law Journal* 35 (2011–12), p 835. See Chapter 2, Section III.3, pp 81–3.
[27] *Prosecutor v Tadić*, Case No IT–94–1, Appeals Chamber Judgment, 15 July 1999, paras 120 ff. For the reasons discussed in Chapter 3, Section II.3.2, pp 138–9, it is the overall control standard that is an element of the primary rule defining an internationalized armed conflict, not effective control, which is a standard of attribution under the law of state responsibility.
[28] See Chapter 3, Section II.1.
[29] Schindler, 'Transformations', pp 375–6. Some commentators have argued that, while in a war in the legal sense third states are obliged to comply with the law of neutrality so that, if they do not, they expose themselves to the risk of countermeasures by the aggrieved belligerent, in an international armed conflict not qualifying as 'war' states remain free to comply with the law of neutrality or not, with the consequence that the aggrieved belligerent will not be entitled to resort to countermeasures, as the non-neutral service is not an internationally wrongful act (Gioia, 'Neutrality', p 213). See comments on this view by Heintschel von Heinegg, '"Benevolent" Third States', pp 558–9.

of the law of neutrality to international 'armed conflicts' is confirmed by the fact that the Geneva Conventions contain several provisions referring to 'neutral states', which therefore apply to all cases covered by Article 2 Common to the Geneva Conventions, not only to declared wars.[30]

In Chapter 3, it has been seen that an international armed conflict is any 'resort to armed force between States',[31] be it by kinetic or cyber means.[32] Unlike in non-international armed conflicts, the *jus in bello* does not require a minimum threshold of intensity for international armed conflicts to occur: according to the ICRC, international humanitarian law applies to any shot fired between states, which prevents specious claims that the minimum threshold has not been reached.[33] The Commentary to Article 2 Common to the Geneva Conventions clearly states that '[i]t makes no difference how long the conflict lasts, or how much slaughter takes place, or how numerous are the participating forces'.[34] Even the total absence of hostilities does not preclude, in certain cases, the application of the Conventions, in particular in the case of a declared war not followed by armed clashes or an occupation that meets with no armed resistance. The question is whether the threshold for the application of the law of neutrality is as low as that of the law on the conduct of hostilities: they both imply the existence of an armed conflict (in the case of the law of neutrality, only of an international character), but does the resort to armed force between states have to be more severe in relation to the former? Little research has been done on this problem, but it has been suggested that there exists a threshold of application for the law of neutrality higher than that for the law on the conduct of hostilities. The reason is that the law of neutrality considerably modifies the relationship between the neutrals and the belligerents, with the neutrals having to tolerate certain limitations to the rights and freedoms that they normally enjoy under the law of peace. Such limitations would be justified only if the conflict is serious enough, not in the case of any armed incident: in other words, 'the law of neutrality must be applied in any conflict which has reached a scope which renders its legal limitation by the application of the law of neutrality meaningful and necessary'.[35] For the same reasons, unlike in the law on the conduct of hostilities,[36] it would be necessary that a state react against the initial resort to armed force by another state for the law of neutrality to apply. The 1992 German Military Manual, currently under revision, supports the view that 'the neutrality of a state begins with the outbreak of an armed conflict of

---

[30] Gioia, 'Neutrality', p 213.

[31] ICTY, *Tadić*, Case No IT-94-1, Decision on the Defence Motion for Interlocutory Appeals on Jurisdiction, 2 October 1995, para 70.

[32] See Chapter 3, Section II.3.

[33] ICRC, *International Humanitarian Law and the Challenges of Contemporary Armed Conflicts*, October 2011, p 7, <http://www.icrc.org/eng/assets/files/red-cross-crescent-movement/31st-int ernational-conference/31-int-conference-ihl-challenges-report-11-5-1-2-en.pdf>.

[34] Jean S Pictet (ed), *Commentary on the Geneva Conventions of 12 August 1949* (Geneva: International Committee of the Red Cross, 1952–60), Vol 3, p 23. See also Yves Sandoz, Christophe Swinarski, and Bruno Zimmermann (eds), *Commentary on the Additional Protocols of 8 June 1977 to the Geneva Conventions of 12 August 1949* (Dordrecht: Nijhoff, 1987), para 62.

[35] Bothe, 'The Law of Neutrality', p 556.   [36] See Chapter 3, Section II.3.2, pp 136–7.

considerable size between other states'.[37] As Michael Bothe concedes, however, '[i]t is...impossible to establish this threshold in a general way' and '[o]ne can only say that there must be a conflict of a certain duration and intensity'.[38]

Should one disagree with the above views and consider the threshold for the application of the law of neutrality identical to that of the *jus in bello*, the considerations made in Chapter 3 with regard to the latter would also apply here. If the above opinions are considered correct, however, the law of neutrality would be potentially applicable to cyber operations associated with an international armed conflict only when this is 'of significant scope': should a belligerent nexus be established, for instance, this would be the case of the 2008 armed conflict between the Russian Federation and Georgia, where the cyber operations against the latter accompanied a territorial invasion by the former. Standalone cyber operations will have to cause a very significant amount of physical damage to property or injury or death of persons, as well as a kinetic or cyber reaction by the target state, in order to trigger the application of the law of neutrality. As it is only cyber operations that exceed mere inconvenience and significantly disrupt the functioning of critical infrastructure, be it military or civilian, that can potentially qualify as 'resort to armed force',[39] and as (if one accepts Bothe's view) the 'resort to armed force' has to be 'of significant scope' for the law of neutrality to become relevant, this double threshold would make it highly unlikely that cyber operations causing mere loss of functionality without physical consequences in the analogue world would, in themselves, ever trigger the application of the law of neutrality.

Neutrality ceases when the neutral state enters the conflict by declaring war on a belligerent and/or by resorting to armed force against it,[40] or when the armed conflict ends (apart from the case of permanent neutrality). A few caveats in relation to these general statements are, however, necessary. According to Bothe, '[w]ithin the meaning of the law of neutrality, a conflict must be considered as terminated where, after the cessation of active hostilities, there is a certain degree of normalization of the relations between the parties to a conflict'.[41] Furthermore, 'limited actions of armed defence of neutrality' by the neutral state against a belligerent

---

[37] German Ministry of Defence, *Humanitarian Law in Armed Conflicts*, Section 1106 (emphasis added). In international armed conflicts that have not reached the minimum threshold for the application of the law of neutrality, then, non-belligerent states would not be 'neutrals' in the technical sense of the expression, but more generally 'other states not parties to a conflict', an expression used in Additional Protocol I (Bothe, 'The Law of Neutrality', p 551). See in particular Art 2(c), 9(2)(a), 19, 22(2)(a), 31, 39(1), and 64 of Additional Protocol I. Article 122 of Geneva Convention III also mentions 'neutral and non-belligerent powers'.

[38] Bothe, 'The Law of Neutrality', p 556. Ziolkowski also suggests that the law of neutrality applies 'to an armed conflict of significant scope' (Katharina Ziolkowski, 'Computer Network Operations and the Law of Armed Conflict', *Military Law and Law of War Review* 49 (2010), p 85).

[39] See Chapter 3, Section II.3.1.1, pp 130–2, 135–6.

[40] Action short of resort to armed force does not end neutral status, even if it amounts to a violation of the law of neutrality. The German Military Manual confirms that 'breaches of single duties of neutrality' by the neutral state are not sufficient, in themselves, to terminate its neutrality (German Ministry of Defence, *Humanitarian Law in Armed Conflicts*, Section 1107).

[41] Bothe, 'The Law of Neutrality', p 557.

are not sufficient, in themselves, to end its neutral status.[42] Indeed, an act that is consistent with the law of neutrality is not a hostile act, even though it might favour one of the belligerents.[43] A neutral state that uses active cyber defences, including those amounting to a use of force, to react against violations of its neutrality by a belligerent, therefore, does not become *ipso facto* a party to the conflict.

III. The Law of Neutrality and its Consequences on the Conduct of Cyber Operations

Having established that the law of neutrality applies, the critical issue to be addressed is how the cyber context can be brought within the applicable principles. The traditional law of neutrality, as contained in the 1907 Hague Conventions V and XIII and the corresponding customary international law, is articulated around three main duties of neutral states: the duty to abstain from committing any acts of hostilities against the belligerents, from providing them with military assistance and from allowing the use of their territory for the conduct of hostilities; the duty to terminate violations of their neutrality, including by force if necessary; and the duty not to discriminate between belligerents in allowing or prohibiting certain conduct that does not amount to acts of hostilities.[44] As to the belligerents, they are under a general duty to respect the inviolability of neutral territory and under more specific obligations not to use it for certain hostile conduct.

In the following sections, these duties will be tested in the cyber context by examining whether cyber operations from, through, and with effects on neutral territory conducted by the belligerents, the neutral states or neutral nationals are a violation of the law of neutrality.[45] Other cyber-related activities, including the use of cyber infrastructure located in neutral territory by the belligerents for communication purposes, will also be discussed in order to establish whether they amount to non-neutral service.

---

[42] German Ministry of Defence, *Humanitarian Law in Armed Conflicts*, Section 1107. See also Dietrich Schindler, 'L'emploi de la force par un Etat belligérant sur le territoire d'un Etat non belligérant', in *Estudios de derecho internacional. Homenaje al Professor Miaja de la Muela* (Madrid: Tecnos, 1979), Vol II, p 863. Rule 169 of the HPCR Manual specifies that 'the use of force by the Neutral must not exceed the degree required to repel the incursion and maintain its neutrality' (p 390).

[43] Article 26 of Hague Convention XIII concerning the Rights and Duties of Neutral Powers in Naval War.

[44] Non-discrimination does not mean identical treatment: neutral states are, for instance, free to allow the continuation of commerce with the belligerent states according to the principle of the *courant normal*, ie under the conditions existing when the hostilities broke out, even if the commercial positions of the belligerents were not analogous (Bothe, 'The Law of Neutrality', p 550). Humanitarian assistance may be provided discriminatorily providing that the difference is motivated by humanitarian reasons and has not the purpose of benefiting one party to the conflict to the detriment of another (William H Boothby, *The Law of Targeting* (Oxford: Oxford University Press, 2012), p 520).

[45] Neutral territory is considered as including the cyber infrastructure located therein.

1. Cyber operations from neutral territory

Cyber operations from neutral territory against a belligerent could be conducted by an opposing belligerent, by the neutral state, or by neutral nationals or other individuals located on neutral territory. A prohibition on belligerents to conduct cyber operations from cyber infrastructure located on a neutral state's territory derives first and foremost from Article 1 of Hague Convention V, which provides that '[t]he territory of neutral Powers is inviolable'.[46] As the ICJ has recognized, this provision reflects customary international law.[47] Article 1 of Hague Convention V applies to all 'public or private cyber infrastructure that is located within neutral territory (including civilian cyber infrastructure owned by a party to the conflict or nationals of that party)'.[48] Article 1 prohibits both the remote taking over of computer systems located in neutral territory to conduct cyber operations and the execution of cyber operations by organs or agents of the belligerent state physically located in the territory of the neutral state.[49]

Article 1 of Hague Convention V, however, is not the only legal basis of the prohibition for belligerents to conduct cyber operations from neutral territory. Article 2 of Hague Convention V provides that '[b]elligerents are forbidden to move troops or convoys of either munitions of war or supplies across the territory of a neutral Power': as has been suggested, 'a cyber attack moves a weapon across the territory of the neutral state' and is therefore a violation of its neutrality.[50] All in all, the norm should be interpreted in a manner that takes into account the digitalization of essential services in modern information societies, the dependency of the military on computerized systems and networks and the potentially severe harmful effects of at least certain cyber operations.[51]

Even though it does not result expressly from Hague Convention V, belligerents are prohibited not only from conducting cyber operations qualifying as 'attacks' in the sense of Article 49(1) of Additional Protocol I from the neutral state's territory, but all cyber operations amounting to acts of hostilities.[52] This is confirmed

---

[46] See also Art 40 of the Hague Rules of Aerial Warfare. Article 1 of Hague Convention XIII provides that '[b]elligerents are bound to respect the sovereign rights of neutral Powers'. The neutral state's territory covers not only its land territory, but also the waters under its sovereignty, including internal, territorial, and archipelagic waters, as well as the airspace above them (Tallinn Manual, p 248; HPCR Manual, p 384; German Ministry of Defence, *Humanitarian Law in Armed Conflicts*, Section 1118). Article 1 of Hague Convention V, therefore, protects all cyber infrastructures located in those areas. It has been argued that the belligerents are also prohibited from using neutral cyber infrastructure located outside neutral territory when such infrastructure is governmental in character and enjoys sovereign immunity, or when it has the nationality of a neutral state, providing it is not located on a belligerent's territory (Jensen, 'Sovereignty', pp 824–5; Tallinn Manual, p 248).

[47] *Nuclear Weapons*, para 88–89.   [48] Tallinn Manual, p 248.

[49] Commentary to Rule 92, in Tallinn Manual, p 251.

[50] Jeffery TG Kelsey, 'Hacking into International Humanitarian Law: The Principles of Distinction and Neutrality in the Age of Cyber Warfare', *Michigan Law Review* 106 (2007–08), p 1443. Rule 92 of the Tallinn Manual, that prohibits '[t]he exercise of belligerent rights by cyber means in neutral territory', founds this obligation, inter alia, on Art 2 of Hague Convention V (Tallinn Manual, p 251).

[51] As to cyber operations routed through neutral cyber infrastructure but originating elsewhere, see below, Section 2 of this Chapter.

[52] The Tallinn Manual prefers the use of the expression 'exercise of belligerent rights', intended as 'actions that a party to the conflict is entitled to take in connection with the conflict' (Tallinn Manual, p 249).

by Article 1 of Hague Convention XIII, which explicitly provides that belligerents must 'abstain, in neutral territory or neutral waters, from *any act* which would, if knowingly permitted by any Power, constitute a violation of neutrality',[53] while Article 2 of the same Convention states that '[a]ny act of hostility' is a violation of neutrality and is thus forbidden.[54] Article 39 of the Hague Rules of Aerial Warfare provides that '[b]elligerent aircraft are bound to respect the rights of neutral Powers and to abstain within the jurisdiction of a neutral state from the commission of *any act* which it is the duty of that state to prevent'.[55] Rule 166 of the HPCR Manual also states that '*[h]ostilities*... must not be conducted within neutral territory', while Rule 167(a) stresses that belligerents 'are prohibited in neutral territory to conduct *any hostile actions*'.[56] Rule 171(d) also prohibits the belligerents to conduct '[a]ny other activity involving the use of military force or contributing to the war-fighting effort, including transmission of data or combat search-and-rescue operations in neutral territory'.[57] The UK *Manual of the Law of Armed Conflict* declares that '[n]eutral states must refrain from allowing their territory to be used by belligerent states for the purposes of *military operations*',[58] while according to Section 1108 of the German Military Manual belligerents are prohibited from conducting 'any act of war' on neutral territory.[59] It is therefore not only cyber attacks amounting to 'attack' in the sense of Article 49(1) of Additional Protocol I, but also cyber attacks not resulting in violent consequences that the belligerents are forbidden from conducting from neutral territory. The references in the above documents to acts of hostilities, military operations, hostile actions, and any activity that contributes to the 'warfighting effort' suggest that cyber exploitation is also prohibited on neutral's territory, at least when it aims to obtain tactical intelligence.[60] Article 47 of the Hague Rules of Aerial Warfare confirms that '[a] neutral state is bound to take such steps as the means at its disposal permit to prevent within its jurisdiction aerial observation of the movements, operations or defenses of one belligerent, with the intention of informing the other belligerent'. Rule 171(b) of the HPCR Manual also prohibits the belligerents from using 'neutral territory or airspace as a base of operations—for attack, targeting, *or intelligence purposes*—against enemy targets in the air, on land or on water outside that territory'.[61]

---

[53] Emphasis added.
[54] Article 5 of Hague Convention XIII also explicitly forbids the belligerents 'to use neutral ports and waters as a base of naval operations against their adversaries'.
[55] Emphasis added.   [56] HPCR Manual, pp 383, 385 (emphasis added).
[57] HPCR Manual, p 395.
[58] UK Ministry of Defence, *The Manual of the Law of Armed Conflict* (Oxford: Oxford University Press, 2004), p 20 (emphasis added).
[59] German Ministry of Defence, *Humanitarian Law in Armed Conflicts*, Section 1108.
[60] See Rule 171(d) of the HPCR Manual, p 395. The Commentary states that the 'transmission of military data from neutral territory by a Belligerent Party must be considered a violation of neutral territory and airspace even if it is not performed for attack, targeting or other purposes', unless this occurs through 'a public, internationally and openly accessible network' (HPCR Manual, p 396). It is also generally accepted that intelligence offices may not be established in neutral territory (Walker, 'Information Warfare', p 1149). See also Heller, 'The Law of Neutrality', p 131 (citing Erik Castrén, *The Present Law of War and Neutrality* (Helsinki: Suomalaisen Kirjallisuuden Seuran Kirjapainon, 1954), p 479).
[61] HPCR Manual, p 394 (emphasis added).

If the belligerent state launches a cyber operation from neutral territory feigning that it is conducted by the neutral state, it may commit a perfidious act.[62] As has been seen, however, for a cyber operation to be perfidious, deception is not sufficient: the operation has to 'invite the confidence of an adversary with respect to protection' under international humanitarian law and betray it, and also result in the death, injury, or capture of the adversary. Sending malware attached to an email appearing to be from a neutral state or national which, once executed, results in infrastructure malfunction causing death or injury would amount to perfidy. If the cyber operation leads to destruction or incapacitation of infrastructure but not to loss of life, injury, or capture of the adversary, however, it would not amount to a violation of Article 37 of Additional Protocol I, although it may fall under the prohibitions contained in Article 39 of Additional Protocol I on the misuse of emblems of nationality.

If the belligerent states have to abstain from conducting cyber operations amounting to hostilities from neutral cyber infrastructure, all the more this obligation of abstention applies to the neutral states themselves. Under Article 5 of Hague Convention V, the neutral states also have an obligation not to allow a belligerent's cyber operations amounting to acts of hostilities from cyber infrastructure located on their territory. According to the Dutch government, for instance, '[i]n an armed conflict involving third parties, the Netherlands can protect its neutrality by impeding the use by such parties of [cyber] infrastructure and systems (e.g. botnets) on Dutch territory. Constant vigilance, as well as sound intelligence and a permanent scanning capability, are required here'.[63] Rule 93 of the Tallinn Manual provides that '[a] neutral State may not knowingly allow the exercise of belligerent rights by the parties to the conflict from cyber infrastructure located in its territory or under its exclusive control'.[64] The rule introduces a mental element ('knowingly') that is not present in Article 5 of the Hague Convention.[65]

Although it has been suggested that Article 8 of Hague Convention V, according to which '[a] neutral Power is not called upon to forbid or restrict the use on behalf of the belligerents of telegraph or telephone cables or of wireless telegraphy apparatus belonging to it or to companies or private individuals', might be construed as an exception to Article 2,[66] the situations covered in the two provisions are different: the former refers to the use of 'telegraph or telephone cables or of wireless telegraphy apparatus' for communications, as this was the only known use of such installations when the Convention was drafted, while the latter concerns the

---

[62] See Chapter 4, Section III.1.5.
[63] Dutch Government Response to the AIV/CAVV Report on Cyber Warfare, p 6, <http://www.rijksoverheid.nl/bestanden/documenten-en-publicaties/rapporten/2012/04/26/cavv-advies-nr-22-bijlage-regeringsreactie-en/cavv-advies-22-bijlage-regeringsreactie-en.pdf>.
[64] Tallinn Manual, p 252. The Commentary explains that cyber infrastructure 'under the exclusive control' of a neutral state refers to 'non-commercial government cyber infrastructure' (p 253).
[65] See the critical comments in Rain Liivoja and Tim McCormack, 'Law in the Virtual Battlespace: The Tallinn Manual and the *Jus in Bello*', Melbourne Legal Studies Research Paper No 650, 23 July 2013, p 12, <http://papers.ssrn.com/sol3/papers.cfm?abstract_id=2297159>.
[66] See the position of the minority of the Group of Experts that drafted the Tallinn Manual (Tallinn Manual, p 252).

movement of troops and munitions of war. Article 8 was never meant to allow activities, such as cyber attacks, that could qualify as acts of hostilities against a belligerent. Furthermore, Article 8 was included to take into account the fact that a state cannot control the 'extraterritorially initiated use of publicly accessible transnational communications networks'.[67] As will be seen, this consideration might apply to the routing of cyber operations through neutral infrastructure, but not to cyber operations launched *from* the neutral's territory, as in this case they are not 'extraterritorially initiated' and the neutral state may be able to terminate them.

While the neutrals' duty not to tolerate certain belligerent activities on their territory is absolute on land,[68] in maritime neutrality the obligation is one of due diligence and only requires the neutral state to use 'the means at its disposal' to prevent them.[69] Similarly, in air neutrality, Articles 42, 46, and 47 of the Hague Rules of Aerial Warfare require a neutral state to use 'the means at its disposal' to prevent violations of its neutrality by belligerent aircraft.[70] The neutral state is therefore not required to possess or acquire the latest technology to enforce its neutrality.[71] This lower standard is due to the different characteristics of the respective domains of warfare: while it is presumed that a state has sufficient forces to prevent and terminate violations of its neutrality on land, not all states have the naval or aerial means to fully control access to, or certain uses of, their sea and airspace.[72] Because of its evanescent characteristics and the difficulty of exercising effective jurisdiction over it, the situation in cyberspace seems to have more in common with the maritime and air domains than with land warfare.[73] In the cyber domain too, then, the neutrals' duty not to allow certain activities on their territory should be an obligation of conduct, not of result. This is the position adopted in the Tallinn Manual, whose Commentary maintains that there is 'a duty on the part of neutral States to take all feasible measures to terminate any exercise of belligerent rights employing cyber infrastructure falling within the scope of' Rule 93.[74] The Group of Experts, however, was divided on whether the neutral state has an obligation to use all feasible measures not only to terminate, but also to prevent the

---

[67] Melzer, *Cyberwarfare*, p 20.

[68] In his Separate Opinion, however, Judge Ammoun argues that 'governments must show due diligence in preventing any individual or collective act contrary to neutrality' (*Namibia*, Separate Opinion of Vice-President Ammoun, p 95).

[69] See Art 8 ('A neutral Government is bound to employ the means at its disposal to prevent the fitting out or arming of any vessel within its jurisdiction which it has reason to believe is intended to cruise, or engage in hostile operations, against a Power with which that Government is at peace. It is also bound to display the same vigilance to prevent the departure from its jurisdiction of any vessel intended to cruise, or engage in hostile operations, which had been adapted entirely or partly within the said jurisdiction for use in war') and 25 ('A neutral Power is bound to exercise such surveillance as the means at its disposal allow to prevent any violation of the provisions of the above Articles occurring in its ports or roadsteads or in its waters') of Hague Convention XIII. See also Rule 15 of the San Remo Manual, in Roberts and Guelff, *Documents*, p 578.

[70] See also Rules 168(a) and 170(c) in HPCR Manual, pp 387, 393, respectively.

[71] Bothe, 'The Law of Neutrality', p 577. During the 1991 Gulf War, for instance, Jordan affirmed its incapacity to intercept the missiles launched by Iraq against Israel in spite of its neutrality because it lacked the necessary technology (Ronzitti, *Diritto internazionale*, p 330).

[72] Walker, 'Information Warfare', p 1148.

[73] Walker, 'Information Warfare', pp 1186, 1191–2.   [74] Tallinn Manual, p 253.

exercise of belligerent rights by the parties to the conflict from cyber infrastructure located on its territory.[75] As a duty of prevention would probably require states to continuously monitor cyber traffic, which is not only costly but also hard to reconcile with internet freedoms, a negative answer seems preferable.[76] Be that as it may, feasible measures that a neutral state could adopt to defend its neutrality include using passive cyber defences and standard measures of cyber hygiene, establishing national public CERTs and setting up formal frameworks of cooperation with the CERTs of other states, adopting legislative measures providing for penalties for cyber criminal activities and cooperating with other governments, including the belligerent ones, in the investigations and prosecutions, working with national ISPs to block the suspicious systems, and distributing protective software to users.[77] In extreme scenarios, the neutral state could at least in theory 'disconnect' itself from the internet to stop the continuation of cyber attacks originating from its territory. However, the repercussions on global internet communications that this could entail, in particular if more than one neutral state adopts this drastic measure or if the neutral state has key gatepoints for transatlantic cables, make it an unlikely option.[78]

Cyber operations amounting to acts of hostilities from neutral territory could also be conducted by neutral nationals or other individuals whose conduct cannot be attributed to the neutral state under the law of state responsibility. The combined effect of Articles 4 and 5 of Hague Convention V imposes an obligation on neutral states not to allow the formation of 'corps of combatants' on their territory to assist the belligerents. This situation, however, only covers the 'formation' of groups of individuals that are organized in a military structure. Article 6 states that the responsibility of the neutral state is not involved in case of 'persons crossing the frontier separately to offer their services to one of the belligerents', but does not envisage the case of hostile acts carried out by volunteers from the neutral's territory without their crossing the frontier. The situation of individuals that are not members of the armed forces ('corps of combatants') and operate from neutral territory, therefore, is not expressly regulated by Hague Convention V. Nonetheless, Article 5 requires the neutral states not to allow certain conduct, including acts of hostilities, on their territory without specifying the author of the conduct: the provision could therefore be extended to acts of volunteers operating from neutral territory. In any case, a neutral state has no obligation to prevent its subjects from conducting propaganda in favour of a belligerent or from providing information to it.[79] If it prohibits such activities, it must do so without discrimination between belligerents.

---

[75] Tallinn Manual, pp 253–4.
[76] Wolff Heintschel von Heinegg, 'Territorial Sovereignty and Neutrality in Cyberspace', *International Law Studies* 89 (2013), pp 151–2.
[77] The notion of due diligence in the cyber context has been discussed in Chapter 2, Section III.3.
[78] Davis Brown, 'A Proposal for an International Convention to Regulate the Use of Information Systems in Armed Conflict', *Harvard International Law Journal* 47 (2006), p 210; Kelsey, 'Hacking', p 1445.
[79] Heller, 'The Law of Neutrality', p 133.

Article 16 of Hague Convention V provides that '[t]he nationals of a State which is not taking part in the war are considered as neutrals'. According to Article 17, however,

[a] neutral cannot avail himself of his neutrality
  (a) [i]f he commits hostile acts against a belligerent;
  (b) [i]f he commits acts in favor of a belligerent, particularly if he voluntarily enlists in the ranks of the armed force of one of the parties. In such a case, the neutral shall not be more severely treated by the belligerent as against whom he has abandoned his neutrality than a national of the other belligerent State could be for the same act.

Hackers who are neutral nationals and who commit hostile acts against or in favour of belligerents, then, lose their neutral status. The question is whether this also applies to an unwilling or even unaware owner of a 'zombie' computer 'enlisted' in a DDoS attack. While Article 17(b) refers to 'voluntary' enlistment, this is offered only as an example of an act in favour of a belligerent. Furthermore, Article 17(a) provides that the neutral national cannot invoke his neutrality if he commits hostile acts against a belligerent, without further specification on whether he does so by his own will or under coercion. By participating in the commission of hostile acts against a belligerent, then, zombie computers located on neutral territory forfeit their neutral status: whether or not they may be attacked, however, depends on whether they are military objectives under the *jus in bello*.[80]

2. Cyber operations through neutral territory

Because of the interconnectivity of networks and the use of packet switching as a method of data transmission for the internet and most local area networks, data and malware are highly likely to transit through cyber infrastructure located in neutral territory before reaching their destination. The question is whether a belligerent is under an obligation to abstain from routing cyber operations against an opposing belligerent through neutral cyber infrastructure and whether the neutral states are under an obligation not to allow it.

As has been seen, Article 1 declares the territory of neutral states 'inviolable' and Article 2 of Hague Convention V provides that '[b]elligerents are forbidden to move troops or convoys of either munitions of war or supplies across the territory of a neutral Power'. Neutral states are under an obligation not to allow such activities in their territory (Article 5), unless otherwise required by the UN Security Council (Article 43(1) of the UN Charter).[81] The US DoD Cyberspace Policy Report mentions '*[t]he issue of the legality of transporting cyber "weapons" across the Internet through the infrastructure owned and/or located in neutral third*

---

[80] See Chapter 4, Section III.1.2.
[81] The export and transport on behalf of a belligerent of arms, munitions of war or 'anything which can be of use to an army or a fleet' by private companies or individuals for commercial purposes, however, do not have to be prevented by the neutral state (Art 7 of Hague Convention V).

countries without obtaining the equivalent of "overflight rights",[82] and thus implicitly acknowledges that, at least potentially, Article 2 of Hague Convention V applies to the routing of cyber operations through neutral cyber infrastructure. Most experts of the International Group that drafted the Tallinn Manual also considered the transmission of cyber weapons across neutral territory incompatible with Article 2.[83]

Nonetheless, it should be recalled that neutral states have only an obligation to use the means at their disposal to prevent violations of their neutrality. Unlike the situation of cyber operations originating from neutral cyber infrastructure, it is difficult to see how a neutral state could have the means to prevent the routing of data and malware through its territory, or even in most cases be aware of it. Indeed, data are divided by the communications protocol into packets that are then sent independently from each other, and take different paths to reach their destination depending on availability and traffic volume. Neither the transited states nor the senders can influence the route that the packets take.[84] Even if the transited neutral state were able to detect the routing, the data transiting its cyber infrastructure would probably look entirely innocuous, as only when all packets are reassembled at destination they would cause harm.[85] All in all, the risk of the neutral state being harmed and thus becoming involved in the conflict because of routing is much lower than in the case of troop movements through its territory.[86] The US DoD's *Assessment of International Legal Issues in Information Operations* confirms that 'use of a nation's communications networks as a conduit for an electronic attack would not be a violation of its sovereignty in the same way that would be a flight through its airspace by a military aircraft', providing that the routing does not cause damage to the neutral state.[87] The fact that mere routing is not a violation of the law of neutrality also finds implicit support in the US DoD *Cyberspace Policy Report* itself, which only focuses on the responses that the United States could adopt against 'cyber activity *originating* from within' the borders of a neutral third country.[88]

---

[82] US DoD, *Cyberspace Policy Report*, p 8 (italics in the original).
[83] Tallinn Manual, p 252. [84] Kelsey, 'Hacking', p 1433.
[85] Heintschel von Heinegg, 'Territorial Sovereignty', pp 137–8.
[86] Johann-Christoph Woltag, 'Cyber Warfare', in *Max Planck Encyclopedia of Public International Law* (2012), Vol II, pp 992–3. In maritime warfare, certain forms of transit are not inconsistent with neutrality: Art 10 of Hague Convention XIII provides that '[t]he neutrality of a Power is not affected by the mere passage through its territorial waters of war-ships or prizes belonging to belligerents'. See also Rule 20(a) of the San Remo Manual, in Roberts and Guelff, *Documents*, p 579; On the other hand, belligerent military aircraft may not penetrate a neutral's airspace, as there is no equivalent for aircraft of the right of innocent passage through territorial waters for ships (Art 40 of the Hague Rules of Aerial Warfare; Rules 167(a) and 170(a) in HPCR Manual, pp 385, 390, respectively; German Ministry of Defence, *Humanitarian Law in Armed Conflicts*, Section 1126). Unlike in peacetime, then, neutral states are not only entitled to prevent the overflight of military aircraft, but have a duty to do so (Gioia, 'Neutrality', p 190). Overflight by civilian aircraft, however, is not a violation of neutrality and depends on the consent of the territorial state (Bothe, 'The Law of Neutrality', p 576). Similarly, overflight by ballistic missiles and satellites does not constitute a violation of neutrality as long as they transit through outer space (p 576).
[87] US Department of Defense, *An Assessment of International Legal Issues in Information Operations*, May 1999, p 23, <http://www.au.af.mil/au/awc/awcgate/dod-io-legal/dod-io-legal.pdf>.
[88] US DoD, *Cyberspace Policy Report*, p 8.

Similarly, the Commentary to Rule 167(b) of the HPCR Manual explains that 'the mere fact that military communications, including CNAs, have been transmitted via a router situated in the territory of a Neutral is not to be considered a violation of neutrality'.[89] This is so because 'it is impossible for any State to effectively control or interfere with communications over such a network'.[90]

According to some commentators, however, a belligerent violates the neutrality of a state when it *intentionally* directs cyber weapons through the neutral's internet nodes.[91] Unintentional routing, on the other hand, would not be a violation of neutrality and the neutral state would not be under an obligation not to allow it.[92] This view cannot be accepted for two reasons. First, identifying the origin of a cyber operation and the identity of the perpetrator, and even more its intention, is notoriously difficult in the cyber context.[93] Secondly, as has been noted, the path followed by the data packages is outside the control of the sender: it is therefore impossible to speak of an 'intention' to send data through a certain state's cyber infrastructure. Of course, should preventing the routing of cyber attacks through a state's cyber infrastructure become technically possible in the future, the neutral state would be under un obligation to use the means at its disposal to do so.

3. Cyber operations against or with incidental harmful effects on neutral territory

The inviolability of neutral territory also entails that the belligerents are prohibited from conducting any cyber operation against targets located therein, regardless of their governmental or private character or of whether they belong to the neutral or other belligerents. The prohibition covers not only cyber operations amounting to attack, but more in general those qualifying as acts of hostilities.[94] The inviolability of neutral territory rules out both cyber operations that target cyber infrastructure and those against physical infrastructure that, being operated by a computer system, can be attacked by cyber means.[95] It also precludes not only those cyber operations that cause or are expected to cause material damage to property or death/injuries of persons, but also those resulting in loss of functionality of infrastructure, although not mere inconvenience.[96] But the language of Article 1

---

[89] HPCR Manual, p 387. The affirmation that CNAs are an example of military 'communications', however, is questionable.

[90] HPCR Manual, p 387. See also Melzer, *Cyberwarfare*, p 20; Jensen, 'Sovereignty', p 827; Heintschel von Heinegg, 'Territorial Sovereignty', p 149; Jenny Döge, 'Cyber Warfare—Challenges for the Applicability of the Traditional Laws of War Regime', *Archiv des Völkerrechts* 48 (2010), p 497; Woltag, 'Cyber Warfare', p 992; Ziolkowski, 'Computer Network Operations', p 88. *Contra*, Kelsey, 'Hacking', pp 1443–4; Joshua E Kastenberg, 'Non-intervention and Neutrality in Cyberspace: An Emerging Principle in the National Practice of International Law', *Air Force Law Review* 64 (2009), p 53.

[91] Kelsey, 'Hacking', p 1448; Brown, 'A Proposal', pp 210–11.

[92] Kelsey, 'Hacking', p 1449.   [93] See Chapter 1, Section IV.

[94] According to the Commentary of the Tallinn Manual, however, espionage operations against the neutral states are not prohibited (Tallinn Manual, p 249).

[95] See Rule 91 of the Tallinn Manual, p 250.

[96] Heintschel von Heinegg, 'Territorial Sovereignty', pp 145–6.

is probably broad enough to also prohibit cyber operations intended to merely destroy data or software contained in cyber infrastructure located in a neutral state or having the nationality of a neutral state without consequences in the analogue world. The 2012 cyber attack on Saudi Aramco, that wiped off three quarters of the data stored in the company's corporate computers, may be an example of an operation affecting data that, if conducted against a neutral state, would violate the law of neutrality.[97]

The prohibition of cyber operations by a belligerent against the territory of a neutral state may be lifted only in case the neutral state has breached its neutrality or when the neutral state is unable or unwilling to prevent the use of its territory by another belligerent. As will be seen, cyber operations amounting to a use of force against the neutral state will be possible only within the limits of the *jus ad bellum* provisions contained in the UN Charter.[98] If they amount to 'attacks', the cyber operations against the neutral state will also have to comply with the relevant *jus in bello* rules, in particular with the law of targeting, and could therefore be directed only against military objectives.

Belligerents are prohibited from conducting not only cyber operations against a neutral state, but also those against another belligerent that have prejudicial incidental effects on neutral territory. Indeed, malicious software may well spread uncontrollably as a consequence of its own characteristics and the interconnectivity of networks: for instance, although most of the affected computers were located in Iran, Stuxnet also hit computers in Azerbaijan, India, Indonesia, Pakistan, United States, and, to a lesser degree, other states.[99] Cyber attacks that destroy or disrupt a belligerent's cyber infrastructure, then, could also affect connectivity services in neutral states,[100] while cyber attacks on a belligerent's critical infrastructure, such as electrical power stations, may disrupt the provision of essential services also to neutrals.[101] For instance, electricity is shared between Laos and Thailand, Venezuela and Brazil, Canada and the United States, Indonesia and Singapore, while Malaysia provides half of Singapore's water.[102] Telecommunication infrastructure is also shared by several states or entire regions.[103] Unlike the law on the conduct of hostilities, that tolerates a certain level of expected incidental damage on civilians and civilian objects so long as it is not excessive with respect to the military advantage anticipated,[104] there is no counterpart in the law of

---

[97] Nicole Perlroth, 'In Cyberattack on Saudi Firm, U.S. Sees Iran Firing Back', *The New York Times*, 23 October 2012, <http://www.nytimes.com/2012/10/24/business/global/cyberattack-on-saudi-oil-firm-disquiets-us.html?pagewanted=all&_r=0>.

[98] See Chapter 5, Section VI.2.

[99] Nicolas Falliere, Liam O Murchu, and Eric Chien, 'W32.Stuxnet Dossier', Version 1.4, February 2011, pp 5–6 <http://www.symantec.com/content/en/us/enterprise/media/security_response/whitepapers/w32_stuxnet_dossier.pdf>.

[100] Commentary to Rule 91, in Tallinn Manual, p 250. An example are satellites used for both civilian and military purposes and owned by consortia formed by both neutrals and belligerents (US DoD, *An Assessment*, pp 10–11).

[101] Kiolkowski, 'Computer Network Operations', p 87.

[102] William Church, 'Information Warfare', *International Review of the Red Cross* 82 (2000), p 210.

[103] Church, 'Information Warfare', p 210.   [104] See Chapter 4, Section III.2.1.

The Law of Neutrality and its Consequences

neutrality to Article 51(5)(b) of Additional Protocol I. It may be, then, that *any* incidental damage on neutral states and their nationals caused by a kinetic or cyber operation by the belligerents is a violation of the inviolability of the neutral's territory, whether or not this is proportionate to the military advantage anticipated from the attack on a military objective.[105] As the Commentary of Rule 91 of the Tallinn Manual sensibly explains, however, 'States would be unlikely to regard *de minimis* effects as precluding the prosecution of an otherwise legitimate attack'.[106]

4. Use of cyber infrastructure for communications

There is an obvious ontological difference between the use of cyber infrastructure to conduct cyber attacks that 'disrupt, deny, degrade, or destroy information resident in computers and computer networks, or the computers and networks themselves'[107] or cyber exploitation activities, and the use of cyber infrastructure for communications. In other words, it is one thing to spread malware, penetrate a system to exfiltrate information, or create a botnet to conduct a DDoS attack and shut down a system, and quite another to send emails, use social networks, or post information on a website (even when this is for propaganda purposes).

The use of cyber infrastructure exclusively for communication purposes, as opposed to the conduct of cyber attacks and cyber exploitation, does not fall under the scope of Articles 1 and 2 of Hague Convention V, but rather of its Articles 3 and 8. Under Article 3, the belligerents are forbidden to erect their own 'wireless telegraphy station or other apparatus' on the territory of a neutral state 'for the purposes of communicating with belligerent forces on land or sea'; and to use any communication installation established by them before the war on neutral territory 'for purely military purposes' if it 'has not been opened for the service of public messages'.[108] The German Military Manual confirms that a neutral state is not

---

[105] Bothe, 'The Law of Neutrality', p 560; Louise Doswald-Beck, 'Some Thoughts on Computer Network Attack and the International Law of Armed Conflict', *International Law Studies* 76 (2002), pp 177–8. During the Second World War, for instance, the Allies paid compensation for damage caused in Switzerland as a result of attacks on military objectives in Germany whose effects reached Swiss territory (Bothe, 'The Law of Neutrality', p 560). See, however, Art 22 of the Harvard Draft Convention on the Rights and Duties of Neutral States in Naval and Aerial War, according to which '[a] belligerent has no duty to pay compensation for damage to a neutral vessel or other neutral property or persons, when such damage is incidental to a belligerent's act of war against the armed forces of its enemy and not in violation of the provisions of this Convention or the law of war' (*American Journal of International Law* 33 (1939), Supplement, p 179); and Rule 1.4 of the Helsinki Principles on the Law of Maritime Neutrality, according to which 'belligerents must exercise due regard to prevent to the maximum extent possible collateral damage on neutral territory, neutral waters or the airspace over these areas' (the text of the Principles is in ILA, Report of the Sixty-eighth Conference (Taipei, 1998), pp 497 ff).

[106] Tallinn Manual, p 250. According to the Commentary of Rule 91, it is necessary to balance competing rights.

[107] Chairman of the Joint Chiefs of Staff, *The National Military Strategy for Cyberspace Operations*, December 2006, p GL–1, <http://www.dod.mil/pubs/foi/joint_staff/jointStaff_jointOperations/07-F-2105doc1.pdf>.

[108] See also Art 5 of Hague Convention XIII, which however refers only to 'erecting' communication installations for military communications.

required to prevent 'the use by a party to the conflict of generally accessible means of Communications on its territory'.[109] Rule 167(b) of the HPCR Manual also states that 'when Belligerent Parties use for military purposes a public, internationally and openly accessible network such as the Internet, the fact that part of this infrastructure is situated within the jurisdiction of a Neutral does not constitute a violation of neutrality'.[110] Belligerents may thus be allowed to (a) 'erect' a new cyber communication installation on the territory of the neutral state, such as servers, routers, and networks, as long as it is exclusively for non-military communications; (b) use an existing one established by them before the war, even for military communications, provided that it is open 'for the service of public messages'; (c) use an existing communication installation established by them before the war and which is not open 'for the service of public messages', provided it is for non-military communications.

If Article 3 applies to communication installations established by a belligerent, Article 8 of Hague Convention V deals with 'telegraph or telephone cables or wireless telegraphy apparatus' belonging to the neutral state or its companies or private individuals: unlike in the situations covered by Article 3, in this case the neutral state is not called upon to forbid or restrict their use for communication purposes by the belligerents, providing it does so without discrimination under the principle of *courant normal* (Article 9).[111] The neutral state could then either prohibit or allow all belligerents to use its cyber communication infrastructure, even if it is for military communications. During the 2008 armed conflict with the Russian Federation, which was accompanied by cyber operations that disrupted access to certain governmental websites, Georgia relocated the Presidential website to a private US web hosting company, while the Ministry of Foreign Affairs' press dispatches were issued through Google's Blogspot. Google's servers in California were used to restore cyber communication with Georgian citizens and forces.[112] The relocation of internet capabilities to US-based servers apparently occurred without informing the US government. In any case, such activities did not amount to a violation of the law of neutrality by either Georgia or the United States. Indeed, the situation fell under the scope of application of Article 8, not Article 3 of Hague Convention V, as the cyber infrastructure had not been erected by Georgia before or during the conflict: the use of communication installations belonging to a neutral state or its companies or private individuals is allowed by Article 8, even if it is for military communications (providing that the neutral does not discriminate between belligerents). Of course, the conduct of cyber operations amounting to hostilities from neutral territory would have been a violation of the law of neutrality

---

[109] German Ministry of Defence, *Humanitarian Law in Armed Conflicts*, Section 1116.
[110] HPCR Manual, p 386.
[111] According to Art 9(2), '[a] neutral Power must see to the same obligation [of non-discrimination] being observed by companies or private individuals owning telegraph or telephone cables or wireless telegraphy apparatus'.
[112] Kastenberg, 'Non-intervention', pp 46–7, 60–1; Daniel J Ryan, Maeve Dion, Eneken Tikk, and Julie JCH Ryan, 'International Cyberlaw: A Normative Approach', *Georgetown Journal of International Law* 42 (2011), p 1190.

both by the belligerent and, if it allows it, by the neutral state as well: it does not seem, however, that Georgia conducted cyber operations against Russia from US cyber infrastructure.

5. Other cyber-related activities: the recruitment of hackers and the supply of cyber weapons

Support to a belligerent constitutes non-neutral service that a neutral state must not allow on its territory (Article 5 of Hague Convention V).[113] The prohibited support covers all military assistance, including supply of troops, arms, and war *matériel* and military intelligence.[114] As to the provision of troops, according to Article 4 of Hague Convention V '[c]orps of combatants cannot be formed nor recruiting agencies opened on the territory of a neutral Power to assist the belligerents'. This would arguably include use of the internet for recruiting purposes on neutral territory. Establishing the degree of government exhortation that qualifies as recruitment, however, is a difficult task.[115] This is particularly true in the cyber context. In 2001, for example, after a US Navy spy plane collided with a Chinese jet fighter in the South China Sea, websites appeared offering instructions to hackers on how to incapacitate US government computers.[116] It also appears that the Russian government might have encouraged 'patriotic hackers' to conduct the 2007 cyber attacks against Estonia,[117] and Russian blogs, forums, and websites published instructions on how to overwhelm Georgian government websites as well as a target list of vulnerable Georgian websites.[118] As to the recruitment of contractors, such as professional hackers, by a belligerent in neutral territory, whether or not this amounts to a violation of Article 4 of Hague Convention V depends on whether they have been hired to conduct activity that amounts to direct participation in hostilities: if this is not the case, the situation falls under the rules of the law of neutrality regulating the commercial relations between the neutrals and the belligerents.[119]

The language of Article 4 of Hague Convention V, which refers to the formation of corps of combatants and to recruitment, is also broad enough to prohibit the taking over of 'zombie' computers located in a neutral state by a belligerent to

---

[113] In his Separate Opinion attached to the ICJ's Advisory Opinion on *Namibia*, Judge Ammoun provides a list of cases of support that constitutes non-neutral service (Separate Opinion of Vice-President Ammoun, pp 94–5).

[114] Schindler, 'Transformations', pp 379–80.

[115] Ian Brownlie, 'Volunteers and the Law of War and Neutrality', *International and Comparative Law Quarterly* 5 (1956), p 572.

[116] Noah Weisbord, 'Conceptualizing Aggression', *Duke Journal of Comparative and International Law* 20 (2009), p 20.

[117] Catherine Lotrionte, 'Active Defense for Cyber: A Legal Framework for Covert Countermeasures', in *Inside Cyber Warfare*, edited by Jeffery Carr, 2nd edn (Sebastopol, CA: O'Reilly, 2012), p 282.

[118] Eneken Tikk, Kadri Kaska, Kristel Rünnimeri, Mari Kert, Anna-Maria Talihärm, and Liis Vihul, *Cyber Attacks Against Georgia: Legal Lessons Identified* (CCDCOE, November 2008), pp 9–10, <http://www.carlisle.army.mil/DIME/documents/Georgia%201%200.pdf>.

[119] Ronzitti, *Diritto internazionale*, p 316.

form botnets employed in a DDoS attack. The neutral state would therefore be obliged under Article 5 to dismantle them. On the other hand, a neutral state is not required to prevent volunteers from joining the belligerents' forces from its territory: according to Article 6, 'the fact of persons crossing the frontier separately to offer their services to one of the belligerents' does not entail the responsibility of the neutral state. The difference between armed forces ('corps of combatants') and volunteers is the element of organization, which characterizes the former but not the latter: '[a]s long as the volunteering proceeds on a purely individual basis, it is not hindered by international law (even if the overall number of volunteers is considerable)'.[120] In most cases, however, it will be unlikely that the hacktivists crossing the frontier to conduct operations in support of a belligerent will be 'organized', or that they will cross the frontier at all.[121]

A neutral state may not directly or indirectly supply belligerents with 'war material of any kind whatever' (Article 6 of Hague Convention XIII), which is an expression broad enough to include cyber weaponry.[122] Nonetheless, the neutral 'is not called upon to prevent the export or transport, on behalf of one or other of the belligerents, of arms, munitions of war, or, in general, of anything which can be of use to an army or a fleet' by private persons (Article 7 of Hague Convention V), provided it does not discriminate between belligerents (Article 9 of the same Convention).[123] Cyber crime firms like the RBN, for instance, are suspected of developing malicious codes and setting up botnets and then sell them to interested parties.[124] It has been claimed that Article 7 reflects 'a neat separation between the civil society and the state, based upon the conditions prevailing in the nineteenth century', while today 'arms and ammunitions industries...are subject to state control'.[125] According to Section 1112 of the German Military Manual, then, '[t]o the extent to which arms export is subject to control by the state, the permission of such export [by private persons] is to be considered as unneutral Service'.[126]

---

[120] Dinstein, *War, Aggression and Self-Defence*, p 27. It is well-known that, during the Korean War, hundreds of thousands of Chinese 'volunteers' participated in the conflict against the United Nations (Norton, 'Between the Ideology', p 279). The UN General Assembly found that the People's Republic of China 'engaged in aggression in Korea' (GA Res 498 (V), 1 February 1951).

[121] It has, however, been suggested that, as 'all governments exercise substantial control over the activities of citizens affecting the national interest, especially in the area of foreign relations', the 'state-private dichotomy' has been repudiated 'to the extent that a neutral state is under a duty to use reasonable efforts to prevent its citizens or others subject to its control from joining either belligerent' (Walter L Williams, Jr, 'Neutrality in Modern Armed Conflicts: A Survey of the Developing Law', *Military Law Review* 90 (1980), p 30).

[122] The rule also applies to land warfare (Dinstein, *War, Aggression and Self-Defence*, p 28). According to Bothe, the prohibition covers 'weapons *stricto sensu*, that is material which is capable of being used for killing enemy soldiers or destroying enemy goods' (Bothe, 'The Law of Neutrality', p 563). See also Art 44 of the Hague Rules of Aerial Warfare.

[123] See also Art 45 of the Hague Rules of Aerial Warfare. It could, however, be asked whether the authorization to sell cyber tools that might be used for military purposes to any belligerent, while apparently non-discriminatory, may actually be non-neutral service if only one of the belligerents has the capacity to use them, or has wired infrastructures that could be targeted.

[124] Tikk, Kaska, Rünnimeri, Kert, Talihärm, and Vihul, *Cyber Attacks*, p 11; Alexander Klimburg, 'Mobilising Cyber Power', *Survival* 53(1), pp 49–50.

[125] Kolb and Hyde, *An Introduction*, p 280.

[126] German Ministry of Defence, *Humanitarian Law in Armed Conflicts*, Section 1112.

If this is correct, 'a neutral state is under a duty to take all reasonable measures to prevent provision of materials and other assistance to a belligerent by individuals and associations under its control'.[127] It is, however, far from sure that this is an accurate view of customary international law, or even that this is the prevailing view in literature:[128] in any case, it is difficult to justify in the cyber context, where states would hardly be able to verify and ensure compliance.[129]

IV. Non-Belligerency

It has been claimed that there is a middle ground between belligerency and traditional neutrality, ie 'non-belligerency', or 'qualified', or 'benevolent', neutrality' that allows states not participating in an armed conflict to support indirectly (for instance through the provision of financial or material assistance, including weapons, or the consent to the overflight of their airspace) the belligerent that is the victim of a violation of the *jus ad bellum*.[130] Non-belligerents, however, may not actively participate in hostilities by kinetic or cyber means, or allow the use of their territory, including cyber infrastructure, as a base to conduct hostilities.[131]

Those that maintain the existence of this intermediate status between neutrality and belligerency rely on state practice during and after the Second World War, where states rarely fully complied with the strict rules on neutrality contained in the Hague Conventions,[132] and invoke, as evidence of the existence of such status, Article 4(B)(2) of the Geneva Convention III and Article 2(c) of Additional Protocol I, that refer to 'neutral *or* non-belligerent Powers' and 'neutral *or* other State not a Party to the conflict', respectively.[133] The prevailing view, however, denies

---

[127] Williams, 'Neutrality', p 33.

[128] See the Commentary of Rule 173 of the HPCR Manual, according to which the 'increasing control of exports of arms and other military equipment by States...gives no evidence that States consider themselves obliged by the law of neutrality to exercise such control' (HPCR Manual, p 399).

[129] Of course, if the Security Council has decided sanctions against a belligerent, the sanction regime takes precedence over the law of neutrality: the neutral states must implement those sanctions and prevent individuals on their territory from providing the prohibited items to the belligerent in question.

[130] The Budapest Articles of Interpretation of the Pact of Paris, adopted by the ILA in 1934, provided that, in case of violation of the Pact by one signatory against another, the other states may '[d]ecline to observe towards the State violating the Pact the duties prescribed by International Law, apart from the Pact, for a neutral in relation to a belligerent' and thus '[s]upply the State attacked with financial or material assistance, including munitions of war' and '[a]ssist with armed forces the State attacked' (*American Journal of International Law* 33 (1939), Supplement, pp 825–6). Examples of non-belligerency are the policies of Italy during the Spanish Civil War and during the Second World War before its entry into the conflict, Spain and Ireland during the Second World War and the United States in favour of the United Kingdom before it became a belligerent itself (Stephen C Neff, *The Rights and Duties of Neutrals. A General History* (Manchester: Manchester University Press, 2000), pp 188–9).

[131] UK Ministry of Defence, *The Manual of the Law of Armed Conflict*, p 19; Roberts and Guelff, *Documents*, p 85; Ronzitti, *Diritto internazionale*, p 322.

[132] This practice is surveyed in Heintschel von Heinegg, '"Benevolent" Third States', pp 544–51.

[133] Emphasis added. See Ronzitti, *Diritto internazionale*, p 322; Gioia, 'Neutrality', pp 214–19; Schindler, 'Transformations', p 372. Meyrowitz accepts non-belligerency but only attaches political

that non-belligerency has become part of customary international law and sees instances of non-belligerency as violations of the law of neutrality:[134] the fact that 'non-belligerents' have often tried to conceal their assistance to the belligerents proves the lack of *opinio juris*.[135] Non-belligerent states that provide material or financial assistance to a belligerent, then, expose themselves to the risk of countermeasures not involving the use of force.[136] The Helsinki Principles on Maritime Neutrality, the San Remo Manual, the HPCR Manual, and the Tallinn Manual do not incorporate non-belligerency status and only refer to neutrality. The German Federal Administrative Tribunal also concluded that Germany had violated its obligations under the law of neutrality during the 2003 US/UK armed operations against Iraq by allowing the use of the US bases and installations situated in German territory, the overflight of US aeroplanes, the transport of US armed forces, supplies, weapons, and military equipment, and the escort of US vessels transporting troops and military equipment by the German Navy.[137] As to the above-mentioned references in the Geneva Convention III and Additional Protocol I, their inclusion can be explained in the light of the uncertainties that emerged during the drafting process on whether the law of neutrality had survived the entry into force of the UN Charter and, if so, whether it applied only in cases of declared war.[138]

Specific instances of support by a neutral state in favour of a belligerent, however, may be seen as countermeasures in case the other belligerent has breached that state's neutrality. The view according to which non-neutral service may be justified as a countermeasure against a state that has breached the prohibition of the use of force in international relations, however, is questionable, apart from the case of collective self-defence.[139] Indeed, even admitting that Article 2(4) is an *erga omnes* obligation, Article 42(b)(i) of the ILC Articles prescribes that, although all states have a legal interest in the fulfilment of this type of obligations, only those 'specially affected' by the breach are 'injured states' and are thus entitled to adopt countermeasures under Article 49. Article 54 of the ILC Articles notoriously leaves the matter of whether non-injured states may adopt countermeasures

---

consequences to it (Henri Meyrowitz, *Le principe de l'égalité des belligérants devant le droit de la guerre* (Paris: Pedone, 1970), pp 336 ff).

[134] Heintschel von Heinegg, '"Benevolent" Third States', p 553; Dinstein, *War, Aggression and Self Defence*, p 180; Bothe, 'The Law of Neutrality', p 550 ('there is no sufficiently uniform general practice which would justify the conclusion that non-belligerency has become a notion recognized by customary international law'); Edwin Borchard, 'War, Neutrality and Non-Belligerency', *American Journal of International Law* 35 (1941), p 624; Doswald-Beck, 'Some Thoughts', p 173; Jensen, 'Sovereignty', p 821.

[135] Heintschel von Heinegg, '"Benevolent" Third States', p 553.

[136] See Section VI.1 in this Chapter.

[137] Bundesverwaltungsgericht (BVerwG), Judgment, 21 June 2005, 2 WD 12.04, <http://www.bverwg.de/entscheidungen/entscheidung.php?ent=210605U2WD12.04.0>. See Heintschel von Heinegg, '"Benevolent" Third States', pp 543–4.

[138] Heintschel von Heinegg, '"Benevolent" Third States', p 554.

[139] This argument was, for instance, invoked to justify the United States' non-belligerency policy during the first years of the Second World War (Quincy Wright, 'The Transfer of Destroyers to Great Britain', *American Journal of International Law* 34 (1940), pp 686–9). See also *Namibia*, Separate Opinion of Vice-President Ammoun, p 93.

unresolved by providing that, in case of obligations of a collective character, any states 'other than the injured States', to which the collective obligation is owed, may take 'lawful measures' against the wrongdoing state 'to ensure cessation of the breach and reparation in the interest of the injured State or of the beneficiaries of the obligation breached', but without specifying whether 'lawful measures' include countermeasures.

V. The Law of Neutrality and the UN Charter

In the *Nuclear Weapons* Advisory Opinion, the ICJ held that 'the principle of neutrality' applies 'subject to the relevant provisions of the United Nations Charter'.[140] The entry into force of the Charter and the provisions on the use of force that it contains have affected the application of the traditional law of neutrality codified in the 1907 Hague Conventions V and XIII in a twofold manner: they have limited the coercive remedies at the disposal of the neutral and belligerent states in case of violations of neutrality; and they have undermined the neutrals' duties to abstain from supporting the belligerents and not to discriminate between them.[141] The former aspect will be discussed in Section VI.2 of this Chapter. As to the latter, Rule 95 of the Tallinn Manual provides that '[a] State may not rely upon the law of neutrality to justify conduct, including cyber operations, that would be incompatible with preventive or enforcement measures decided upon by the Security Council under Chapter VII of the Charter of the United Nations'.[142] Whenever the Security Council exercises its powers under Chapter VII of the UN Charter to react against an act of aggression or a breach of the peace,[143] then, the law of neutrality is superseded because of the combined effects of Article 2(5), which requires member states to provide the United Nations with 'every assistance in any action it takes in accordance with the present Charter', Article 25, the provisions in Chapter VII on the Security Council's powers to adopt binding measures, and Article 103, according to which the Charter obligations prevail over other international law.[144] As the UK Military Manual explains, as a consequence of the above provisions and in spite of the traditional law of neutrality, the UN member states have to assist the Security Council in its enforcement actions and

---

[140] *Nuclear Weapons*, para 89.

[141] The neutrals' duty of impartiality was founded on the assumption that war was legal and that the belligerents could not be discriminated from a *jus ad bellum* perspective, an assumption rejected by the Pact of Paris first and the UN Charter later (Robert H Jackson, 'Address', *American Journal of International Law* 35 (1941), p 354).

[142] Tallinn Manual, pp 255–6.

[143] The Security Council could of course also react against a threat to the peace (Art 39 of the UN Charter), but in case of situations short of an (international) armed conflict the law of neutrality would not come into consideration.

[144] See Rule 165, HPCR Manual, p 382; Rule 1.2 of the Helsinki Principles on the Law of Maritime Neutrality. Although it only mentions international agreements, according to the preferable interpretation Art 103 ensures the prevalence of the Charter obligations not only over other treaty law, but also over customary international law (Marco Roscini, 'The United Nations Security Council and the Enforcement of International Humanitarian Law', *Israel Law Review* 43 (2010), p 357).

have to refrain from assisting a state against which such actions are undertaken.[145] It might be, for instance, that a state commits an act of aggression by cyber means, or conducts cyber operations in connection with an act of aggression by kinetic means, and that the Security Council adopts a resolution requiring the member states to adopt cyber sanctions against the aggressor: the states that implement the resolution are not responsible for the violation of their duties under the law of neutrality.[146] The prevalence of the Charter's obligations over the law of neutrality, which relieves the neutrals from their duties of abstention and non-discrimination, however, does not determine a non-belligerency status intermediate between belligerency and neutrality, but is rather a case of conflict of norms settled by Article 103 of the UN Charter.[147]

It is important to distinguish measures decided by the Security Council from those that are merely recommended: only if the Council adopts decisions are the UN member states required to disregard their obligations under the law of neutrality and discriminate against the aggressor state, while in case of recommendations they are allowed but not obliged to do so. In both cases, however, 'any [UN] member which has been condemned by the Council for breach of the Charter and subjected to any type of enforcement measure is precluded from maintaining that other members assisting the victim of its aggression are in breach of their neutrality'.[148] If the Security Council has identified the aggressor state and, in spite of this, a neutral state supports it, then, the neutral state breaches not only the law of neutrality but also the UN Charter. If the neutral state supports the victim, it breaches the law of neutrality but not the UN Charter (providing that, if the support amounts to a use of kinetic or cyber force, the conditions for collective self-defence are met or the use of force is authorized by the Security Council). If the neutral state abstains and there is no binding Security Council resolution requiring it to adopt measures against the aggressor, it breaches neither the law of neutrality nor the UN Charter. If it abstains and the Security Council has adopted a decision imposing sanctions, the neutral state breaches the UN Charter but not the law of neutrality.

It is well-known that measures short of armed force under Article 41 of the UN Charter can be either decided or recommended by the Security Council.[149] Neutral states would thus be allowed, or obliged, to disregard the principle of

---

[145] UK Ministry of Defence, *The Manual of the Law of Armed Conflict*, p 20. See also Arts 3–4 of the Institut de droit international's Resolution on the Conditions of Application of Humanitarian Rules of Armed Conflict to Hostilities in Which United Nations Forces May Be Engaged, *Annuaire de l'Institut de droit international* 56 (1975), p 543. Non-members of the United Nations also have a right to discriminate between the aggressor and the victim of the aggression, although it is uncertain whether they are under an obligation to do so (Dinstein, *War, Aggression and Self-Defence*, p 177).

[146] Commentary to Rule 95, in Tallinn Manual, p 256.

[147] 'Fragmentation of International Law: Difficulties Arising from the Diversification and Expansion of International Law', Report of the Study Group of the International Law Commission, UN Doc A/CN.4/L.682, 13 April 2006, pp 168 ff.

[148] Green, *The Contemporary Law*, p 298.

[149] Benedetto Conforti, *The Law and Practice of the United Nations*, 3rd edn (Leiden and Boston: Nijhoff, 2005), pp 186, 192.

courant normal without breaching their duties under the law of neutrality. As far as military measures are concerned, the Security Council can in theory conduct 'action by air, sea, or land forces as may be necessary to maintain or restore international peace and security' as provided in Article 42. Lacking implementation of Articles 43 ff of the Charter, however, the Council has only been able to authorize, but not to decide, military action. States remain, therefore, free not to participate in a cyber operation amounting to a use of force against an aggressor state that has been authorized by the Council, although if they do so the breach of the law of neutrality would be excluded by the authorization.[150] Even if they do not participate in the operation, neutral states may discriminate between belligerents and provide logistic and material assistance to the states authorized by the Security Council to use force against the aggressor, without breaching the law of neutrality.[151]

States would also not be prevented by the law of neutrality from acting in collective self-defence under Article 51 of the Charter upon request 'by the State which regards itself as the victim of an armed attack':[152] the illegality of the action under the law of neutrality is excluded in this case not only by the UN Charter, but also by the law of state responsibility, which includes self-defence as a circumstance precluding wrongfulness.[153] Unless otherwise provided in a mutual defence treaty, collective self-defence, however, is a right and not a duty, and, therefore, it does not deprive the neutral state of the option to remain impartial.[154]

In case the Security Council remains inactive, or until it takes enforcement action, the traditional duties under the law of neutrality continue to apply.[155] Even if the Security Council decides non-forcible measures against one or more belligerents, it could require their implementation by some UN members only, and not by all.[156]

---

[150] Michael Bothe, 'Neutrality, Concept and General Rules', in *Max Planck Encyclopedia of Public International Law* (2012), Vol VII, p 620. Article 103 only refers to '*obligations* of the Members of the United Nations under the present Charter' (emphasis added), and would thus seem to apply only to binding resolutions, ie decisions, of the Security Council, not mere authorizations. Nonetheless, '[b]ecause authorizations by the Council to member states have effectively taken over the role of armed forces under UN command, as was originally envisaged in the Charter, and thus have a central place in the system of collective security, Article 103 has generally been interpreted to extend to Council authorizations as well as to its commands' (Marko Milanović, 'Norm Conflict in International Law: Whither Human Rights?', *Duke Journal of Comparative and International Law* 20 (2009–10), p 78 (footnote omitted). See also Vera Gowlland-Debbas, 'The Limits of Unilateral Enforcement of Community Objectives in the Framework of UN Peace Maintenance', *European Journal of International Law* 11 (2000), p 371; Robert Kolb, 'Does Article 103 of the Charter of the United Nations Apply only to Decisions or also to Authorizations Adopted by the Security Council?', *Zeitschrift für ausländisches öffentliches Recht und Völkerrecht* 64 (2004), pp 31–5. The point was also made by the UK House of Lords in the 2007 *Al-Jedda* Judgment (*R (Al-Jedda) v Secretary of State for Defence* [2007] UKHL 58, para 33 (Lord Bingham)).

[151] Gioia, 'Neutrality', p 216.

[152] *Case Concerning Oil Platforms (Iran v US)*, Merits, Judgment, 6 November 2003, ICJ Reports 2003, para 51. See Ronzitti, *Diritto internazionale*, p 314; Schindler, 'Transformations', p 374.

[153] See Art 21 of the ILC Articles on Responsibility of States for Internationally Wrongful Acts, with commentaries ('ILC Commentary'), *Yearbook of the International Law Commission*, 2001, Vol II, Part Two, p 74.

[154] Bothe, 'The Law of Neutrality', p 553. See also US DoD, *An Assessment*, p 7.

[155] See *Namibia*, Separate Opinion of Vice-President Ammoun, p 92.

[156] Article 48(1) of the UN Charter.

It could also decide an embargo only on selected goods, or, in case of a total embargo, exempt certain items, such as medical supplies and foodstuff. The traditional law of neutrality would continue to apply to residual situations and states not covered by the Security Council's resolutions.[157]

VI. Remedies Against the Violations of the Law of Neutrality

Violations of the law of neutrality by cyber means could be committed by the neutral states or by the belligerents. In both cases, the injured state (be it the neutral or another belligerent, or both)[158] can use the remedies provided under the law of state responsibility to react against internationally wrongful acts and, if the violation amounts to a use of force, by the *jus ad bellum*.

1. Acts of retorsion and non-forcible countermeasures

A state injured by a violation of the law of neutrality could first and foremost react by adopting acts of retorsion and countermeasures. If the former, which are unfriendly acts not involving any breach of international law, can be adopted at any time, countermeasures are 'measures that would otherwise be contrary to the international obligations of an injured State *vis-à-vis* the responsible State, if they were not taken by the former in response to an internationally wrongful act by the latter in order to procure cessation and reparation'.[159] As has been observed, '[t]he right to employ economic sanctions against neutral states supplying or permitting the supply of war matériel to an enemy belligerent is quite firmly established'.[160] Countermeasures against the violation of the law of neutrality can be adopted under the same conditions already described in Chapter 2.[161] In any case, no cyber operation amounting to a use of force, ie with destructive effects or seriously disrupting essential services, could be carried out in countermeasure by a neutral against a belligerent or vice versa: Article 50 of the ILC Articles, which reflects customary international law, provides that countermeasures cannot affect the prohibition of the threat and use of force.

It should be emphasized again that countermeasures require the commission of a previous internationally wrongful act that affects the 'injured state', although, contrary to what is stated in the Tallinn Manual,[162] the violation does not necessarily

---

[157] Walker, 'Information Warfare', pp 1129–30.
[158] According to Art 42 of the ILC Articles on State Responsibility, '[a] State is entitled as an injured State to invoke the responsibility of another State if the obligation breached is owed to: (*a*) that State individually; or (*b*) a group of States including that State, or the international community as a whole, and the breach of the obligation: (i) specially affects that State; or (ii) is of such a character as radically to change the position of all the other States to which the obligation is owed with respect to the further performance of the obligation' (ILC Commentary, p 117).
[159] ILC Commentary, p 128. [160] Norton, 'Between the Ideology', p 296.
[161] See Chapter 2, Section IV.
[162] Commentary to Rule 94, in Tallinn Manual, pp 254–5. The Commentary refers to Rule 22 of the San Remo Manual, but such provision requires the threat to the security of the opposing

have to be serious in order to trigger the aggrieved belligerent's right to respond. A party to the conflict, then, may adopt a countermeasure against the failure by the neutral state to terminate a violation of its neutrality exclusively if it is 'injured' by the violation in the sense of Article 42 of the ILC Articles on State Responsibility; otherwise, it is only the neutral state that is entitled to respond.[163] If the neutral state has used the means at its disposal to prevent violations of its neutrality but has not succeeded, it must tolerate the aggrieved belligerent's action on its territory to terminate the violation. Such action, however, could not be justified as a countermeasure against the neutral state in the absence of an internationally wrongful act attributable to it,[164] and can only be founded on a state of necessity under Article 25 of the ILC Articles on State Responsibility.[165]

2. Use of kinetic or cyber force

The law of neutrality requires a neutral state to terminate violations of its neutrality by force, if necessary.[166] This is implied in Article 10 of Hague Convention V, according to which '[t]he fact of a neutral Power resisting, even by force, attempts to violate its neutrality cannot be regarded as a hostile act', and in the similarly phrased Article 48 of the Hague Rules of Aerial Warfare.[167]

However, it is now the UN Charter that determines the legality of forcible reactions by the neutrals against violations of their neutrality by the belligerents: whether such use of force is lawful will have to be ascertained according to the provisions on the use of force contained therein.[168] In Chapter 2, it has been seen that Article 2(4) of the UN Charter contains a general prohibition on the use of force in international relations, with two exceptions: the right of individual and collective self-defence and the use of force in the framework of Chapter VII. Neutrals,

belligerent to be 'serious and immediate' exclusively for the adoption of forcible reactions, not any countermeasures.

[163] Commentary to Rule 94, in Tallinn Manual, p 254.

[164] This is clearly expressed in Art 24 of the 1939 Harvard Draft Convention on the Rights and Duties of Neutral States in Naval and Aerial War, according to which '[a] belligerent may not resort to acts of reprisal or retaliation against a neutral State except for illegal acts of the latter, and a State is not to be charged with failure to perform its duties as a neutral State because it has not succeeded in inducing a belligerent to respect its rights as a neutral State' (*American Journal of International Law* 33 (1939), Supplement, p 179).

[165] According to this provision, '1. Necessity may not be invoked by a State as a ground for precluding the wrongfulness of an act not in conformity with an international obligation of that State unless the act: (*a*) is the only way for the State to safeguard an essential interest against a grave and imminent peril; and (*b*) does not seriously impair an essential interest of the State or States towards which the obligation exists, or of the international community as a whole. 2. In any case, necessity may not be invoked by a State as a ground for precluding wrongfulness if: (*a*) the international obligation in question excludes the possibility of invoking necessity; or (*b*) the State has contributed to the situation of necessity'.

[166] German Ministry of Defence, *Humanitarian Law in Armed Conflicts*, Section 1109. Such measures could be justified as belligerent reprisals or self-help measures (Schindler, 'L'emploi', p 851).

[167] See also Rules 168(b) and 169 in HPCR Manual, p 53.

[168] Of course, in case of an armed conflict, any act of hostilities will also have to comply with the relevant *jus in bello* norms.

therefore, could use force by kinetic or cyber means in reaction to a violation of the law of neutrality only if such violation amounts to an armed attack or the forcible reaction is authorized by the UN Security Council. A belligerent's violation of the law of neutrality by cyber means can amount to an armed attack on the neutral state only when the operation causes or is reasonably likely to cause either material damage to persons or property or significant disruption of the functioning of critical infrastructures, and if its destructive or disruptive character meets the 'scale and effects' standard identified by the ICJ in the *Nicaragua* Judgment.[169] Unlike a belligerent, that is free to decide whether to react or not, the neutral is obliged not to allow violations of its neutrality by force, if necessary: as has been effectively observed, 'the Charter of the United Nations grants a right to use counter-force; the law of neutrality may, under certain circumstances, impose an obligation to exercise this right'.[170]

The above considerations also apply, *mutatis mutandis*, to the use of force by a belligerent to react against a violation of the law of neutrality by a neutral state. If a neutral state conducts cyber operations short of an armed attack against a belligerent, this could not use force against the neutral, but would only be able to resort to non-forcible countermeasures, as armed reprisals are unlawful under the current *jus ad bellum* regime.[171] Similarly, non-neutral service such as the transfer of arms or military *matériel* can be a use of force, but is not necessarily an armed attack:[172] an aggrieved belligerent could not react forcibly to terminate those violations of the law of neutrality.[173] The same conclusion applies, *a maiore ad minus*, to other violations of the law of neutrality that do not amount to a use of force, such as providing intelligence to a belligerent, allowing recruitment and movement of troops on neutral territory, or discriminating with regard to commerce and communications in favour of a belligerent. The US DoD study on the legal aspects of information operations, then, goes too far where it seems to suggests that the belligerent has a 'limited right of self-defence' if the neutral state provides satellite imagery of the belligerent's armed forces, weather information or precision navigation services to an adversary.[174]

It should be recalled that neither the fact that a cyber operation originates from a neutral state's governmental cyber infrastructure nor that it has been routed through the cyber infrastructure located in a neutral state are sufficient evidence for attributing the operation to those states.[175] Rule 7 of the Tallinn Manual, however, ambiguously specifies that the origin from governmental cyber infrastructure

---

[169] See Chapter 2, Section III.1.   [170] Bothe, 'The Law of Neutrality', p 561.
[171] Michael Bothe, 'Neutrality in Naval Warfare: What is Left of Traditional International Law', in *Humanitarian Law*, edited by Delissen and Tanja, p 396.
[172] *Military and Paramilitary Activities in and against Nicaragua (Nicaragua v US)*, Merits, Judgment, 27 June 1986, ICJ Reports 1986, para 228. See Schindler, 'L'emploi', p 860.
[173] Ronzitti, *Diritto internazionale*, p 319; Bothe, 'Neutrality in Naval Warfare', p 396.
[174] US DoD, *An Assessment*, p 10.
[175] Tallinn Manual, pp 34–6. According to the ICJ, 'it cannot be concluded from the mere fact of the control exercised...over its territory...that State necessarily knew, or ought to have known, of any unlawful act perpetrated therein' (*Corfu Channel (United Kingdom v Albania)*, Merits, Judgment, 9 April 1949, ICJ Reports 1949, p 18).

might be an 'indication' that the state was associated with the operation.[176] Such unnecessary specification may open the way to abuses.[177]

According to the law of neutrality, a belligerent may take action to terminate violations of neutrality also 'when the neutral Power is unable to prevent belligerent use of its territory and when the action is necessary and proportional to lawful defensive objectives.'[178] Rule 22 of the San Remo Manual states that

> [s]hould a belligerent State be in violation of the regime of neutral waters,… the neutral State is under an obligation to take the measures necessary to terminate the violation. If the neutral State fails to terminate the violation of its neutral waters by a belligerent, the opposing belligerent must so notify the neutral State and give that neutral State a reasonable time to terminate the violation by the belligerent. If the violation of the neutrality of the State by the belligerent constitutes a serious and immediate threat to the security of the opposing belligerent and the violation is not terminated, then that belligerent may, in the absence of any feasible and timely alternative, use such force as is strictly necessary to respond to the threat posed by the violation.[179]

The 1956 US Manual on Land Warfare provides that '[s]hould the neutral State be unable, or fail for any reason, to prevent violations of its neutrality by the troops of one belligerent entering or passing through its territory, the other belligerent may be justified in attacking the enemy forces on this territory'.[180] The US *Commander's Handbook on the Law of Naval Operation* also states that '[i]f the neutral nation is unable or unwilling to enforce effectively its right of inviolability, an aggrieved belligerent may take such acts as are necessary in neutral territory to counter the activities of enemy forces, including warships and military aircraft, making unlawful use of that territory'.[181] A similar position is adopted in relation to cyber operations in the US DoD's *Assessment of International Legal Issues in Information Operations*, which however subordinates the right to attack the enemy in neutral territory to the fact that such territory is used by the enemy 'in a manner that gives it a military advantage'.[182] The US DoD *Cyberspace Policy Report* also refers to '[t]he ability and willingness of the third country to respond effectively to the malicious cyber activity' as one of the factors to determine the nature of the DoD response to a hostile act or threat originating from a neutral state.[183] Rule 94 of the Tallinn Manual provides that '[i]f a neutral State fails to terminate the exercise of belligerent rights on its territory, the aggrieved party to the conflict

---

[176] Tallinn Manual, p 34.
[177] Dieter Fleck, 'Searching for International Rules Applicable to Cyber Warfare—A Critical First Assessment of the New *Tallinn Manual*', *Journal of Conflict and Security Law* 18 (2013), p 339.
[178] John N Moore, 'Legal Dimensions of the Decision to Intercede in Cambodia', *American Journal of International Law* 65 (1971), p 51; Ashley Deeks, '"Unwilling or Unable": Toward a Normative Framework for Extra-Territorial Self-Defense', *Virginia Journal of International Law* 52 (2011–12), p 499.
[179] Roberts and Guelff, *Documents*, pp 579–80.
[180] US Department of the Army, *The Law of Land Warfare*, Field Manual No 27-10, 1956, Rule 520, <http://www.loc.gov/rr/frd/Military_Law/pdf/law_warfare-1956.pdf>.
[181] *The Commander's Handbook on the Law of Naval Operations*, July 2007, para 7.3, <http://www.usnwc.edu/getattachment/a9b8e92d-2c8d-4779-9925-0defea93325c/>.
[182] US DoD, *An Assessment*, p 7.   [183] US DoD, *Cyberspace Policy Report*, p 8.

may take such steps, including by cyber operations, as are necessary to counter that conduct', although it does not specify what kind of cyber operations may be conducted.[184]

Although the issue is not uncontroversial, it is submitted that the right under the law of neutrality to react forcibly on the territory of the neutral state if this is unable or unwilling to terminate violations of its neutrality must be reconciled with the *jus ad bellum* provisions contained in the UN Charter, that prevail under Article 103 of the Charter over conflicting norms. As correctly explained in the UK Military Manual, '[i]f a neutral state is unable or unwilling to prevent the use of its territory for the purposes of... military operations, a belligerent state may become entitled to use force in self-defence against enemy forces operating from the territory of that neutral state': this, however, 'will depend on the ordinary rules of the *jus ad bellum*'.[185] An armed reaction, therefore, would be allowed only if it is the aggressor state that uses the territory of the neutral state, including the cyber infrastructure located therein, to conduct kinetic or cyber hostilities against the victim of the armed attack, and the forcible reaction on neutral territory is necessary and proportionate to repel the armed attack;[186] or if the reaction has been authorized by the Security Council. In any case, the territory of the neutral may be attacked only to the extent that it is used to conduct hostilities by the enemy.[187]

Whether cyber operations amounting to the use of force may be undertaken against a violation of the law of neutrality that has yet to occur depends on whether the violation would amount to an armed attack and on whether it is imminent enough to justify anticipatory self-defence: the Nuremberg Tribunal referred to the *Caroline* incident to hold that 'preventive action in foreign territory is justified only in case of "an instant and overwhelming necessity for self-defence, leaving no choice of means, and no moment of deliberation"' and concluded that Germany's invasion of Denmark and Norway in order to prevent their occupation by the Allies were 'acts of aggressive war'.[188] As has been seen in Chapter 2, pre-emptive action against non-imminent armed attacks is unlawful under existing international law.[189] An armed reaction against the neutral state or on its territory is therefore possible 'only when the use of [its] territory by the enemy is imminent; it is not sufficient that a belligerent should merely fear that his enemy might perhaps attempt so to use it'.[190]

---

[184] Tallinn Manual, p 254.
[185] UK Ministry of Defence, *The Manual of the Law of Armed Conflict*, p 20.
[186] Bothe, 'Neutrality in Naval Warfare', p 397. According to Castrén, an analogous situation is that of the '*continual* passage of enemy military transports through neutral territory' in order to conduct an armed attack (Castrén, *The Present Law*, p 463; emphasis in the original). See also Schindler, 'L'emploi', pp 860–1. Whether or not this is correct in a traditional scenario, however, such conclusion is not warranted in the cyber context, where routing a cyber attack through neutral cyber infrastructure cannot be considered a violation of the law of neutrality (see Section III.2 in this Chapter).
[187] Schindler, 'L'emploi', p 860.
[188] Nuremberg International Military Tribunal, Judgment, 1 October 1946, reprinted in *American Journal of International Law* 41 (1947), pp 205, 207.
[189] See Chapter 2, Section III.2, p 78.
[190] Lassa Oppenheim, *International Law. A Treatise*, edited by Hersch Lauterpacht, 7th edn (London, New York, and Toronto: Longmans, Green and Co, 1952), Vol II, p 698.

VII. Conclusions

The above analysis has demonstrated that the law of neutrality may extend to cyber operations whenever they are conducted in the context of an international armed conflict and have a nexus with it or when they amount themselves to such a conflict, whether or not there is a declaration of war or a state has declared its neutrality. The traditional law codified in the 1907 Hague Conventions V and XIII, however, must be reconciled with the provisions on the use of force contained in the UN Charter and in customary international law.

In summary, we have determined the following normative points in relation to cyber operations:

(1) Belligerent and, even more, neutral states are prohibited from conducting any cyber operations amounting to acts of hostilities against other belligerents from cyber infrastructure situated in neutral territory or under the exclusive control of neutral states.

(2) If conducted by a belligerent or by private individuals, the neutral state from whose territory the cyber operations are conducted has an obligation to use all the means at its disposal to terminate them.

(3) Unlike cyber operations originating from neutral territory, the routing of cyber operations through neutral cyber infrastructure is not a violation of that state's neutrality, as neither would the belligerent be able to control the pathway taken by the malware, nor would the neutral have the means to prevent the routing.

(4) Belligerents are prohibited from conducting any cyber operation against neutral territory or neutral cyber infrastructure and from conducting cyber operations against other belligerents that have more than nominal prejudicial incidental effects on neutral territory.

(5) As to the use of cyber infrastructure for communications, belligerents are allowed to 'erect' a new cyber communication installation on the territory of the neutral state as long as it is exclusively for non-military communications; use an existing one established by them before the war, even for military communications, provided that it is open 'for the service of public messages'; and use an existing communication installation established by them before the war and which is not open 'for the service of public messages', provided it is for non-military communications.

(6) The neutral state is not called upon to forbid or restrict the use by the belligerents of its cyber infrastructure for communications, providing it does not discriminate between them.

(7) Corps of cyber combatants may not be formed on neutral territory nor hackers recruitment agencies opened to assist the belligerents, but the neutral state is not required to prevent volunteers from crossing the frontier to conduct cyber operations on behalf of a belligerent.

(8) The neutral state may not supply a belligerent with malware that may be used in the conduct of hostilities, although it is not required to prevent its companies or private individuals from supplying it.

(9) A neutral state may not invoke the law of neutrality to justify cyber operations that are incompatible with the UN Charter or resolutions adopted by the UN Security Council under Chapter VII. Similarly, a neutral state may not invoke the law of neutrality to avoid adopting cyber sanctions decided by the Council against a belligerent.

(10) The state injured by a violation of the law of neutrality (be it a belligerent or a neutral state) by cyber means may adopt acts of retorsion or countermeasures, in-kind or not, to respond to the violation. Cyber operations amounting to a use of force may also be conducted by the neutral state in response to a violation of the law of neutrality that amounts to an armed attack and providing that the forcible reaction is necessary and proportionate to repel the attack, and by the belligerent state acting in self-defence if the aggressor state is using the neutral's territory, including its cyber infrastructure, to continue its armed attack, or if the cyber operation has been authorized by the UN Security Council.

The violations of the law of neutrality in the cyber context have been summarized in the following table.

Table 5.1 Violations of the law of neutrality in the cyber context

	Prohibition on the belligerents	Prohibition on the neutral states	Obligation on the neutral states not to allow if conducted by the belligerent states	Obligation on the neutral states not to allow if conducted by individuals
Cyber attacks from neutral territory	✓	✓	✓	✓
Cyber exploitation from neutral territory	✓	✓	✓	
Provision of malware for the conduct of hostilities		✓		
Formation of corps of cyber combatants and recruitment of hackers in neutral territory	✓		✓	
Routing of cyber operations through a neutral state				

Table 5.1 (Continued)

	Prohibition on the belligerents	Prohibition on the neutral states	Obligation on the neutral states not to allow if conducted by the belligerent states	Obligation on the neutral states not to allow if conducted by individuals
Cyber operations against or with incidental harmful effects on neutral territory	✓			
Erection of new cyber infrastructure by the belligerent on neutral territory for military communications	✓		✓	
Use of existing cyber infrastructure established by the belligerent on neutral territory for military communications	✓ (unless it is open 'for the service of public messages')		✓ (unless it is open 'for the service of public messages')	
Use of cyber infrastructure belonging to the neutral state, companies or nationals for communications				

General Conclusions

The militarization of cyberspace is not a risk, it is already a fact, with the armed forces of several states establishing cyber units and including cyber operations in their military doctrines and strategies. States have also been the object of cyber attacks of which other states were suspected, in some cases in connection with a traditional military operation or an armed conflict. It has been this book's submission that international law is well equipped to face these challenges. In particular, the present book has demonstrated that existing *jus ad bellum* and *jus in bello* provisions apply to cyber operations, even though the rules were adopted well before the advent of cyber technologies: the lack of ad hoc provisions does not mean that cyber operations can be conducted by states without restrictions. In order to demonstrate this thesis, resort has been made throughout the book to the notion of the evolutionary interpretation of treaties. Indeed, as recalled by the former President of the Israeli Supreme Court, Aharon Barak, in another context, 'new reality at times requires new interpretation. Rules developed against the background of a reality which has changed must take on a dynamic interpretation which adapts them, in the framework of accepted interpretational rules, to the new reality'.[1] Similarly, the ICJ has emphasized that

> where parties have used generic terms in a treaty, the parties necessarily having been aware that the meaning of the terms was likely to evolve over time, and where the treaty has been entered into for a very long period or is 'of continuing duration', the parties must be presumed, as a general rule, to have intended those terms to have an evolving meaning.[2]

Notions such as 'force', 'armed conflict' and 'attack', therefore, need to be interpreted taking into account the dependency of modern societies on computers, computer systems and networks. Indeed, there is ample verbal state practice, expressed in cyber strategies and doctrines, official statements and, if only exceptionally, in military manuals, which has been examined in-depth in the book, demonstrating that existing *jus ad bellum* and *jus in bello* rules are considered applicable to at least certain cyber operations by the end-users of these rules, ie states and international organizations, as well as by the ICRC. The forward-looking character of the law of armed conflict is also demonstrated by the inclusion in Protocol I additional to the Geneva Conventions on the Protection of Victims of War of provisions like

---

[1] *Public Committee Against Torture in Israel et al v The Government of Israel et al*, Israel's Supreme Court, HCJ 769/02, 11 December 2005, para 28 (Barak). Vice President Rivlin also stated that 'international law must adapt itself to the era in which we are living' (para 2 (Rivlin)).

[2] *Dispute Regarding Navigational and Related Rights (Costa Rica v Nicaragua)*, Judgment, 13 July 2009, ICJ Reports 2009, para 66.

Article 36 on the study, development, acquisition or adoption of a new weapon, means, or method of warfare and by the so-called Martens Clause.

The apparent lack of territoriality of 'cyberspace' is not necessarily an obstacle to the application of existing rules, which can be territorialized, in a Westphalian sense, by focusing on where the prejudicial activity is undertaken and where the effects occur. Similarly, the argument that existing rules are inadequate to regulate a phenomenon, like cyber operations, characterized by anonymity and deception cannot be accepted. Difficulties associated with identifying those responsible for the operations may be technical obstacles, but not legal issues affecting the relevance of the rules. It has also been demonstrated that not only the primary rules, but also the secondary rules on state responsibility, including those on attribution of conduct to states, are flexible enough to be adjusted to the new cyber scenario. After all, difficulties in identification and attribution are not unique to cyber operations, as they are a well-known problem also with regard to international terrorism and asymmetric warfare in general.

After demonstrating that existing international law can extend to the cyber context, the present book has taken the most relevant rules on the use of force, the conduct of hostilities and neutrality, whether based on treaty or custom, and has explained how these rules apply to cyber operations, identifying some potential problems. These rules have been discussed in relation to both cyber attacks and cyber exploitation operations. All in all, it is safe to conclude that the *jus ad bellum* rules are more flexible, and thus easier to apply, to a new phenomenon such as cyber operations, than those on the conduct of hostilities and neutrality. The reason for this is simple: the *jus ad bellum* contains significantly fewer and far less detailed rules than the *jus in bello*. What is evident is that, while states have been prepared to express views in more detail on cyber operations in relation to *jus ad bellum* issues, in particular the right to self-defence, they have been more cautious about doing so in respect of *jus in bello* aspects.

With regard to the *jus ad bellum*, in particular, Chapter 2 has demonstrated the under-inclusive character of the doctrine of kinetic equivalence, which limits the application of Articles 2(4) and 51 of the UN Charter to cyber attacks that cause, or are reasonably likely to cause, physical damage to property, loss of life or injury to persons. If one applies this doctrine, a state's cyber attack that shuts down another state's national grid or stock exchange for a significant period of time would not be a 'use of force', regardless of the severity of its non-physical consequences. This view is contradicted by the statements of several states, that see such situations as a 'new form of violence'.[3] This book has argued, therefore, that a 'new interpretation' of existing rules justifies their extension to cyber attacks that cause *significant* disruption of the functioning of *critical* infrastructures even if they do not materially damage them. Indeed, the increasing digitalization of today's societies has made it possible to cause significant prejudicial consequences on states through non-destructive means: cyber technologies can produce results

---

[3] Comments submitted by Panama to the UN Secretary-General, UN Doc A/57/166/Add.1 29 August 2002, p 5. See the practice examined in Chapter 2, Section II.1.2.

comparable to those of kinetic weapons but without the need of physical damage. Certain disruptive cyber operations, therefore, can, at least potentially, be a use of force prohibited by Article 2(4) and, if they reach the high 'scale and effects' threshold identified by the ICJ, also an 'armed attack' triggering the right of the victim state to individual and collective self-defence.

As to cyber attacks conducted by states but falling below the level of the use of force, ie those that do not result in more than minimal material damage to property, loss of life or injury to persons, or significant disruption of the functioning of critical infrastructures, they may be violations of the customary principle of non-intervention in the internal affairs of other states if they are 'the manifestation of a policy of force',[4] ie if they are accompanied by an intention to coerce the target state to do or not to do something 'on matters in which each State is permitted, by the principle of State sovereignty, to decide freely', such as 'the choice of a political, economic, social and cultural system, and the formulation of foreign policy'.[5] On the other hand, cyber exploitation in order to obtain information could be a violation of the sovereignty of the targeted state when it entails an unauthorized intrusion into cyber infrastructure located in another state (be it governmental or private), but not intervention and even less a use of force, as it lacks the coercive element and, on its own, does not result in physical damage to property, loss of life or injury of persons, or malfunction of infrastructure. The remedies against cyber operations short of 'armed attack' conducted by states include the resort to international courts, the adoption of acts of retorsion and countermeasures, but not forcible measures, unless the cyber operation is an integral part of an imminent armed attack by kinetic or cyber means that justifies the invocation of anticipatory self-defence, or unless the effects of the low-intensity cyber attack can be accumulated with those of others to form a composite armed attack.

The increasing role played by non-state actors is not unique to the cyber context and is a phenomenon that characterizes almost every sector of international law. Chapter 2 has maintained that the right of self-defence may be invoked against an armed attack, by cyber or kinetic means, whoever—state or armed group—is responsible for it. In this context, as the US *International Strategy for Cyberspace* recalls, 'cybersecurity due diligence' is an 'emerging norm' essential in cyberspace that involves the states' 'responsibility to protect information infrastructures and secure national systems from damage or misuse'.[6] With regard to anticipatory self-defence against an imminent cyber armed attack, it has been seen that 'imminence' could be interpreted either in its traditional temporal meaning in order to prevent possible abuses but at the cost of restricting the defensive options of the victim state in the absence of visible indicators of the attack, or taking into account

---

[4] *Corfu Channel (United Kingdom v Albania)*, Merits, Judgment, 9 April 1949, ICJ Reports 1949, p 35.
[5] *Military and Paramilitary Activities in and against Nicaragua (Nicaragua v US)*, Merits, Judgment, 27 June 1986, ICJ Reports 1986, para 205.
[6] White House, *International Strategy for Cyberspace. Prosperity, Security, and Openness in a Networked World*, May 2011, p 10, <http://www.whitehouse.gov/sites/default/files/rss_viewer/international_strategy_for_cyberspace.pdf>.

not only the time factor but also the specific characteristics of the cyber context, including the speed of the attack, its covert character, and the criticality of the target. While states that pursue aggressive policies or, vice versa, states that are the frequent target of cyber attacks, like China, the Russian Federation and the United States, are likely to favour the more flexible approach to imminence, states that do not play an active role in the cyber arena and fear possible abuses by more powerful states will probably go for the stricter temporal notion of imminence.

Moving to the *jus in bello*, Chapter 3 has determined that the international law of armed conflict applies to cyber operations when they are preceded by a declaration of war or are carried out in the context of a traditional armed conflict (providing they are conducted in support of a party to the conflict to the detriment of another and cause, or are reasonably likely to cause, the required threshold of harm), or if the cyber operations are conducted by the occupying state in the exercise of its policing and governance powers in occupied territory or are part of the mounted resistance by the local population to the exercise of such powers. The *jus in bello*, however, also applies when the exchange of cyber operations between states amounts in itself to 'resort to armed force', ie the operations entail the use of cyber means or methods of warfare resulting in material damage to property, loss of life or bodily injury, or serious disruption of critical infrastructures, or when an organized armed group conducts cyber operations amounting to 'protracted armed violence' against a state or against another organized armed group.[7] Having said that, it is possible that, in the future, 'cyber conflicts' may come to be seen as a 'lesser form of international conflict'[8] intermediate between peacetime and armed conflict. According to the *Finnish Security and Defence Policy 2009* Government Report, for instance, '[t]he state between war and peace is increasingly nebulous.... This phase also includes...various means of information warfare and asymmetric warfare, such as cyber attacks, with the intention of disturbing the normal functions of society'.[9] Whether or not introducing a *status mixtus* would be beneficial remains to be seen: the analysis conducted in Chapter 3 seems to reinforce the view that, at least for the time being, in the cyber context *inter bellum et pacem nihil est medium*.[10]

As the United States has observed, the application in the cyber context of the law of armed conflict, conceived with kinetic weaponry in mind, presents 'new and unique challenges that will require consultation and cooperation among nations'.[11] Several characteristics of cyber operations are likely to affect, in particular, the

---

[7] Whether the conflict falls under Common Art 3 of the Geneva Conventions or Additional Protocol II depends on their respective thresholds.

[8] James P Terry, 'Responding to Attacks on Critical Computer Infrastructure. What Targets? What Rules of Engagement?', *International Law Studies* 76 (2002), p 434.

[9] *Finnish Security and Defence Policy 2009*, Government Report, Prime Minister's Office Publications, 13/2009, p 18, <http://vnk.fi/julkaisukansio/2009/j11-turvallisuus-j12-sakerhets-j13-finnish/pdf/en.pdf>.

[10] The expression was famously used by Grotius, *De jure belli ac pacis*, Book III, § XXI, I; and Cicero, *Philippics*, § VIII, I, 4.

[11] Comments submitted by the United States to the UN Secretary-General, UN Doc A/66/152, 15 July 2011, p 19.

application of the law on the conduct of hostilities: they are carried out remotely, they may produce effects almost instantaneously, they use essentially dual-use infrastructures and programs, their effects on the infrastructures controlled by the information systems are often more relevant than the direct effects on the information itself, and their technology is easily accessible to anyone, not just to the military. On the basis of these characteristics, not all of which are necessarily unique to cyber operations, Chapter 4 has examined how the existing law may regulate acts of cyber hostilities. It has determined that cyber capabilities are so diverse and their effects so dependent on the circumstances, including the characteristics of the targeted system, that a legal review of their legality as a means or method of warfare can only be conducted on each individual capability and that, in most cases, it will be *how* the means or method is used, more than the means or method itself, that may be incompatible with the law of armed conflict.

From this perspective, it should be recalled that the main provisions of the law of targeting only apply to 'attacks', defined in Article 49(1) of Additional Protocol I as 'acts of violence against the adversary, whether in offence or in defence'. Chapter 4 has determined that cyber operations can be qualified as such when they employ means or methods of cyber warfare that have or are reasonably likely to have 'violent' effects in the form of loss of life or injury of persons, more than minimal material damage to property or loss of functionality of infrastructures. In particular, it has been this book's contention that Article 49(1) of Additional Protocol I must be interpreted in a manner that takes into account the radically increasing reliance of modern societies on information technologies and that the concept of 'violence' should be expanded to include not only material damage to objects, but also incapacitation of infrastructures without destruction. Whenever a belligerent's cyber attack in the context of an armed conflict causes loss of functionality of infrastructure that goes beyond mere inconvenience, then, it qualifies as an 'attack' and it must comply with the law of targeting, regardless of whether concomitant physical damage to the infrastructure occurs. The view according to which merely disruptive cyber operations may be 'attacks' only 'if restoration of functionality requires replacement of physical components'[12] cannot be accepted: the attacker may not be able to know in advance whether the restoration of functionality will require replacement of physical components or mere reinstallation of the operating system and, therefore, it could claim that it was not aware that it was conducting an 'attack' to which the law of targeting applied.

Chapter 4 has then analysed the main provisions of the law on the conduct of hostilities and identified areas where the unique characteristics of cyber operations may create interpretive problems. The overall conclusion is that these problems are often overestimated. Take, for instance, the case of the application of the principle of proportionality in attacks. The potentially less damaging character of cyber operations may offer a more effective means to minimize incidental damage on civilians

---

[12] This view was expressed by some of the experts that drafted the Tallinn Manual (*Tallinn Manual on the International Law Applicable to Cyber Warfare* (Cambridge: Cambridge University Press, 2013), p 108).

and civilian property. Cyber operations also present advantages for the attacking state, as they virtually entail no risk for its forces thanks to their remote character and the difficulties with regard to identification and attribution. True, the problem with calculating proportionality in the cyber context resides in the speed and covert nature of cyber attacks: it may be difficult for the parties to the conflict to readily establish their magnitude and consequences. Furthermore, as with biological weapons, some kinds of malware sent through cyberspace might spread uncontrollably because of the malware's characteristics and the interconnectivity of information systems. All in all, however, meeting the proportionality criterion is essentially a technical issue: customized proportionate cyber reactions are possible if the software is written with this purpose in mind and the targeted system is sufficiently known.[13] The code could, for instance, be designed in a way as to be activated only by the presence of certain characteristics, as in the case of Stuxnet. This requires a high degree of information on the targeted systems, which may be obtained through traditional intelligence collection and/or cyber exploitation.

Another overestimated problem concerns the dual-use nature of most cyber infrastructures, ie the fact that they are at the same time used by civilians and the military. This is not unique to the cyber context. The fact that an object is *also* used for civilian purposes does not affect its qualification under the principle of distinction: if the two requirements provided in Article 52(2) of Additional Protocol I are present, the object is a military objective and is thus targetable, but the neutralization of its civilian component needs to be taken into account when assessing the incidental damage on civilians and civilian property under the principle of proportionality. What is prohibited is to attack the dual-use cyber infrastructure *because* of its civilian function or to attack a dual-use facility where the incidental civilian damage expected from the attack is excessive in relation to the anticipated concrete and direct military advantage.

Finally, while it is correct that the automated character of certain cyber active defences may increase the risk of violations of the principles of distinction and proportionality, this is not necessarily inconsistent with the duty to take precautions in attack: in automated processes, such duty applies before the attack, ie at the time that the targeting software is programmed and the data designed and uploaded, or up to any time the attack can be called off. With regard to passive precautions against the effects of cyber attacks, if the separation of civilian and military networks and information infrastructure is at present not feasible, either technically or financially, there are plenty of 'other necessary precautions'[14] that the belligerents in control of civilians and civilian objects should, as far as possible, adopt to protect them from the dangers of military cyber operations.

Chapter 5 has examined how the law of neutrality affects the conduct of cyber operations by neutrals and belligerents. In spite of the fact that it has been codified

---

[13] The same considerations apply to proportionality as a requirement of the reaction in self-defence under the *jus ad bellum* and of countermeasures and belligerent reprisals.
[14] Article 58(c) of Additional Protocol I.

at the beginning of the twentieth century, the law of neutrality may extend to cyber operations whenever they are conducted in the context of an international armed conflict and have a nexus with it, or when they amount themselves to such a conflict, whether or not there is a declaration of war or a state has declared its neutrality. If belligerent and, even more, neutral states are prohibited from conducting any cyber operations amounting to acts of hostilities against other belligerents from cyber infrastructure situated in neutral territory and under the neutral's exclusive control, and the neutral state has an obligation to use all the means at its disposal to terminate them, routing cyber operations through neutral cyber infrastructure is not a violation of that state's neutrality: indeed, neither would the belligerent be able to control the pathway taken by the malware, nor would the neutral have the means to effectively prevent such transit. Belligerents are of course also prohibited from conducting any cyber operation against neutral territory or neutral cyber infrastructure and from conducting cyber operations against other belligerents that have more than nominal prejudicial incidental effects on neutral territory, in the same way as they would in a traditional operation, although they may use force, by cyber or kinetic means, where the neutral state is unable or unwilling to enforce its neutrality and terminate the use of its territory by another belligerent. The use of cyber infrastructure to conduct cyber attacks or cyber exploitation activities should, however, be distinguished from the use of cyber infrastructure for communications. Indeed, it is one thing to spread malware or create a botnet to conduct a DDoS attack and shut down a system, but quite another to send emails, use social networks, or post information on a website (even when this is for propaganda purposes). The use of cyber infrastructure exclusively for communication purposes, as opposed to the conduct of cyber attacks and cyber exploitation, does not fall under the scope of Articles 1 and 2 of Hague Convention V, but rather of its Articles 3 and 8, that provide for the more permissible legal framework examined in Section III.4 of Chapter 5.

To conclude. While it appears likely that cyber operations against states will increase both in frequency and in gravity, in the near future they will probably supplement, not replace, traditional warfare. Their potentially less lethal effects and covert character make them particularly appealing means and methods of warfare not only in non-international armed conflicts, but also in international ones, given the present trend towards effects-based warfare. With regard to their legal regime, possible developments include the conclusion of specific treaties regulating cyber operations conducted by states from a *jus ad bellum, jus in bello*, or arms control perspective, the formation of ad hoc customary rules, or the adoption of Confidence Building Measures (CBMs).[15] The possibility of requesting an advisory opinion on the legality of cyber operations to the ICJ could also be explored.

If it is correct—as the present book has submitted—that existing international law on the use of force is flexible enough to regulate, with a sufficient degree of efficiency, cyber operations, none of the above developments is however urgently

---

[15] On CBMs, see Katharina Ziolkowski, *Confidence Building Measures for Cyberspace—Legal Implications* (CCDCOE, 2013).

needed. It is known, for instance, that Russia and China have long supported the conclusion of a convention to regulate the offensive use of cyber technologies by states and to ban attacks on computer networks.[16] Russia has even drafted a proposal for a Convention on International Information Security.[17] In contrast, a majority of states, including the United States,[18] argue that there is no need for such a treaty, as existing rules and law enforcement mechanisms suffice. Although the increasing frequency and gravity of cyber operations might determine a rapprochement of these two opposite positions, the chances of the adoption of a treaty on cyber warfare in the foreseeable future remain slim. The rushed adoption of such a treaty may actually be counterproductive: as Montesquieu argued well before the Information Age, 'les lois inutiles affaiblissent les lois nécessaires'.[19] Yoram Dinstein has observed in another context that the sporadic character of treaty provisions for air warfare may be due 'to over-hasty and unrealistic endeavours to cope with air warfare by treaty in an earlier period'.[20] The present author agrees: drafting a treaty on cyber warfare today may be prejudicial to future efforts. We still have to fully understand the realities and potentialities of cyber capabilities, and the developments in these technologies occur at such a speed that any treaty would potentially be outdated the day after it has been opened for signature. Existing rules are capable of adequately regulating the phenomenon and of limiting the conduct of states in the cyber context: let us start by correctly interpreting and applying them.

---

[16] John Markoff and Andrew E Kramer, 'U.S. and Russia Differ on a Treaty for Cyberspace', *The New York Times*, 27 June 2009, <http://www.nytimes.com/2009/06/28/world/28cyber.html?pagewanted=all&_r=0>. On 12 September 2011, Russia, China, Tajikistan and Uzbekistan submitted a draft resolution to the UN General Assembly on an International Code of Conduct for Information Security (UN Doc A/66/359, 14 September 2011). See also China's and Brazil's statements in the UN General Assembly's First Committee (UN Doc A/C.1/66/PV.17, 20 October 2011, p 9, and UN Doc A/C.1/65/PV.16, 21 October 2010, p 3, respectively). On the prospects for a treaty on cyber warfare, see eg Phillip A Johnson, 'Is It Time for a Treaty on Information Warfare?', *International Law Studies* 76 (2002), pp 439–53; Louise Arimatsu, 'A Treaty for Governing Cyber-Weapons: Potential Benefits and Practical Limitations', in *2012 4th International Conference on Cyber Conflict* (2012), edited by Christian Czosseck, Rain Ottis, and Katharina Ziolkowski, pp 91 ff.

[17] See Russian Foreign Ministry and Security Council, Convention on International Information Security (Concept), 2011, <http://www.mid.ru/bdomp/ns-osndoc.nsf/1e5f0de28fe77fdcc32575d900298676/7b17ead7244e2064c3257925003bcbcc!OpenDocument>.

[18] John Markoff, 'Before the Gunfire, Cyberattacks', *The New York Times*, 12 August 2008, <http://www.nytimes.com/2008/08/13/technology/13cyber.html>.

[19] Montesquieu, *De l'Esprit des lois* (1748), Livre XXIX, Chapitre XVII, in *Œuvres complètes de Montesquieu* (Paris: Firmin Didot Frères, 1838), p 477 ('unnecessary laws weaken those that are necessary'; the translation is mine).

[20] Yoram Dinstein, 'Air Warfare', in *Max Planck Encyclopedia of Public International Law* (2012), Vol I, p 252.

Select Bibliography[1]

BOOKS (INCLUDING EDITED BOOKS)

Boothby, William H. *The Law of Targeting* (Oxford: Oxford University Press, 2012).
Brenner, Susan W. *Cyberthreats. The Emerging Fault Lines of the Nation State* (New York: Oxford University Press, 2009).
Byström Karin, ed. *Proceedings of the Conference: International Expert Conference on Computer Network Attacks and the Applicability of International Humanitarian Law, 17–19 September 2004, Stockholm, Sweden* (Stockholm: Swedish National Defence College, 2004).
Delibasis, Dimitrios. *The Right to National Self-Defense in Information Warfare Operations* (Suffolk: Arena Books, 2007).
Greenberg, Lawrence T, Seymour E Goodman, and Kevin Soo Hoo. *Information Warfare and International Law* (Washington: National Defense University Press, 1998).
Harrison Dinniss, Heather. *Cyber Warfare and the Laws of War* (Cambridge: Cambridge University Press, 2012).
Owens, William A, Kenneth W Dam, and Herbert S Lin, eds. *Technology, Policy, Law, and Ethics Regarding U.S. Acquisition and Use of Cyberattack Capabilities* (Washington: The National Academies Press, 2009).
Palojärvi, Pia. *A Battle in Bits and Bytes: Computer Network Attacks and the Law of Armed Conflict* (Helsinki: The Erik Castrén Research Reports 27/2009).
Saxon, Dan, ed. *International Humanitarian Law and the Changing Technology of War*, (Leiden: Martinus Nijhoff Publishers, 2013).
Sharp Sr, Walter G. *Cyberspace and the Use of Force* (Falls Church, VA: Aegis Research Corporation, 1999).
Schmitt, Michael N, ed. *Tallinn Manual on the International Law Applicable to Cyber Warfare* (Cambridge: Cambridge University Press, 2013).

ARTICLES

Aldrich, Richard W. 'The International Legal Implications of Information Warfare'. *Airpower Journal* 10, no 3 (1996), pp 99–110.
Aldrich, Richard W. 'How Do You Know You Are at War in the Information Age'. *Houston Journal of International Law* 22 (2000), pp 224–63.
Anderson, Douglas S and Christopher R Dooley. 'Information Operations in the Space Law Arena: Science Fiction Becomes Reality'. *International Law Studies* 76 (2002), pp 265–311.
Antolin-Jenkins, Vida M. 'Defining The Parameters of Cyberwar Operations: Looking For Law in All the Wrong Places?'. *Naval Law Review* 51 (2005), pp 132–75.

---

[1] The bibliography includes works that analyse the international law aspects of military cyber operations. Therefore, it includes neither works on cyber crime nor those that discuss cyber operations conducted by states from the perspective of other disciplines.

Bachmann, Sascha-Dominik. 'Hybrid Threats, Cyber Warfare and NATO's Comprehensive Approach for Countering 21st Century Threats—Mapping the New Frontier of Global Risk and Security Management'. *Amicus Curiae* 88 (2011), pp 14–17.

Backstrom, Alan and Ian Henderson. 'New Capabilities in Warfare: An Overview of Contemporary Technological Developments and the Associated Legal and Engineering Issues in Article 36 Weapons Reviews'. *International Review of the Red Cross* 94 (2012), pp 483–514.

Banks, William. 'The Role of Counterterrorism Law in Shaping *ad Bellum* Norms for Cyber Warfare'. *International Law Studies* 89 (2013), pp 156–97.

Barkham, Jason. 'Information Warfare and International Law on the Use of Force'. *New York University Journal of International Law and Politics* 34 (2001), pp 57–114.

Beckett, Jason. 'New War, Old Law: Can the Geneva Paradigm Comprehend Computers?'. *Leiden Journal of International Law* 13 (2000), pp 33–51.

Benatar, Marco. 'The Use of Cyber Force: Need for Legal Justification?'. *Goettingen Journal of International Law* 3 (2009), pp 375–96.

Blake, Duncan and Joseph S Imburgia. '"Bloodless Weapons"? The Need to Conduct Legal Reviews of Certain Capabilities and the Implications of Defining Them as "Weapons"'. *Air Force Law Review* 66 (2010), pp 159–200.

Blank, Laurie R. 'International Law and Cyber Threats from Non-State Actors'. *International Law Studies* 89 (2013), pp 406–37.

Blount, PJ. 'The Preoperational Legal Review of Cyber Capabilities: Ensuring the Legality of Cyber Weapons'. *Northern Kentucky Law Review* 39 (2012), pp 211–20.

Boothby, William. 'Some Legal Challenges Posed By Remote Attack'. *International Review of the Red Cross* 94 (2012), pp 579–95.

Boothby, William H. 'Methods and Means of Cyber Warfare'. *International Law Studies* 89 (2013), pp 387–405.

Bowman, ME. 'Is International Law Ready for the Information Age?'. *Fordham International Law Journal* 19 (1995), pp 1935–46.

Bradbury, Steven G. 'Keynote Address: The Developing Legal Framework for Defensive and Offensive Cyber Operations'. *Harvard National Security Journal* 2 (2011), pp 633–51.

Brenner, Susan W. '"At Light Speed": Attribution and Response to Cybercrime/Terrorism/Warfare'. *The Journal of Criminal Law and Criminology* 97 (2007), pp 379–475.

Brown, Davis. 'A Proposal for an International Convention to Regulate the Use of Information Systems in Armed Conflict'. *Harvard International Law Journal* 47 (2006), pp 179–221.

Brown, Gary and Keira Poellet. 'The Customary International Law of Cyberspace'. *Strategic Studies Quarterly* 6, no 3 (2012), pp 126–45.

Buchan, Russell. 'Cyber Attacks: Unlawful Uses of Force or Prohibited Interventions?'. *Journal of Conflict & Security Law* 17 (2012), pp 212–27.

Cammack, Chance. 'The Stuxnet Worm and Potential Prosecution by the International Criminal Court Under the Newly Defined Crime of Aggression'. *Tulane Journal of International and Comparative Law* 20 (2011), pp 303–26.

Chainoglou, Kalliopi. 'An Assessment of *Jus In Bello* Issues Concerning Computer Network Attacks: A Threat Reflected in National Security Agendas'. *Romanian Journal of International Law* 12 (2010), pp 25–63.

Church, William. 'Information Warfare'. *International Review of the Red Cross* 82 (2000), pp 205–15.

Condron, Sean M. 'Getting It Right: Protecting American Critical Infrastructure in Cyberspace'. *Harvard Journal of Law & Technology* 20 (2007), pp 404–22.

Cox, Stephen J. 'Confronting Threats Through Unconventional Means: Offensive Information Warfare as a Covert Alternative to Preemptive War'. *Houston Law Review* 42 (2005), pp 881–910.

Crane, David M. 'Fourth Dimensional Intelligence: Thoughts on Espionage, Law, and Cyberspace'. *International Law Studies* 76 (2002), pp 311–21.

Crawford, Emily. 'Virtual Battlegrounds: Direct Participation in Cyber Warfare'. Sydney Law School Research Paper No 12/10 (2012), <http://papers.ssrn.com/sol3/papers.cfm?abstract_id=2001794>.

Creekman, Daniel M. 'A Helpless America? An Examination of the Legal Options Available to the United States in Response to Varying Types of Cyber-Attacks from China'. *American University International Law Review* 17 (2002), pp 641–81.

Cronin, Audrey K. 'Cyber-Mobilization: The New *Levée En Masse*'. *Parameters* 36, No 2 (2006), pp 77–87.

D'Amato, Anthony. 'International Law, Cybernetics, and Cyberspace'. *International Law Studies* 76 (2002), pp 59–73.

Deeks, Ashley. 'The Geography of Cyber Conflict: Through a Glass Darkly'. *International Law Studies* 89 (2013), pp 1–20.

Dinstein, Yoram. 'Computer Network Attacks and Self-Defense'. *International Law Studies* 76 (2002), pp 99–121.

Dinstein, Yoram. 'The Principle of Distinction and Cyber War in International Armed Conflicts'. *Journal of Conflict & Security Law* 17 (2012), pp 261–77.

Dinstein, Yoram. 'Cyber War and International Law: Concluding Remarks at the 2012 Naval War College International Law Conference'. *International Law Studies* 89 (2013), pp 275–87.

Döge, Jenny. 'Cyber Warfare'. *Archiv des Völkerrechts* 48 (2010), pp 486–501.

Doswald-Beck, Louise. 'Some Thoughts on Computer Network Attack and the International Law of Armed Conflict'. *International Law Studies* 76 (2002), pp 163–87.

Doyle Jr, James H. 'Computer Networks, Proportionality, and Military Operations'. *International Law Studies* 76 (2002), pp 147–63.

Droege, Cordula. 'Get Off My Cloud: Cyber Warfare, International Humanitarian Law, and the Protection of Civilians'. *International Review of the Red Cross* 94 (2012), pp 533–78.

Dunlap, Charles J. 'Meeting the Challenge of Cyberterrorism: Defining the Military Role in a Democracy'. *International Law Studies* 76 (2002), pp 353–75.

Dunlap, Charles J. 'Perspectives for Cyber Strategists on Law for Cyberwar'. *Strategic Studies Quarterly* 5, no 1 (2011), pp 81–99.

Fidler, David P. 'International Law and the Future of Cyberspace: The Obama Administration's International Strategy for Cyberspace'. *ASIL Insights* 15(15) (2011). <http://www.asil.org/insights/volume/15/issue/15/international-law-and-future-cyberspace-obama-administration%E2%80%99s>.

Fidler, David P. 'Was Stuxnet an Act of War? Decoding a Cyberattack'. *IEEE Security & Privacy* 9, no 4 (2011), pp 72–5.

Fidler, David P. 'Recent Developments and Revelations Concerning Cybersecurity and Cyberspace: Implications for International Law'. *ASIL Insights* 16(22) (2012). <http://www.asil.org/insights/volume/16/issue/22/recent-developments- and-revelations-concerning-cybersecurity-and>.

Fidler, David P. 'Tinker, Tailor, Soldier, Duqu: Why Cyberespionage is More Dangerous Than You Think'. *International Journal of Critical Infrastructure Protection* 5 (2012), pp 28–9.

Fleck, Dieter. 'Searching for International Rules Applicable to Cyber Warfare—A Critical First Assessment of the New Tallinn Manual'. *Journal of Conflict & Security Law* 18 (2013), pp 1–21.
Geiss, Robin. 'War and Law in Cyberspace: The Conduct of Hostilities in and via Cyber-Space'. *American Society of International Law Proceedings* 104 (2010), pp 371–74.
Geiss, Robin. 'Cyber Warfare: Implications for Non-international Armed Conflicts'. *International Law Studies* 89 (2013), pp 626–45.
Geiss, Robin and Henning Lahmann. 'Cyber Warfare: Applying the Principle of Distinction in an Interconnected Space'. *Israel Law Review* 45 (2012), pp 381–99.
Gervais, Michael. 'Cyber Attacks and the Laws of War'. *Berkeley Journal of International Law* 30 (2012), pp 525–79.
Gill, Terry D and Paul A Ducheine. 'Anticipatory Self-Defense in the Cyber Context'. *International Law Studies* 89 (2013), pp 438–71.
Glennon, Michael J. 'The Dark Future of International Cybersecurity Regulation'. *Journal of National Security Law & Policy* 89 (2013), pp 563–70.
Glennon, Michael J. 'The Road Ahead: Gaps, Leaks and Drips'. *International Law Studies* 89 (2013), pp 361–86.
Goldsmith, Jack. 'How Cyber Changes the Laws of War'. *European Journal of International Law* 24 (2013), pp 129–38.
Gosnell, Stephanie H. 'The New Cyber Face of Battle: Developing a Legal Approach to Accommodate Emerging Trends in Warfare'. *Stanford Journal of International Law* 48 (2012), pp 209–37.
Graham, David E. 'Cyber Threats and the Law of War'. *Journal of National Security Law & Policy* 4 (2010), pp 87–102.
Hanseman, Robert G. 'The Realities and Legalities of Information Warfare'. *Air Force Law Review* 42 (1997), pp 176–200.
Haslam, Emily. 'Information Warfare: Technological Changes and International Law'. *Journal of Conflict and Security Law* 5 (2000), pp 157–75.
Hathaway, Oona A et al. 'The Law of Cyber-Attack'. *California Law Review* 100 (2012), pp 817–86.
Heintschel von Heinegg, Wolff. 'Territorial Sovereignty and Neutrality in Cyberspace'. *International Law Studies* 89 (2013), pp 122–56.
Hinkle, Katharine C. 'Countermeasures in the Cyber Context: One More Thing to Worry About'. *Yale Journal of International Law Online* (Fall 2011), pp 11–21. <http://www.yjil.org/docs/pub/o-37-hinkle-countermeasures-in-the-cyber-context.pdf>.
Hoffman, Michael H. 'The Legal Status and Responsibilities of Private Internet Users under the Law of Armed Conflict: A Primer for the Unwary on the Shape of Law to Come'. *Washington University Global Studies Law Review* 2 (2003), pp 415–26.
Hoisington, Matthew. 'Cyberwarfare and the Use of Force Giving Rise to the Right of Self-Defense'. *Boston College International and Comparative Law Review* 32 (2009), pp 439–54.
Hollis, Duncan B. 'Why States Need an International Law for Information Operations'. *Lewis and Clark Law Review* 11 (2007), pp 1023–62.
Hollis, Duncan B. 'An e-SOS for Cyberspace'. *Harvard International Law Journal* 52 (2011), pp 374–432.
Hughes, Rex. 'A Treaty for Cyberspace'. *International Affairs* 86 (2010), pp 523–41.
Huntley, Todd C. 'Controlling the Use of Force in Cyber Space: The Application of the Law of Armed Conflict During a Time of Fundamental Change in the Nature of Warfare'. *Naval Law Review* 60 (2010), pp 1–40.

Intoccia, Gregory F and Joe W Moore. 'Communications Technology, Warfare, and the Law: Is the Network a Weapon System?'. *Houston Journal of International Law* 28 (2006), pp 470–90.

Jacobson, Mark R. 'War in the Information Age: International Law, Self-Defense, and the Problem of "Non-Armed" Attacks'. *Journal of Strategic Studies* 21, no 3 (1998), pp 1–23.

Jensen, Eric Talbot. 'Computer Attacks on Critical National Infrastructure: A Use of Force Invoking the Right of Self-Defense'. *Stanford Journal of International Law* 38 (2002), pp 207–40.

Jensen, Eric Talbot. 'Unexpected Consequences from Knock-On Effects: A Different Standard for Computer Network Operations?'. *American University International Law Review* 18 (2003), pp 1145–88.

Jensen, Eric Talbot. 'Cyber Warfare and Precautions Against the Effects of Attacks'. *Texas Law Review* 88 (2010), pp 1533–69.

Jensen, Eric Talbot. 'President Obama and the Changing Cyber Paradigm'. *William Mitchell Law Review* 37 (2011), pp 5049–60.

Jensen, Eric Talbot. 'Sovereignty and Neutrality in Cyber Conflict'. *Fordham International Law Journal* 35 (2011–12), pp 815–41.

Jensen, Eric Talbot. 'Cyber Attacks: Proportionality and Precautions in Attack'. *International Law Studies* 89 (2013), pp 198–217.

Johnson, Phillip A. 'Is It Time for a Treaty on Information Warfare?'. *International Law Studies* 76 (2002), pp 439–531.

Joyner, Christopher C and Catherine Lotrionte. 'Information Warfare as International Coercion: Elements of a Legal Framework'. *European Journal of International Law* 12 (2001), pp 825–65.

Jurich, Jon P. 'Cyberwar and Customary International Law: The Potential of a "Bottom-Up" Approach to an International Law of Information Operations'. *Chicago Journal of International Law* 9 (2008), pp 275–95.

Kanuck, Sean. 'Information Warfare: New Challenges for Public International Law'. *Harvard International Law Journal* 37 (1996), pp 272–92.

Kanuck, Sean. 'Sovereign Discourse on Cyber Conflict Under International Law'. *Texas Law Review* 88 (2010), pp 1571–97.

Kastenberg, Joshua E. 'Non-intervention and Neutrality in Cyberspace: An Emerging Principle in the National Practice of International Law'. *Air Force Law Review* 64 (2009), pp 44–64.

Keber, Tobias O and Przemysław N Roguski. 'Ius ad bellum electronicum? Cyberangriffe im Lichte der UN-Charta und aktueller Staatenpraxis'. *Archiv des Völkerrechts* 49 (2011), pp 399–434.

Kelsey, Jeffrey TG. 'Hacking Into International Humanitarian Law: The Principles of Distinction and Neutrality in the Age of Cyber Warfare'. *Michigan Law Review* 106 (2008), pp 1427–52.

Kesan, Jay P, and Carol M Hayes. 'Mitigative Counterstriking: Self-Defense and Deterrence in Cyberspace'. *Harvard Journal of Law and Technology* 25 (2012), pp 417–527.

Kim, Jasper. 'Law of War 2.0: Cyberwar and the Limits of the UN Charter'. *Global Policy* 2 (2011), pp 322–28.

Kirchner, Stefan. 'Distributed Denial-of-Service Attacks Under Public International Law: State Responsibility in Cyberwar'. *The IUP Journal of Cyber Law* 8, nos 3–4 (2009), pp 10–23.

Kleffner, Jann K and Heather A Harrison Dinniss. 'Keeping the Cyber Peace: International Legal Aspects of Cyber Activities in Peace Operations'. *International Law Studies* 89 (2013), pp 511–35.

Kodar, Erki. 'Computer Network Attacks in the Grey Areas of *Jus ad Bellum* and *Jus in Bello*'. *Baltic Yearbook of International Law* 9 (2009), pp 133–55.

Kritsiotis, Dino. 'Enforced Equations'. *European Journal of International Law* 24 (2013), pp 139–49.

Kuehl, Daniel T. 'Information Operations, Information Warfare, and Computer Network Attack: Their Relationship to National Security in the Information Age'. *International Law Studies* 76 (2002), pp 35–58.

Kulesza, Joanna. 'State Responsibility for Cyber-Attacks on International Peace and Security'. *Polish Yearbook of International Law* 29 (2009), pp 139–52.

Lentz, Christopher E. 'A State's Duty to Prevent and Respond to Cyberterrorist Acts'. *Chicago Journal of International Law* 10 (2010), pp 799–823.

Lichtenbaum, Peter and Melanie Schneck. 'The Response to Cyberattacks: Balancing Security and Cost'. *The International Lawyer* 36 (2002), pp 39–48.

Lin, Herbert S. 'Offensive Cyber Operations and the Use of Force'. *Journal of National Security Law & Policy* 4 (2010), pp 63–86.

Lin, Herbert. 'Cyber Conflict and International Humanitarian Law'. *International Review of the Red Cross* 94 (2012), pp 515–31.

Liivoja, Rain and Tim McCormack. 'Law in the Virtual Battlespace: The Tallinn Manual and the *Jus in Bello*'. Melbourne Legal Studies Research Papers no 650 (2013), <http://papers.ssrn.com/sol3/papers.cfm?abstract_id=2297159>.

Lubell, Noam. 'Lawful Targets in Cyber Operations: Does the Principle of Distinction Apply?'. *International Law Studies* 89 (2013), pp 251–75.

Malawer, Stuart S. 'Cyber Warfare: Law and Policy Proposals for U.S. and Global Governance'. *Virginia Lawyer* 58 (2010), pp 27–31.

Margulies, Peter. 'Networks in Non-International Armed Conflicts: Crossing Borders and Defining "Organized Armed Group"'. *International Law Studies* 89 (2013), pp 53–76.

McGavran, Wolfgang. 'Intended Consequences: Regulating Cyber Attacks'. *The Tulane Journal of Technology and Intellectual Property* 12 (2009), pp 259–75.

Melnitzky, Alexander. 'Defending America Against Chinese Cyber Espionage Through the Use of Active Defenses'. *Cardozo Journal of International and Comparative Law* 20 (2012), pp 537–70.

Melzer, Nils. 'Cyber Operations and *Jus In Bello*'. *Disarmament Forum* 2011, no 4 (2011), pp 3–17.

Morth, Todd A. 'Considering Our Position: Viewing Information Warfare as a Use of Force Prohibited by Article 2(4) of the U.N. Charter'. *Case Western Reserve Journal of International Law* 30 (1998), pp 567–600.

Murphy, John F. 'Computer Network Attacks by Terrorists: Some Legal Dimensions'. *International Law Studies* 76 (2002), pp 321–53.

Murphy, John F. 'Cyber War and International Law: Does the International Legal Process Constitute a Threat to U.S. Vital Interests?'. *International Law Studies* 89 (2013), pp 308–40.

O'Connell, Mary Ellen. 'Cyber Security Without Cyber War'. *Journal of Conflict and Security Law* 17 (2012), pp 187–209.

O'Donnell, Brian T and James C Kraska. 'International Law of Armed Conflict and Computer Network Attack: Developing the Rules of Engagement'. *International Law Studies* 76 (2002), pp 359–421.

O'Donnell, Brian T and James C Kraska. 'Humanitarian Law: Developing International Rules for the Digital Battlefield'. *Journal of Conflict and Security Law* 8 (2003), pp 133–60.

Ophardt, Jonathan A. 'Cyber Warfare and the Crime of Aggression: The Need for Individual Accountability on Tomorrow's Battlefield'. *Duke Law and Technology Review* 9 (2010), pp 1–28.

Padmanabhan, Vijay M. 'Cyber Warriors and the *Jus in Bello*'. *International Law Studies* 89 (2013), pp 287–308.

Prescott, Jody. 'War by Analogy. US Cyberspace Strategy and International Humanitarian Law'. *The RUSI Journal* 156, no 6 (December 2011), pp 32–9.

Raboin, Bradley. 'Corresponding Evolution: International Law and the Emergence of Cyber Warfare'. *Journal of the National Association of Administrative Law Judiciary* 31 (2011), pp 603–68.

Richardson, John. 'Stuxnet as Cyberwarfare: Applying the Law of War to the Virtual Battlefield'. *The John Marshall Journal of Computer & Information Law* 29 (2011), pp 1–39.

Richmond, Jeremy. 'Evolving Battlefields: Does Stuxnet Demonstrate a Need for Modifications to the Law of Armed Conflict?'. *Fordham International Law Journal* 35 (2012), pp 842–94.

Ricou Heaton, John. 'Civilians at War: Reexamining the Status of Civilians Accompanying the Armed Forces'. *Air Force Law Review* 57 (2005), pp 155–208.

Robbat, Michael J. 'Resolving the Legal Issues Concerning the Use of Information Warfare in the International Forum: The Reach of the Existing Legal Framework, and the Creation of a New Paradigm'. *Journal of Science & Technology Law* 6 (2000), pp 264–89.

Robertson Jr, Horace B. 'Self-Defense against Computer Network Attack under International Law'. *International Law Studies* 76 (2002), pp 121–47.

Roscini, Marco. 'World Wide Warfare: *Jus ad bellum* and the Use of Cyber Force', *Max Planck Yearbook of United Nations Law* 14 (2010), pp 85–130.

Ryan, Daniel J, Maeve Dion, Eneken Tikk, and Julie JCH Ryan 'International Cyberlaw: A Normative Approach'. *Georgetown Journal of International Law* 42 (2011), pp 1161–97.

Schaap, Arie J. 'Cyber Warfare Operations: Development and Use Under International Law'. *Air Force Law Review* 64 (2009), pp 121–74.

Schmitt, Michael N. 'Computer Network Attack and the Use of Force in International Law: Thoughts on a Normative Framework'. *Columbia Journal of Transnational Law* 37 (1999), pp 885–937.

Schmitt, Michael N. 'Computer Network Attack: The Normative Software'. *Yearbook of International Humanitarian Law* 4 (2001), pp 53–85.

Schmitt, Michael N. 'Wired Warfare: Computer Network Attack and the *Jus in Bello*'. *International Law Studies* 76 (2002), pp 187–219. Also in *International Review of the Red Cross* 84 (2002), pp 365–98.

Schmitt, Michael N. 'Cyber Operations and the *Jus Ad Bellum* Revisited'. *Villanova Law Review* 56 (2011), pp 569–606.

Schmitt, Michael N. 'Cyber Operations and the *Jus in Bello*: Key Issues'. *International Law Studies* 87 (2011), pp 89–110.

Schmitt, Michael N. 'Classification of Cyber Conflict'. *International Law Studies* 89 (2013), pp 233–51. Also in *Journal of Conflict and Security Law* 17 (2012), pp 245–60.

Schmitt, Michael N. 'International Law in Cyberspace: The Koh Speech and Tallinn Manual Juxtaposed'. *Harvard International Law Journal Online* 54 (2012), pp 13–37. <http://www.harvardilj.org/2012/12/online-articles-online_54_schmitt>.

Scott, Roger D. 'Legal Aspects of Information Warfare: Military Disruption of Telecommunications'. *Naval Law Review* 45 (1998), pp 57–76.

Segura-Serrano, Antonio. 'Internet Regulation and the Role of International Law'. *Max Planck Yearbook of United Nations Law* 10 (2006), pp 191–272.
Shackelford, Scott J. 'From Nuclear War to Net War: Analogizing Cyber Attacks in International Law'. *Berkeley Journal of International Law* 27 (2009), pp 192–252.
Shackelford, Scott J and Richard B Andres. 'State Responsibility for Cyber Attacks: Competing Standards for a Growing Problem'. *Georgetown Journal of International Law* 42 (2011), pp 971–1016.
Sharp Sr, Walter Gary. 'The Past, Present, and Future of Cybersecurity'. *Journal of National Security Law and Policy* 4 (2010), pp 13–26.
Sherman, Adam. 'Forward unto the Digital Breach: Exploring the Legal Status of Tomorrow's High-Tech Warriors'. *Chicago Journal of International Law* 5 (2004), pp 335–41.
Shulman, Mark R. 'Discrimination in the Laws of Information Warfare'. *Columbia Journal of Transnational Law* 37 (1999), pp 939–68.
Shulman, Mark R. 'Legal Constraints on Information Warfare', Occasional Paper no 7, Center for Strategy and Technology Air War College, Alabama (1999).
Silver, Daniel B. 'Computer Network Attack as a Use of Force under Article 2(4) of the United Nations Charter'. *International Law Studies* 76 (2002), pp 73–99.
Sinisalu, Arnold. 'Propaganda, Information War and the Estonian-Russian Treaty Relations: Some Aspects of International Law'. *Juridica International* 15 (2008), pp 154–62.
Sklerov, Matthew J. 'Solving the Dilemma of State Responses to Cyberattacks: A Justification for the Use of Active Defenses Against States Who Neglect Their Duty to Prevent'. *Military Law Review* 201 (2009), pp 1–85.
Smith, Bruce P. 'Hacking, Poaching, and Counterattacking: Digital Counterstrikes and the Contours of Self-Help'. *Journal of Law, Economics and Policy* 1 (2005), pp 171–97.
Smith, Jeffrey H and Gordon N Lederman. '"Weapons like to Lightning": US Information Operations and US Treaty Obligations'. *International Law Studies* 76 (2002), pp 375–59.
Solce, Natasha. 'The Battlefield of Cyberspace: The Inevitable New Military Branch—The Cyber Force'. *Albany Law Journal of Science and Technology* 18 (2008), pp 293–324.
Stein, Torsten and Thilo Marauhn. 'Völkerrechtliche Aspekte von Informationsoperationen'. *Zeitschrift für ausländisches öffentliches Recht und Völkerrecht* 60 (2000), pp 1–40.
Swanson, Lesley. 'The Era of Cyber Warfare: Applying International Humanitarian Law to the 2008 Russian-Georgian Cyber Conflict'. *Loyola of Los Angeles International and Comparative Law Review* 32 (2010), pp 303–33.
Tappero Merlo, Germana. 'Il dominio degli spazi: il cosmo, la cyberwar e l'urgenza di una dottrina operativa per la guerra futura'. *La Comunità internazionale* 65 (2010), pp 535–59.
Terry, James P. 'Responding to Attacks on Critical Computer Infrastructure: What Targets? What Rules of Engagement?'. *International Law Studies* 76 (2002), pp 421–39.
Tikk, Eneken. 'Ten Rules for Cyber Security'. *Survival* 53, no 3 (2011), pp 119–32.
Todd, Graham H. 'Armed Attack in Cyberspace: Deterring Asymmetric Warfare with an Asymmetric Definition'. *Air Force Law Review* 64 (2009), pp 65–103.
Tsagourias, Nicholas. 'Cyber Attacks, Self-Defence and the Problem of Attribution'. *Journal of Conflict and Security Law* 17 (2012), pp 229–44.
Turns, David. 'Cyber Warfare and the Notion of Direct Participation in Hostilities'. *Journal of Conflict and Security Law* 17 (2012), pp 279–97.
Walker, George K. 'Information Warfare and Neutrality'. *Vanderbilt Journal of Transnational Law* 33 (2000), pp 1079–202.
Walker, Paul A. 'Organizing for Cyberspace Operations: Selected Issues'. *International Law Studies* 89 (2013), pp 340–61.

Wallace, David and Shane R Reeves. 'The Law of Armed Conflict's "Wicked" Problem: *Levée en Masse* in Cyber Warfare'. *International Law Studies* 89 (2013), pp 645–68.
Watkin, Kenneth. 'The Cyber Road Ahead: Merging Lanes and Legal Challenges'. *International Law Studies* 89 (2013), pp 472–511.
Watts, Sean. 'Combatant Status and Computer Network Attack'. *Virginia Journal of International Law* 50 (2010), pp 391–447.
Watts, Sean. 'Low-Intensity Computer Network Attack and Self-Defense'. *International Law Studies* 87 (2011), pp 59–87.
Waxman, Matthew C. 'Cyber-Attacks and the Use of Force: Back to the Future of Article 2(4)'. *Yale Journal of International Law* 36 (2011), pp 420–58.
Waxman, Matthew C. 'Self-Defensive Force against Cyber Attacks: Legal, Strategic and Political Dimensions'. *International Law Studies* 89 (2013), pp 108–22.
Wedgwood, Ruth G. 'Proportionality, Cyberwar, and the Law of War'. *International Law Studies* 76 (2002), pp 219–33.
Weisbord, Noah. 'Conceptualizing Aggression'. *Duke Journal of Comparative and International Law* 20 (2009), pp 1–68.
Williams, Robert D. '(Spy) Game Change: Cyber Networks, Intelligence Collection, and Covert Action'. *The George Washington Law Review* 79 (2011), pp 1162–200.
Woltag, Johann-Christoph. 'Computer Network Operations Below the Level of Armed Force'. *European Society of International Law Conference Paper Series* 1 (2011), pp 1–16.
Ziolkowski, Katharina. 'Computer Network Operations and the Law of Armed Conflict'. *Military Law and the Law of War Review* 49 (2010), pp 47–94.

CHAPTERS IN EDITED BOOKS

Arimatsu, Louise. 'A Treaty for Governing Cyber-Weapons: Potential Benefits and Practical Limitations'. In *2012 4th International Conference on Cyber Conflict—Proceedings*, edited by Christian Czosseck, Rain Ottis, and Katharina Ziolkowski (Tallinn: CCDCOE, 2012), pp 91–109.
Boothby, William. 'Cyber Deception and Autonomous Attack—Is There a Legal Problem?'. In *2013 5th International Conference on Cyber Conflict*, edited by Karlis Podins, Jan Stinissen, and Markus Maybaum (Tallinn: CCDCOE, 2013), pp 245–61.
Brenner, Susan W and Leo L Clarke. 'Conscription and Cyber Conflict: Legal Issues'. In *2011 3rd International Conference on Cyber Conflict*, edited by Christian Czosseck, Enn Tyugu, and Thomas Wingfield (Tallinn: CCDCOE, 2011), pp 1–13.
Brown, Gary. 'Law at Cyberspeed: Answering Military Cyber Operators' Legal Questions'. In *International Humanitarian Law and New Weapon Technologies*, edited by Wolff Heintschel von Heinegg and Gian Luca Beruto (Milano: Franco Angeli, 2012), pp 166–70.
Bufalini, Alessandro. 'Uso della forza, legittima difesa e problemi di attribuzione in situazioni di attacco informatico'. In *Uso della forza e legittima difesa nel diritto internazionale contemporaneo*, edited by Attila Tanzi and Alessandra Lanciotti (Napoli: Jovene, 2011), pp 405–36.
Bufalini, Alessandro. 'Les cyber-guerres à la lumière des règles internationales sur l'interdiction du recours à la force'. In *La gouvernance globale face aux défis de la sécurité collective*, edited by Maurizio Arcari and Louis Balmond (Napoli: Editoriale Scientifica, 2012), pp 89–109.

Busuttil, James J. 'A Taste of Armageddon: The Law of Armed Conflict as Applied to Cyberwar'. In *The Reality of International Law. Essays in Honour of Ian Brownlie*, edited by Guy S Goodwin-Gill, Stefan Talmon, and Robert Jennings (Oxford: Clarendon Press, 1999), pp 37–56.

Dion, Maeve. 'Different Legal Constructs for State Responsibility'. In *International Cyber Security Legal & Policy Proceedings*, edited by Eneken Tikk and Anna-Maria Talihärm (Tallinn: CCDCOE, 2010), pp 67–75.

Fidler, David P. '*Inter Arma Silent Leges Redux?* The Law of Armed Conflict and Cyber-Conflict'. In *Cyberspace and National Security: Threats, Opportunities and Power in a Virtual World*, edited by Derek S Reveron (Washington: Georgetown University Press, 2012), pp 71–87.

Fidler, David P. 'The Path to Less Lethal and Destructive War? Technological and Doctrinal Developments and International Humanitarian Law after Iraq and Afghanistan'. In *International Humanitarian Law and the Changing Technology of War*, edited by Dan Saxon (Leiden: Martinus Nijhoff Publishers, 2013), pp 315–36.

Garnett, Richard and Paul Clarke. 'Cyberterrorism: A New Challenge for International Law'. In *Enforcing International Law Norms Against Terrorism*, edited by Andrea Bianchi (Oxford and Portland: Hart, 2004), pp 465–88.

Harrison Dinniss, Heather. 'Participants in Conflict—Cyber Warriors, Patriotic Hackers and the Laws of War'. In *International Humanitarian Law and the Changing Technology of War*, edited by Dan Saxon (Leiden: Martinus Nijhoff Publishers, 2013), pp 251–78.

Häussler, Ulf. 'Cyber Security and Defence from the Perspective of Articles 4 and 5 of the NATO Treaty'. In *International Cyber Security Legal & Policy Proceedings*, edited by Eneken Tikk and Anna-Maria Talihärm (Tallinn: CCDCOE, 2010), pp 100–25.

Healey, Jason. 'When "Not My Problem" Isn't Enough: Political Neutrality and National Responsibility in Cyber Conflict'. In *2012 4th International Conference on Cyber Conflict—Proceedings*, edited by Christian Czosseck, Rain Ottis, and Katharina Ziolkowski (Tallinn: CCDCOE, 2012), pp 21–33.

Heinsch, Robert. 'Methodology of Law-Making. Customary International Law and the New Military Technologies'. In *International Humanitarian Law and the Changing Technology of War*, edited by Dan Saxon (Leiden: Martinus Nijhoff Publishers, 2013), pp 17–41.

Heintschel von Heinegg, Wolff. 'Neutrality in Cyberspace'. In *2012 4th International Conference on Cyber Conflict—Proceedings*, edited by Christian Czosseck, Rain Ottis, and Katharina Ziolkowski (Tallinn: CCDCOE, 2012), pp 35–45.

Hiller, Janine S. 'Legal Aspects of a Cyber Immune System'. In *2013 5th International Conference on Cyber Conflict*, edited by Karlis Podins, Jan Stinissen, and Markus Maybaum (Tallinn: CCDCOE, 2013), pp 263–77.

Hollis, Duncan B. 'New Tools, New Rules: International Law and Information Operations'. In *Ideas as Weapons: Influence and Perception in Modern Warfare*, edited by GJ David Jr and TR McKeldin III (Washington: Potomac Books, 2009), pp 59–72.

Lotrionte, Catherine. 'Active Defense for Cyber: A Legal Framework for Covert Counter Measures'. In *Inside Cyber Warfare: Mapping the Cyber Underworld*, 2nd edn, edited by Jeffrey Carr (Sebastopol, CA: O'Reilly Media, 2011), pp 273–85.

Newton, Michael A. 'Proportionality and Precautions in Cyber Attacks'. In *International Humanitarian Law and the Changing Technology of War*, edited by Dan Saxon (Leiden: Martinus Nijhoff Publishers, 2013), pp 229–49.

Prescott, Jody M. 'Direct Participation in Cyber Hostilities: Terms of Reference for Like-Minded States?'. In *2012 4th International Conference on Cyber Conflict—Proceedings*, edited by

Christian Czosseck, Rain Ottis, and Katharina Ziolkowski (Tallinn: CCDCOE, 2012), pp 251–66.

Rattray, Gregory J and Jason Healey. 'Non-State Actors and Cyber Conflict'. In *America's Cyber Future: Security and Prosperity in the Information Age*, edited by Kristin M Lord and Travis Sharp (Washington: Center for a New American Security, 2011), pp 65–87.

Ryan, Julie JCH, Daniel J Ryan, and Eneken Tikk. 'Cybersecurity Regulation: Using Analogies to Develop Frameworks for Regulation'. In *International Cyber Security Legal and Policy Proceedings*, edited by Eneken Tikk and Aanna-Maria Talihärm (Tallinn: CCDCOE, 2010), pp 76–99.

Schmitt, Michael N. 'War, Technology and the Law of Armed Conflict'. In *The Law of War in the 21st Century: Weaponry and the Use of Force*, edited by Anthony M Helm (Rhode Island: Naval War College, 2006), pp 137–83.

Schmitt, Michael N. 'Cyber Operations in International Law: The Use of Force, Collective Security, Self-Defense and Armed Conflicts'. In *Proceedings of a Workshop on Deterring Cyberattacks: Informing Strategies and Developing Options for U.S. Policy*, edited by National Research Council of the National Academies (Washington: National Academies Press, 2010), pp 151–78.

Schmitt, Michael N. '"Attack" as a Term of Art in International Law: The Cyber Operations Context'. In *2012 4th International Conference on Cyber Conflict—Proceedings*, edited by Christian Czosseck, Rain Ottis, and Katharina Ziolkowski (Tallinn: CCDCOE, 2012), pp 283–93.

Schmitt, Michael N. 'The "Use of Force" in Cyberspace: A Reply to Dr Ziolkowski'. In *2012 4th International Conference on Cyber Conflict—Proceedings*, edited by Christian Czosseck, Rain Ottis, and Katharina Ziolkowski (Tallinn: CCDCOE, 2012), pp 311–17.

Turns, David. 'Cyber War and the Concept of "Attack" in International Humanitarian Law'. In *International Humanitarian Law and the Changing Technology of War*, edited by Dan Saxon (Leiden: Martinus Nijhoff Publishers, 2013), pp 209–27.

Watts, Sean. 'The Notion of Combatancy in Cyber Warfare'. In *2012 4th International Conference on Cyber Conflict—Proceedings*, edited by Christian Czosseck, Rain Ottis, and Katharina Ziolkowski, (Tallinn: CCDCOE, 2012), pp 235–49.

REPORTS AND OTHER DOCUMENTS

Advisory Council on International Affairs/Advisory Committee on Issues of Public International Law, *Cyber Warfare*, no 77, AIV/No 22, CAVV, December 2001, <http://www.aiv-advies.nl/ContentSuite/upload/aiv/doc/webversie__AIV77CAVV_22_ENG.pdf>.

Dörmann Knut, *Computer Network Attack and International Humanitarian Law*, 19 May 2001, <http://www.icrc.org/eng/resources/documents/misc/5p2alj.htm>.

Dörmann, Knut. *Applicability of the Additional Protocols to Computer Network Attacks*. International Expert Conference on Computer Network Attacks and the Applicability of International Humanitarian Law, Stockholm, November 2004, <http://www.icrc.org/eng/resources/documents/misc/68lg92.htm>.

International Committee of the Red Cross. *Technological Challenges for the Humanitarian Legal Framework*. Proceedings of the Bruges Colloquium, October 2010, <http://www.coleurope.eu/sites/default/files/uploads/page/collegium_41_0.pdf>.

International Committee of the Red Cross. *International Humanitarian Law and the Challenges of Contemporary Armed Conflicts*. Report, 31st International Conference of the Red Cross and Red Crescent, 31IC/11/5.1.2, Geneva, October 2011, <http://www.icrc.org/eng/assets/files/red-cross-crescent-movement/31st-international-conference/31-int-conference-ihl-challenges-report-11-5-1-2-en.pdf>.

Melzer, Nils. *Cyberwarfare and International Law*, United Nations Institute for Disarmament Research, Geneva, 2011, <http://unidir.org/files/publications/pdfs/cyberwarfare-and-international-law-382.pdf>.

Rauscher, Karl F and Andrey Korotkov. *Working Towards Rules for Governing Cyber Conflict—Rendering the Geneva and Hague Conventions in Cyberspace* (New York: EastWest Institute, 2011), <http://www.ewi.info/working-towards-rules-governing-cyber-conflict>.

Schmitt, Michael N, Heather Harrison Dinniss, and Thomas C Wingfield. *Computers and War: The Legal Battlespace*, Background Paper prepared for Informal High-Level Expert Meeting on Current Challenges to International Humanitarian Law, Cambridge, June 2004, <http://www.hpcrresearch.org/sites/default/files/publications/schmittetal.pdf>.

Schreier, Fred, *On Cyber Warfare*, DCAF Horizon 2015 Working Paper no 7, <http://www.dcaf.ch/Publications/On-Cyberwarfare>.

Tikk, Eneken, Kadri Kaska, Kristel Rünnimeri, Mari Kert, Anna-Maria Talihärm, and Liis Vihul. *Cyber Attacks Against Georgia: Legal Lessons Identified* (CCDCOE, 2008), <http://www.carlisle.army.mil/DIME/documents/Georgia%201%200.pdf>.

Tikk, Eneken, Kadri Kaska, and Liis Vihul. *International Cyber Incidents: Legal Considerations* (Tallinn: CCDCOE, 2010), <http://www.ccdcoe.org/publications/books/legalconsiderations.pdf>.

United States Department of Defense, Office of General Counsel. *An Assessment of International Legal Issues in Information Operations* (Washington, May 1999), <http://www.au.af.mil/au/awc/awcgate/dod-io-legal/dod-io-legal.pdf>.

Vihul, Liis et al. *Legal Implications of Countering Botnets* (CCDCOE, 2012).

Wingfield, Thomas C. *Legal Aspects of Offensive Information Operations in Space*. Report, United States Department of Defense, 2005.

Wingfield, Thomas C and James B Michael. *An Introduction to Legal Aspects of Operations in Cyberspace*. Report, Naval Postgraduate School Homeland Security Leadership Development Program, April 2004.

Ziolkowski, Katharina. *Stuxnet—Legal Considerations* (CCDCOE, 2012).

Ziolkowski, Katharina. *Confidence Building Measures for Cyberspace—Legal Implications* (CCDCOE, 2013).

ENCYCLOPAEDIC ENTRIES

Woltag, Johann-Christoph. 'Cyber Warfare', in *Max Planck Encyclopedia of Public International Law*, edited by Rüdiger Wolfrum (Oxford: Oxford University Press, 2012), Vol II, pp 988–94.

Woltag, Johann-Christoph. 'Internet', in *Max Planck Encyclopedia of Public International Law*, edited by Rüdiger Wolfrum (Oxford: Oxford University Press, 2012), Vol VI, pp 227–38.

Index

Afghanistan 98, 189, 202
African Union 43, 85, 93
Aggression 43, 44, 46, 48, 60, 67, 71–3, 75, 77, 85, 93, 110–14, 116, 129, 149, 250, 266, 269, 270–1
    cyber operations as acts of 67, 71– 2, 75, 111, 114, 116, 270
Al-Qaeda 2, 38, 84, 98, 122, 140, 155
'Anonymous' 8, 9, 38, 147, 155–6
Argentina 10, 187
'Armed attack' 25, 41–2, 88–91, 98–102, 104–10, 111, 116, 271, 274, 276, 278, 282
    cyber operations as 13, 17, 41, 43, 47, 54, 55, 60, 61, 63, 66, 70–7, 92–7, 108–10, 115–16, 282
    cyber operations below the level of 44, 104–10, 116, 274, 282
    imminent 77–80, 91, 107, 282
    by non-state actors 80–8, 282
    scale and effects standard 52, 72–5, 84, 85, 96, 98, 104–5, 108, 109, 115–16, 282
    See also Self-defence
Armenia 39, 65
Association of Southeast Asian Nations (ASEAN) 64
'Attack'
    cyber operations as 17, 165, 166, 167, 178–82, 183–4, 189–90, 222, 245, 262, 284
    cyber operations short of 42, 167, 195, 239–42
    definition 178
    geographical limitations on 213–15, 245
    on objects indispensable to the survival of the civilian population 221, 230–1, 244
    prohibition of indiscriminate 217–19, 235, 245
    on works and installations containing dangerous forces 221, 230, 232–3, 244
Australia 21, 57, 92, 134, 171, 225
Automated warfare 42
Azerbaijan 5, 38–9, 65, 262

Belarus 51, 170
Belgium 10, 171, 228
Belligerency (recognition of) 140, 147, 149–50, 157, 199, 249
Belligerent nexus 123–5, 139, 147, 148, 182, 196, 204, 208, 211, 242, 245, 252, 283, 286

Belligerent occupation 119, 120, 126, 134, 136–7, 139, 140, 168, 211, 214, 244, 251, 276
    cyber operations during 141–7, 161, 283
    difference with 'invasion' 142–4, 213
    and the resumption of hostilities 145–7, 161–3
Belligerent reprisals 243–5, 273, 274, 285
Bolivia 111
Bosnia and Herzegovina 129, 138
Botnets 5, 18–19, 36, 48, 51, 73, 101, 155–6, 169, 204, 215
    and direct participation in hostilities 202, 210–11
    'Mariposa' 18–19, 210
    and neutrality 256, 263, 265–6, 286
    See also Denial of service (DoS)/Distributed denial of service (DDoS) attacks
Brazil 10, 45–6, 262, 287

Canada 10, 32, 54, 57, 89, 137, 171, 225, 262
Capture-rather-than-kill 215
Central Intelligence Agency (CIA) 4, 53, 103, 202
China 3, 5, 10, 17, 21, 24, 31, 39, 59, 67, 68, 80, 111, 186, 188, 195, 265, 266, 283, 287
Civil defence personnel 193, 211–12
Civil war *see* Non-international armed conflict
Civilian objects 1, 76, 129, 131, 135, 164, 166, 171, 173–4, 180, 181, 182, 185, 186, 217–18, 239, 245, 252, 254, 262, 285
    prohibition of attacks on *see* Distinction (principle of)
    See also Proportionality (in the law of targeting), Precautions (duty to take)
Civilians 22, 34, 119, 127, 129, 143, 144, 152, 166, 169, 171, 173–4, 175, 180, 182, 184, 185, 187, 190, 213–15, 217–18, 230–2, 239, 241, 242, 243, 245, 248, 262, 285
    accompanying the armed forces 211
    definition 203
    prohibition of attacks on *see* Distinction (principle of)
    prohibition of reprisals against 244
    See also Direct participation in hostilities, Proportionality (in the law of targeting), Precautions (duty to take)
Cloud computing 2, 185
Colombia 10, 194–5

Index

Combatants 22, 42, 125, 146, 172, 173, 175, 177, 202, 204, 205, 207, 209, 212, 213–17, 220, 225, 231, 241, 258, 265–6, 277–8
  definition 192–3, 199
  members of a belligerent state's armed forces as 193–5
  members of militias and volunteer corps belonging to a belligerent state as 195–9
  members of a non-state actor's armed forces as 199–202
  and POW status 198–9
  See also Distinction (principle of)
Computer Emergency Readiness Team (CERT) 35, 36, 239, 258
Computer network attack (CNA) *see* Cyber attacks
Computer network defence (CND) *see* Cyber defence
Computer network exploitation (CNE) *see* Cyber exploitation
Computer network operation (CNO) *see* Cyber operations
Conference on Security and Co-operation in Europe (CSCE) 64
  See also Organization for Security and Co-operation in Europe (OSCE)
Confidence Building Measures (CBMs) 286
'Continuous combat function' 148, 199, 200–2, 203, 210
Costa Rica 109, 249
Council of Europe 19, 88
Countermeasures 15, 63, 104–6, 115–16, 242–4, 250, 268–9, 272–3, 274, 278, 282
Cuba 21, 51, 111, 114, 170
Cultural property (protection of) 117, 164, 229, 233, 244
Cyber attacks
  definition 13, 15, 17–18, 65, 179, 263
  effects of 52–3, 169–70, 183, 207, 220–3
  by states against other states 4–9
  tools for 18
  See also Cyber operations, Denial of service (DoS)/Distributed denial of service (DDoS) attacks, Operation 'Olympic Games', 'Shamoon', 'Stuxnet'
Cyber crime: 4, 12, 19, 35, 41, 88, 93, 143, 147, 156, 199, 202, 239, 266
Cyber defence 12, 13–14, 15, 92–3, 166, 205, 212, 238
  active 14, 69, 79, 89, 106, 253
  automated 14, 237, 285
  passive 14, 63, 69, 75, 79, 89, 106, 160, 258
Cyber exploitation 12–13, 15, 16–17, 32, 33, 52, 65–6, 71, 91, 110, 111, 115–16, 122, 148, 156, 157, 160, 162, 163, 168, 195, 199, 201, 205, 207, 228, 234, 240, 255, 263, 278, 281, 282, 285, 286
  See also 'DuQu', 'Flame', 'Moonlight Maze', Operation 'Olympic Games', 'Solar Sunrise', 'Titan Rain'
Cyber operations
  as acts of aggression 67, 71– 2, 75, 111, 114, 116, 270
  as acts of hostilities 20, 25, 32, 122–4, 129–32, 148, 161, 165–6, 178, 180–1, 205, 254–5, 256, 257, 258, 261, 277, 286
  and the applicability of existing treaties 20–4, 41
  as 'armed attack' 13, 17, 41, 43, 47, 54, 55, 60, 61, 63, 66, 70–7, 92–7, 108–10, 115–16, 282
  as 'attack' 17, 165, 166, 167, 178–82, 183–4, 189–90, 222, 245, 262, 284
  attribution to states of 4, 7, 8, 26, 30, 32, 34–40, 41, 42, 44, 63, 70, 77, 98, 100–2, 110, 123, 135, 136, 147, 229, 274–5, 281, 285
  below the level of 'armed attack' 44, 104–10, 116, 274, 282
  below the level of 'use of force' 15, 63–6, 75, 79, 110, 242–3, 282
  causing physical damage 2, 6, 17, 47–8, 52–5, 59, 71–2, 75, 106, 116, 131–2, 136, 161, 179, 205–6, 231, 235–6, 252, 261, 281, 284
  and customary international law 19, 21, 24–30, 41
  definition and classification of military 10–18
  disrupting services 2, 52–3, 55–63, 71–2, 74–6, 106, 116, 131–2, 135–6, 161, 179–81, 205–6, 222–3, 231–2, 235–6, 252, 261, 281–2, 284
  identification of the origin of 33, 38, 44, 70, 101, 110, 197, 220, 229, 281, 285
  against or with incidental effects on neutral territory 261–3, 277, 279, 286
  as 'internal disturbances and tensions' 119, 151, 157, 159–61
  in and as international armed conflicts 119–41, 161–3, 283, 286
  from neutral territory 254–9, 264–5, 277–8, 286
  through neutral territory 259–61, 277–8, 286
  in and as non-international armed conflicts 148–59, 161–3, 286
  and the principle of non-intervention 32, 53, 63–5, 66, 115–16, 282
  as remedies against violations of the law of armed conflict 167, 242–5
  as remedies against violations of the law of neutrality 269, 272–6
  and the role of the UN Security Council 110–15

Index

short of 'attack' 42, 167, 195, 239–42
standard of proof for 7, 32, 38, 77, 80, 91, 97–103, 116, 123, 274–5
and territorial borders 23–4, 96–7, 149, 281
and threats of force 41, 67–9
as threats to the peace 113–4
as a use of force 15, 32, 41, 43, 44–67, 52–67, 71, 75, 115–16, 243, 281–2
  effects-based approach 47–8, 50
  instrument-based approach 46–7, 49–50
  target-based approach 47, 58
Cyberspace 3, 9, 10, 11, 14, 15, 17, 21, 23–4, 33, 34, 35, 50, 51, 61, 66, 70, 74, 75, 79, 81, 82, 87, 89, 90, 92, 93, 95, 96, 103, 110, 111, 114, 142, 149, 157, 166, 169, 183, 184, 185, 197, 199, 247, 248, 280, 281, 282, 285
Cyber terrorism 12, 41, 93
Cyber warfare 9, 10–11, 14–15, 25, 30, 32, 61, 67, 68, 69, 164, 165, 166, 167, 176, 213, 228, 287
  means and methods of 168–76, 217–18, 245, 283, 284, 286

Declaration of war 119, 120–2, 127, 134, 161, 250–2, 268, 283, 286
Defacement of websites 5, 8–9, 17, 65, 123, 147, 148, 213, 217, 240
Demilitarized and neutralized zones 231
Democratic Republic of the Congo (DRC) 83–4, 99, 109, 139, 144, 147
Denial of Service (DoS)/Distributed Denial of Service (DDoS) attacks 4–5, 8, 11, 18, 36, 48, 52, 63, 72, 73, 94, 100, 107, 108, 109, 135–6, 151, 155, 160, 169, 176, 188, 202, 204, 210–11, 235, 259, 263, 265–6, 286
Denmark 10, 141, 276
Detention (law of) 119, 159, 164, 167, 193, 198, 199, 204, 213
Direct participation in hostilities 42, 123–4, 146, 148, 177, 182, 192, 193, 195, 196, 198, 199, 200, 201, 202–11, 213–16, 220, 241, 265
  belligerent nexus see Belligerent nexus
  and botnets 210–11
  and cyber operations short of 'attack' 241–2
  direct causation 204, 206–8, 241, 242
  duration of 194, 199, 201–2, 203–4, 209–10, 211, 214
  threshold of harm 123, 124, 147, 161, 204–6, 207, 208, 210, 211, 241, 242, 283
  See also 'Continuous combat function'
Distinction (principle of) 23, 166, 169, 173, 174, 175, 176–8, 179, 180, 182, 185, 192, 193, 194, 195, 198–9, 202–3, 211, 217, 218, 220, 223, 230, 232, 237, 241, 245, 285
Drones 1, 59, 202, 205, 214

Dual-use objects 50, 66, 167, 185–6, 205–6, 220, 222–4, 284, 285
  See also Military objective
Due diligence 40, 87–8, 234, 257–8, 282
'DuQu' 6–7, 52, 240

Economic coercion 45–7, 49, 62, 64
Effects-based warfare 61, 228, 286
Egypt 79, 120, 190
Electronic warfare 11, 52, 60, 69, 130, 170
El Salvador 87, 109
Erga omnes obligations 243, 268
Espionage: 2, 3, 12, 16–7, 66, 114, 122, 127, 192, 240, 241, 261
Estonia 4–5, 35–6, 39, 40, 55, 63, 72, 93, 94–5, 108, 109, 111, 135, 151, 160, 169, 265
Estonian Cyber Defence League 35–6
Ethiopia 122, 229
European Union 21–2, 57–8, 68, 97, 111

Finland 283
'Flame' 6–7, 52, 108, 240
'Flood' attacks see Denial of service (DoS)/Distributed denial of service (DDoS) attacks
France 10, 68, 96, 120, 133, 167, 171, 219, 220, 225, 250

Gaza see Palestinian Territories
Georgia 7–8, 36, 39, 40, 54, 68, 84, 100, 107, 123, 155, 159, 169, 196, 202, 206, 240, 252, 264, 265
Germany 10, 12, 32, 57, 74–5, 95, 101, 121, 141, 171, 228, 263, 268, 276
Google 76, 264
Greece 113

Hamas 8, 38, 147
Hezbollah 8, 38, 84, 108, 139, 155
Honduras 109
Hors de combat 173, 195, 199, 203, 214, 244
Hospital and safety zones 231
Hungary 22

Immediacy
  of self-defence reactions 79, 88, 91, 92, 108
India 10, 56–7, 82, 113, 262
Indonesia 262
Information blockade 72, 168
Information operations 11, 51
Intelligence collection 2, 7, 12, 15, 16, 33, 35, 47, 65–6, 71, 91, 100–1, 108, 127, 131, 160, 166, 168, 201, 207, 228, 234, 239–42, 255, 263, 285
Internal disturbances and tensions 119, 151, 152, 157, 159–61

Index

International armed conflict
  and belligerent occupation 146–7
  'between states' 136–41
  combatants in 193–9
  cyber operations in and as 119–41, 161–3, 283, 286
  definition 126–7
  law applicable to 164–5, 248–52, 277
  minimum level of intensity 132–6
  and 'use of force' 128–32
International Atomic Energy Agency (IAEA) 6, 53, 188
International criminal law 32, 41, 124–5, 191
International Fact-finding Commission 54, 68, 242
International humanitarian law *see* Law of armed conflict
International Telecommunications Union (ITU) 82
Internet 1, 6, 7–8, 9, 10, 11, 18, 24, 33, 40, 51, 58, 67, 72, 76, 86, 88, 114, 121, 122, 175, 180, 184, 188–91, 202, 239, 258, 259, 261, 264, 265
Internet Service Providers (ISPs) 33, 114, 190, 239, 258
IPv6 33
Iran 6–7, 8, 10, 22, 31, 34–5, 36–7, 39, 53, 76, 91, 99, 113–14, 135, 175–6, 179, 183, 188, 202, 206, 227, 232, 240, 262
Iraq 61, 84, 93, 128, 142, 189–90, 202, 228, 240, 257, 268
Ireland 267
Israel 7–8, 10, 31, 79, 84, 107, 108, 128, 139, 143, 146–7, 181, 203, 204–5, 207, 210, 214, 215, 225, 226–7, 228, 241, 257, 280
Italy 12, 22, 70, 95, 101, 122, 133, 226–7, 228, 267

Japan 10, 68, 120
Jus ad bellum
  applicability to cyber operations of 20–30, 43–116, 280–3, 286
  and the *jus in bello* 117–18
  and the law of neutrality 243, 246, 262, 267, 269, 272, 273–6, 278
  and the qualification of armed conflicts 138
  and the use of force by non-state actors 83–5
Jus in bello see Law of armed conflict

Kazakhstan 51, 170
Kenya 84, 108
Kinetic equivalence doctrine 45, 59–60, 96, 180, 281
Kosovo *see* Yugoslavia
Kyrgyzstan 5

Laos 262
Law of armed conflict
  applicability to cyber operations of 20–30, 117–63, 166–8, 243, 280–1, 283, 286

  cyber operations as remedies against the violation of 167, 242–5
  and cyber operations short of 'attack' 239–42
  geographical scope of application of 42, 118, 208, 213–15
  and international human rights law 118, 145, 167–8, 215, 245
  and the *jus ad bellum* 117–18
  temporal scope of application of 118–19
  See also Targeting (law of)
Law enforcement 14, 30, 42, 68, 69, 80, 85, 87, 89, 118, 126, 145, 147, 152, 156, 159–61, 194, 208, 209, 287
League of Nations 43, 120
Lebanon 84, 108, 139
Levée en masse 144, 193, 203, 212–13
Libya 9, 113, 137, 154, 157, 190, 191, 215, 234
Lithuania 5, 92

Malaysia 262
Mali 22, 60, 75
Martens Clause 22–3, 29, 224, 281
Medical units, personnel and means of transportation 136, 145, 192, 194, 197, 202, 229–30, 231, 244
Mercenaries 192, 199
Mexico 111
Military objective
  data as a 183–4
  'definite military advantage' 187–8, 190, 191
  definition 180, 182–8, 223, 224, 226
  'effective contribution to military action' 184–7, 190, 191
  internet as a 188–91
  US definition of 183, 186–7
  See also Distinction (principle of), Dual-use objects
Misuse of emblems 217, 240, 256
Montenegro 5
'Moonlight Maze' 17

National critical infrastructure (NCI) 6, 47, 51, 53, 55–63, 71, 73, 74, 75, 76, 96, 97, 106, 112, 115–16, 131, 132, 135–6, 153, 156, 161–3, 169, 181, 186, 212, 221, 223, 239, 252, 262, 274, 281, 282, 283
National liberation movements *see* War of national liberation
Natural environment (protection of) 172, 174–6, 220, 229–30, 244
Necessity
  and countermeasures 106
  military 145, 150, 172, 174, 187, 209, 220, 230, 232, 237, 275
  and self-defence reactions 69, 75, 77, 78–9, 85–6, 88–90, 91, 92, 102, 106, 107, 276, 278
  state of 273

The Netherlands 10, 22, 74, 84–5, 89, 95,
    101–2, 131–2, 171, 228, 256
Neutrality (law of) 120, 214, 217, 246–8, 281
  applicability of 121, 248–53, 277, 286
  and communication installations 256–7,
    263–5, 277–9, 286
  courant normal (principle) 253, 264, 270–1
  and cyber operations 191, 245, 253–63,
    277–9, 285–6
  and cyber-related activities 265–7, 277–9
  non-belligerency 248, 267–9
  in non-international armed conflicts 249
  remedies against the violation of 272–6, 286
  unable or unwilling standard in
    the 275–6, 286
  and the UN Charter 248, 268, 269–72, 278
New Zealand 92, 225
Nicaragua 37, 66, 68, 87, 103, 109, 137
Non-defended localities 230–1
Non-international armed conflict
  and Article 36 of Additional Protocol I 171
  and belligerent occupation 146–7, 161–3
  combatancy in 192–3, 199–201
  cyber operations in and as 148–59,
    161–3, 286
  definition 150–1
  law applicable to 139–40, 164–5
  and the law of neutrality 249
  minimum threshold of 132, 145, 150–1,
    157–9, 161–3
    organization of the armed group 150–1,
      153–7, 158, 161–3, 195, 283
    'protracted armed violence' 150–1, 152–3,
      158, 161–3, 283
  and the prohibition of perfidy 217
Non-intervention in the domestic affairs
    of other states (principle) 32, 46, 53,
    63, 65, 66, 82, 115–16, 138,
    249–50, 282
North Atlantic Treaty Organization
    (NATO) 3, 7, 30, 31, 61, 68, 84, 92–7,
    99, 189–91, 233–4, 250
  'NATO reservation' 188, 226–7
North Korea 10, 67, 195, 266

Operation 'Olympic Games' 7
Organization of American States (OAS) 93
Organization for Security and Co-operation in
    Europe (OSCE) 3
  See also Conference on Security and
    Co-operation in Europe (CSCE)

Pakistan 113, 214, 262
Palestinian Territories 8, 146–7, 225
Panama 51, 60, 111, 132, 170, 181, 281
Peacekeeping and peace enforcement
    operations 114–15, 231–2, 243
Perfidy (prohibition of) 195, 213, 214,
    215–17, 240, 245, 256
Poland 9–10, 111, 141–2

Precautions (duty to take) 23, 166, 176,
    180, 182, 211, 214, 219, 226, 228, 229,
    232–9, 245
  in attack 176, 214, 219, 226, 229,
    232–7, 285
  against the effects of attack 176, 237–9, 285
Prisoner of War (POW) status 119, 192–3,
    195, 196–9, 204, 211, 212–13, 244
Propaganda 5, 7, 8, 17, 65, 144–5, 147, 157,
    189–91, 201, 239, 241, 258, 263, 286
Proportionality
  and belligerent reprisals 244–5, 285
  and countermeasures 104, 105, 106, 285
  in the law of targeting 23, 166, 169, 173,
    174, 176, 178, 180, 182, 185, 188,
    191, 211, 214, 218, 219–29, 230, 232,
    237, 244, 245, 262–3, 284–5
  and self-defence reactions 69, 75, 78, 79–80,
    84, 85, 88, 90–1, 92, 105, 108, 219,
    276, 278, 285
Protecting Powers 242
Psychological operations 11, 17, 239–42

Qatar 22

Reconnaissance *see* Intelligence collection
Religious personnel and property 192, 194,
    229–30, 233, 244
Retorsion (acts of) 105, 106, 115, 272,
    278, 282
Ruses of war 215–16, 217, 240
  See also Perfidy (prohibition of)
Russian Business Network (RBN) 36, 38, 155,
    202, 266
Russian Federation 3, 5, 7–8, 22, 31, 36,
    39–40, 50–1, 57, 60, 70, 75, 79, 80, 81,
    82, 84, 100, 107, 111–12, 120, 123, 130,
    135, 148, 170, 202, 206, 239–40, 252,
    264–5, 283, 287
Rwanda 124, 139, 191

Sabotage 12, 38, 114, 138, 192
Saudi Arabia 5, 186
Self-defence
  and the accumulation of events
    doctrine 108–110, 115, 282
  anticipatory 42, 77–80, 91, 107–8, 115,
    276, 282–3
  and burden of proof 102–3
  collective 84, 92–7, 104, 268, 271
  against cyber operations 41, 43, 60,
    63, 69–111
  duty to report to the UN Security
    Council 103–4
  interceptive 78
  against non-state actors 40, 42, 80–8,
    282
  pre-emptive 78, 108, 276
  requirements for the exercise of 43, 69, 75,
    77, 78, 85–6, 88–91, 92, 108, 278

Self-defence (*cont.*):
  and Security Council's measures 89–90
  and standard of proof: 7, 80, 91, 97–102, 116
  'unable or unwilling' standard 40, 44, 81, 84, 85–8
  against violations of the law of neutrality 273–4, 276, 278
  See also 'Armed attack', Immediacy, Necessity (and self-defence reactions), Proportionality (and self-defence reactions)
Serbia *see* Yugoslavia
'Shamoon' 5, 55, 169–70, 186, 262
Shanghai Cooperation Organization (SCO) 64, 130
Singapore 262
Slovakia 93
'Solar Sunrise' 33
Somalia 84, 108
South Korea 5, 10, 68, 266
Soviet Union 4, 64,
Spain 51, 54, 170, 228, 267
Sri Lanka 148
State responsibility
  attribution rules 4, 19, 32, 34–40, 42, 44, 80–6, 123–4, 138–9, 148, 195–6, 200, 250, 258, 273, 281
  circumstances precluding wrongfulness 89, 105–6, 271, 273
  complete dependency test 35, 37, 195–6
  composite internationally wrongful act 109–10, 282
  continuing internationally wrongful act 209
  effective control test 35, 37–9, 65, 148, 195–6, 250
  'injured state' 105–6, 242–3, 268–9, 272–3, 278
  overall control test 37–8, 137–9, 146, 195–6, 250
  primary and secondary rules 19, 40, 80–6, 109, 138, 196, 250, 281
  and the qualification of armed conflicts 138–9
  See also Countermeasures, Necessity (state of), Self-defence
'Stuxnet' 6–7, 48, 53, 76, 91, 135, 153, 169, 175–6, 179, 183, 188, 206, 216, 224, 229, 232, 240, 262, 285
Superfluous injury and unnecessary suffering (prohibition of) 171–5, 195, 215, 228, 245
Supervisory Control and Data Acquisition (SCADA) 6, 17–18, 53, 55–6, 185
Sweden 111, 171
Switzerland 5, 10, 233, 248–9, 263
Syria 7, 9, 66, 79, 99, 107, 148, 151, 226–7

Taiwan 5
Tajikistan 3, 287
Targeting (law of) 135, 137, 147, 167, 172, 176–245, 262, 284
  geographical scope of application of 213–15
  and individuals 192–213
  and military objectives 176–91
  See also 'Attack', Civilian objects, Civilians, Combatants, Distinction (principle of), Dual-use objects, Military objective, Precautions (duty to take), Proportionality (in the law of targeting)
Thailand 262
Threat
  to the peace 67, 111–15, 149, 269
  of the use of force 41, 44, 45, 46, 67–9, 105, 107, 121, 129, 138, 272
  of violence against civilians 190, 192, 241
'Titan Rain' 5, 17
Treaties
  application to cyber operations conducted by states 20–4, 280–1
  definition 20
  effects of armed conflict on 121, 127, 150, 151
  interpretation of 19, 20–1, 45, 54, 59, 61, 78, 96, 114, 134, 143–4, 166, 179, 238, 254, 280–1, 284
Turkey 84, 93, 113, 120
Turkmenistan 111, 249

Uganda 77, 83–4, 90, 99, 109, 144, 147
United Arab Emirates 33
United Kingdom 3, 5, 10, 22, 32, 34, 51, 89, 93, 95, 120, 132, 141, 167, 170, 171, 174, 187, 192, 194, 197, 202, 220, 267, 268
UN General Assembly 2, 3, 9, 22, 28, 34, 56, 64, 65, 82, 87, 110, 111, 149, 177, 266, 287
UN Secretary-General 21, 28, 61, 81, 100, 181
UN Security Council 67, 70, 71, 83, 84, 89–90, 103–4, 105, 107, 110–15, 116, 149, 152, 214, 243, 245, 249–50, 259, 267, 269–72, 274, 276, 278
  and regional arrangements or agencies 114–15
United States 1, 3–4, 5, 7, 9, 10, 15, 11–15, 17, 21, 24, 30, 31, 32, 33, 37, 38, 39, 47, 50–1, 53, 55, 56, 58, 60–1, 64, 68, 70, 71, 74, 75, 76, 78, 80, 82, 84, 86–9, 92–3, 96, 98–9, 100, 101, 105, 113, 114, 132–3, 137, 140, 141, 165, 166, 169, 171, 177, 183, 184, 185, 186–7, 189, 192, 194, 199, 214, 221, 225, 229–30, 234–5, 239, 247–8, 260, 262, 264, 265–6, 267, 268, 275, 283, 287
US Cyber Command (CYBERCOM) 10, 21, 28, 31, 38, 47, 58, 70, 75, 81, 100, 170, 184, 186, 194, 221, 239, 248

Use of force
  cyber operations as 15, 32, 41, 43, 44–67, 52–67, 71, 75, 115–16, 243, 281–2
  cyber operations below the level of 15, 63–6, 75, 79, 110, 242–3, 282
  definition 45–50
  prohibition of 41, 43, 44–67, 69, 71, 72, 83, 99–100, 115, 128, 129, 131, 243, 250, 268, 272, 273, 277, 281–2
  and 'resort to 'armed force' 128–32
  See also 'Armed attack', Self-defence
Uzbekistan 3, 287

Venezuela 24, 262

War of national liberation 110, 140–1, 157, 197, 199, 200, 249–50
Weapons
  biological 50, 90, 174, 223, 285
  blinding 28
  chemical 50, 99
  cyber 50–2, 60, 66–7, 71–2, 111, 116, 130, 167, 168–70, 175–6, 208, 221, 254, 259–60, 261, 265, 266–7
  definition 49–50, 60
  indiscriminate 172–4
  legality of 22–3, 25, 164–5, 217–18
  non-lethal 61–2
  nuclear 7, 74, 75, 130, 166, 167, 206, 208
  obligation to review the legality of new 23, 60, 62, 170–2, 280–1
  See also Cyber warfare, (means and methods of)
West Bank *see* Palestinian Territories
World Summit on the Information Security 3

Yugoslavia 7, 61, 138, 189, 190, 191, 228, 234

Printed and bound by CPI Group (UK) Ltd, Croydon, CR0 4YY